Advanced Processing Technologies of Innovative Materials

Advanced Processing Technologies of Innovative Materials

Guest Editors

Sergey N. Grigoriev
Marina A. Volosova
Anna A. Okunkova

Basel • Beijing • Wuhan • Barcelona • Belgrade • Novi Sad • Cluj • Manchester

Guest Editors

Sergey N. Grigoriev
Department of High-efficiency
Processing Technology
Moscow State University
of Technology "Stankin"
Moscow
Russia

Marina A. Volosova
Department of High-efficiency
Processing Technology
Moscow State University
of Technology "Stankin"
Moscow
Russia

Anna A. Okunkova
Department of High-efficiency
Processing Technology
Moscow State University
of Technology "Stankin"
Moscow
Russia

Editorial Office
MDPI AG
Grosspeteranlage 5
4052 Basel, Switzerland

This is a reprint of the Special Issue, published open access by the journal *Technologies* (ISSN 2227-7080), freely accessible at: www.mdpi.com/journal/technologies/special_issues/additive_manufacturing_fundamentals.

For citation purposes, cite each article independently as indicated on the article page online and using the guide below:

Lastname, A.A.; Lastname, B.B. Article Title. *Journal Name* **Year**, *Volume Number*, Page Range.

ISBN 978-3-7258-3004-6 (Hbk)
ISBN 978-3-7258-3003-9 (PDF)
https://doi.org/10.3390/books978-3-7258-3003-9

Cover image courtesy of Moscow State University of Technology "Stankin"

© 2025 by the authors. Articles in this book are Open Access and distributed under the Creative Commons Attribution (CC BY) license. The book as a whole is distributed by MDPI under the terms and conditions of the Creative Commons Attribution-NonCommercial-NoDerivs (CC BY-NC-ND) license (https://creativecommons.org/licenses/by-nc-nd/4.0/).

Contents

About the Editors vii

Preface ix

Sergey N. Grigoriev, Marina A. Volosova and Anna A. Okunkova
Advanced Processing Technologies for Innovative Materials
Reprinted from: *Technologies* **2024**, *12*, 227, https://doi.org/10.3390/technologies12110227 1

Sergey N. Grigoriev, Marina A. Volosova and Anna A. Okunkova
Investigation of Surface Layer Condition of SiAlON Ceramic Inserts and Its Influence on Tool Durability When Turning Nickel-Based Superalloy
Reprinted from: *Technologies* **2023**, *11*, 11, https://doi.org/10.3390/technologies11010011 17

Shwetabh Gupta, Gururaj Parande and Manoj Gupta
Comparison of Shallow (-20 °C) and Deep Cryogenic Treatment (-196 °C) to Enhance the Properties of a Mg/2wt.%CeO$_2$ Nanocomposite
Reprinted from: *Technologies* **2024**, *12*, 14, https://doi.org/10.3390/technologies12020014 39

Jason Daza, Wael Ben Mbarek, Lluisa Escoda, Joan Saurina and Joan-Josep Suñol
Two Fe-Zr-B-Cu Nanocrystalline Magnetic Alloys Produced by Mechanical Alloying Technique
Reprinted from: *Technologies* **2023**, *11*, 78, https://doi.org/10.3390/technologies11030078 51

Viktor Shamakhov, Sergey Slipchenko, Dmitriy Nikolaev, Ilya Soshnikov, Alexander Smirnov and Ilya Eliseyev et al.
Features of Metalorganic Chemical Vapor Deposition Selective Area Epitaxy of Al$_z$Ga$_{1-z}$As ($0 \leq z \leq 0.3$) Layers in Arrays of Ultrawide Windows
Reprinted from: *Technologies* **2023**, *11*, 89, https://doi.org/10.3390/technologies11040089 64

Konstantin Baranov, Ivan Reznik, Sofia Karamysheva, Jacobus W. Swart, Stanislav Moshkalev and Anna Orlova
Optical Properties of $AgInS_2$ Quantum Dots Synthesized in a 3D-Printed Microfluidic Chip
Reprinted from: *Technologies* **2023**, *11*, 93, https://doi.org/10.3390/technologies11040093 75

Muhammad Farooq Saleem, Niaz Ali Khan, Muhammad Javid, Ghulam Abbas Ashraf, Yasir A. Haleem and Muhammad Faisal Iqbal et al.
Moisture Condensation on Epitaxial Graphene upon Cooling
Reprinted from: *Technologies* **2023**, *11*, 30, https://doi.org/10.3390/technologies11010030 88

Yana Suchikova, Sergii Kovachov, Ihor Bohdanov, Artem L. Kozlovskiy, Maxim V. Zdorovets and Anatoli I. Popov
Improvement of β-SiC Synthesis Technology on Silicon Substrate
Reprinted from: *Technologies* **2023**, *11*, 152, https://doi.org/10.3390/technologies11060152 95

Anna A. Okunkova, Marina A. Volosova, Khaled Hamdy and Khasan I. Gkhashim
Electrical Discharge Machining of Alumina Using Cu-Ag and Cu Mono- and Multi-Layer Coatings and ZnO Powder-Mixed Water Medium
Reprinted from: *Technologies* **2022**, *11*, 6, https://doi.org/10.3390/technologies11010006 112

Ilaria Ceccarelli, Luca Filoni, Massimiliano Poli, Ciro Apollonio and Andrea Petroselli
Regenerating Iron-Based Adsorptive Media Used for Removing Arsenic from Water
Reprinted from: *Technologies* **2023**, *11*, 94, https://doi.org/10.3390/technologies11040094 133

Victor Panarin, Eduard Sosnin, Andrey Ryabov, Victor Skakun, Sergey Kudryashov and Dmitry Sorokin
Comparative Effect of the Type of a Pulsed Discharge on the Ionic Speciation of Plasma-Activated Water
Reprinted from: *Technologies* **2023**, *11*, 41, https://doi.org/10.3390/technologies11020041 . . . . **148**

Takafumi Aizawa
Anisotropy Analysis of the Permeation Behavior in Carbon Dioxide-Assisted Polymer Compression Porous Products
Reprinted from: *Technologies* **2023**, *11*, 52, https://doi.org/10.3390/technologies11020052 . . . . **158**

Igor Lebedev, Anastasia Uvarova and Natalia Menshutina
Information-Analytical Software for Developing Digital Models of Porous Structures' Materials Using a Cellular Automata Approach
Reprinted from: *Technologies* **2023**, *12*, 1, https://doi.org/10.3390/technologies12010001 . . . . . **168**

Alexandre Staub, Lucas Brunner, Adriaan B. Spierings and Konrad Wegener
A Machine-Learning-Based Approach to Critical Geometrical Feature Identification and Segmentation in Additive Manufacturing
Reprinted from: *Technologies* **2022**, *10*, 102, https://doi.org/10.3390/technologies10050102 . . . . **188**

Alexander S. Metel, Sergey N. Grigoriev, Tatiana V. Tarasova, Yury A. Melnik, Marina A. Volosova and Anna A. Okunkova et al.
Surface Quality of Metal Parts Produced by Laser Powder Bed Fusion: Ion Polishing in Gas-Discharge Plasma Proposal
Reprinted from: *Technologies* **2021**, *9*, 27, https://doi.org/10.3390/technologies9020027 . . . . . . **197**

About the Editors

Sergey N. Grigoriev

Sergey N. Grigoriev, doctor of engineering science, professor, and head of the Department of High-Efficiency Processing Technologies of the Moscow State University of Technology "Stankin", completed his higher education in 1982. The scientific team of the department have received awards from the most prestigious international scientific competitions and exhibitions under the supervision of Dr. Eng. Sci., Prof. S. N. Grigoriev. Their brightest achievements are advances in surface engineering, coating, and surface-hardening technologies. The researchers and academic staff of the department comprise multiple winners of grants from the President of the Russian Federation for leading scientific schools in the fields of engineering and the technical sciences. The department's research and academic staff develop projects that improve the efficiency of cutting tools, the efficiency of the most advanced and new technologies, adaptive, multi-component, and nanocomposite structures, and the diagnostics of technological processes. Prof. S.N. Grigoriev has initiated and participated in various invited lectures at conferences, he serves as an editor of the most authoritative international scientific journals all over the world, is the author of over 180 patents, and has more than 500 international scientific articles in journals indexed by the Web of Science and Scopus, with H-indexes of 60 and 62, correspondingly.

Marina A. Volosova

Marina A. Volosova is a doctor of engineering science (Dr.Sci.), a docent, and a leading researcher in the Department of High-Efficiency Processing Technologies of the Moscow State University of Technology "Stankin"; she completed her higher education in 2000 at the SMoscow State University of Technology "Stankin" as a chartered engineer of mechanical engineering. She was awarded a Ph.D. in engineering science on 05.02.07, in technology and equipment for mechanical and physical–technical processing in 2003, a docent degree at the department in 2010, and a doctorate in engineering science in 2022. She is a laureate of the Government of the Russian Federation in Science and Engineering for Young Scientists; the winner of many international and national competitions and salons for works that are mostly related to laser, plasma, and sintering processes in the context of ceramic cutting tool coating improvements; and the author of approximately 50 patents and more than 200 international scientific articles in journals indexed in the Web of Science and Scopus (with H-indexes of 28 and 29, correspondingly).

Anna A. Okunkova

Anna A. Okunkova, a Ph.D. candidate in engineering science (Ph.D.) and a senior researcher in the Department of High-Efficiency Processing Technologies of the Stankin Moscow State University of Technology, completed her higher education in 2005 at the Nizhniy Novgorod State Technical University as a chartered engineer in the design of technical and technological complexes. In 2009, she was awarded a Ph.D. in engineering science, specializing in the automation and control of technological processes and productions. In 2015, she was honored as a Laureate of the Government of the Russian Federation in Science and Engineering for Young Scientists (N10904) for her work on mobile laser complexes for the reparation of specialized machines. She has developed 26 patents and published more than 75 international scientific articles in journals indexed in the Web of Science and Scopus (with H-indexes of 22 and 23).

Preface

Innovative processing and synthesizing technologies, and a new class of materials developed in recent decades, have aimed to meet some of the most relevant engineering challenges: the problems of the surface quality of grown 3D objects and the development of new post-processing approaches; the question of the reliable work of cutting tools in machining; the most advanced and heat-resistant alloys with increased cutting speeds and depths; questions of cryogenic treatments for the developed of nanocomposites and the functional ability of 2D objects in the presence of moisture under the cooling conditions of liquid nitrogen; production methods for and the properties of innovative low-dimensional structures (quantum dots, nano-objects, and layers) for the needs of photoelectronics and sensory devices; the question of the serial and mass production of enriched water for growing plants; water purification for the needs of humans; questions of medicament delivery by porous polymer materials; machine learning systems for additive manufacturing; and digital models for medicament delivery. Traditional scientific approaches to improving technologies in order to switch them from the laboratory scale to the production of functional products using the most advanced materials is steadily being replaced by more analytical research that needs to be proved with a set of experiments with an error of 10–15%. The high efficiency of the developed methods has been confirmed by the most advanced experimental demonstrations and investigation of their functional properties, aiming to improve the current state of industries through newly developed, simplified, and replaceable approaches with respect to nature.

Many years of experience with the scientific team of the Moscow State University of Technology "Stankin" in the field of proposing the most advanced technologies and innovative solutions oriented toward the quality and productivity of the final product of certain manufacturing methods have made it possible to present the results of the latest scientific research in the form of a collection of selected scientific articles, a communication, and a review. In addition, the presented achievements contributes to the development of innovative 0D–2D objects. These are in demand by other production branches and present the results of development in the form of patented methods and techniques, developed setups, models, and databases, ready for installation in enterprises for the next technology switch associated with a newly developed class of materials and relevant technologies.

This Special Issue is dedicated to the most recent achievements using the advanced processing technologies of innovative materials, such as metals, ceramics, and polymers, as well as the associated innovative manufacturing methods.

Sergey N. Grigoriev, Marina A. Volosova, and Anna A. Okunkova
Guest Editors

technologies

Editorial

Advanced Processing Technologies for Innovative Materials

Sergey N. Grigoriev, Marina A. Volosova and Anna A. Okunkova *

Department of High-Efficiency Processing Technologies, Moscow State University of Technology STANKIN, Vadkovskiy per., 3A, 127994 Moscow, Russia; s.grigoriev@stankin.ru (S.N.G.); m.volosova@stankin.ru (M.A.V.)
* Correspondence: a.okunkova@stankin.ru; Tel.: +7-499-972-94-29

1. Introduction and Scope

There is a need for further, in-depth research that explores the synthesis of newly developed materials created using advanced technologies. Their potential uses in various applications, as well as their ability to operate under increased thermal and mechanical loads and in the presence of moisture or other contaminants, are of particular interest [1–3]. The latest progressive technology approaches to industrial realities must be adapted for the sixth technological paradigm [4–6]. Many questions have arisen with regard to the quality of products' surfaces post-production [7–9] and repair, particularly with regard to additive manufacturing, 3D printing, or synthesis [10–13], which lend such products exceptional operational properties [14–16]. As such, it is necessary not only to conduct research under various conditions but also to develop new functional coating, surface cleaning, and processing methods [17–19]. Fundamental issues related to the synthesis and processing of multicomponent objects, structures, and composites, the creation of new classes of materials and nanocrystalline alloys, advanced surface treatments and multilayer coatings, and the development of new approaches and technological solutions using the latest achievements in the information sphere are of particular interest to the industry [20–25]. In particular, process monitoring and productivity improvements in the scale of real serial and mass production of newly developed methods and technologies deserve special attention [26–30].

For the last few decades, cutting-edge advances in processing, synthesis, and research methods and technologies relating to developed materials, objects, and coatings have featured in outstanding scientific publications, attracting widespread attention from the research community. In addition, progressive approaches have received multiple awards at the most prestigious scientific conferences, industrial exhibitions, and other events around the world.

This Special Issue is devoted to the latest achievements in the technologies of production and synthesis of innovative 0D–2D, nano-, and functionally gradient objects, nanocomposites, nanoporous structures (aerogels) and coatings, and new processing methods and technologies based on plasma and laser treatment, electrophysical and chemical processing and synthesis, as well as research on their exploitation properties and the development of digital models and machine learning approaches for the needs of the industry.

2. Contributions

Ten scientific articles, three communications, and a review were published in the presented Special Issue, each addressing a critical topic relating to the processing and synthesis of innovative materials. These topics include surface treatment, coating, and film deposition technologies; the synthesis of multicomponent objects and their behavior under various exploitation conditions (cooling, increased thermal and mechanical loads, etc.); and their mechanical and physical properties, including optical and electrical properties, for use in the aviation and tool industry and in different sensory and electronic devices [31–39]. These topics are described in greater detail below.

- Technologies for improving the surface quality of metallic parts produced by laser additive manufacturing and developing of the technology of ion polishing those parts in gas-discharge plasma and subsequent coatings for use in the aviation industry are reviewed. Other finishing techniques, such as mechanical machining, chemical etching, surface plastic deformation, ultrasonic cavitation abrasive finishing, and laser ablation, are also considered [31].
- Comparative research is conducted on the mechanical machining technologies that are used to improve the surface layer condition of the ceramic tools, such as diamond grinding and diamond grinding–lapping–polishing. The development of advanced double-layer (CrAlSi)N/DLC coating deposition technology on SiAlON ceramics used as cutting tools for turning nickel-based Inconel 718-type superalloys is also explored [32].
- Research is conducted on the machinability of insulating material, such as Al_2O_3 oxide ceramics, using the developed mono- and multilayer assistive coatings based on a Cu-tape and Cu-Ag "sandwich" for advanced electrical discharge machining in a ZnO powder–water medium [33].
- One study explores the influence of shallow (at $-20\,°C$) and deep (at $-196\,°C$) cryogenic treatment of a magnesium nanocomposite Mg/2 wt.%CeO_2 produced via the disintegrated melt deposition method followed by hot extrusion, and then investigate its influence on the porosity, grain size, dislocation density, ignition temperature, microhardness, lattice strain, 0.2% compressive yield strength, ultimate strength, and fracture strain of this material when it is operated in sub-zero conditions and strength-based constructions [34].
- Two innovative nanocrystalline soft magnetic Fe-based nanoperm-type alloys are created by mechanical alloying. Their alloys are initially composed of $Fe_{85}Zr_6B_8Cu_1$ (at.%) and $Fe_{80}Zr_5B_{13}Cu_1$ (at.%). Their operating temperature range (thermal stability and Curie point) and magnetic properties depending on the Fe/B ratio were evaluated. Additionally, high-saturation magnetic flux density is determined to be of interest for the development of low-dimensional systems, such as 0D-2D objects [35].
- Metal–organic chemical vapor deposition selective area epitaxy of $Al_zGa_{1-z}As$ ($0 \leq z \leq 0.3$) bulk layers is developed using a passivating mask with ultrawide windows—in other words, a SiO_2-mask/window of alternating stripes of 100 μm—on a GaAs substrate to produce strained quantum wells of AlGaAs/GaAs solid solutions (0D objects) in light-emitting structures for optoelectronic applications [36].
- Researchers explore the physical and optical properties of colloidal nanoparticles, particularly $AgInS_2$ quantum dots (0D objects), produced by a newly developed scalable manufacturing method of additive 3D-printing microfluidic chips for microfluidic synthesis. An increased product yield of 60% is observed [37].
- Research is conducted to evaluate the frequency peaks of epitaxial 2D objects. Graphene is chemically grown on 6H-SiC substrates using a Raman spectrometer and cooled in a range from room temperature to $-180\,°C$ to ensure that the graphene is properly insulated against moisture [38];
- Thermal synthesizing technology for carbide ceramic (β-SiC) film deposition on a mono-silicon substrate is developed, in which porous silicon is used as an intermediate layer for adhesion in the double-layer structure of β-SiC/por-Si/mono-Si composition. This allows the developed carbide ceramic film to meet the needs of electronics and proto-electronics in light of the indicated quantum size effects (0D–2D objects or, in other words, nano-objects) [39].

The process of purifying water using regenerating iron-based adsorptive media and plasma-activated water generation as biocidal agents for growing plants, and the technology of production and properties of porous polymer materials that are used in filters and drug-loading tablets [40–42], are also discussed in depth, as follows:

- An innovative regenerating iron-based adsorptive media and a relevant device for removing arsenic (As) from drinking water are developed [40].

- Research is conducted to determine the effects of barrier and bubble-pulsed discharges on the properties—particularly the ionic speciation and concentrations, pH index, and electrical conductivity—of plasma-activated water produced from distilled and groundwater [41].
- A carbon dioxide-assisted polymer compression method is developed for the creation of porous polymer products with laminated fiber sheets. This method can be used to design filters and drug-loading tablets [42].

Analytical models, software, and interactive databases of production technologies of nanoporous aerogels and metallic parts are also developed [43,44].

- The researchers developed software for digital models of porous and nanoporous structures such as aerogels to create new materials based on SiO_2, silica–resorcinol–formaldehyde, polyamides, carbon, polysaccharides, proteins, etc., to predict their properties (thermal and electrical conductivity, mechanical properties, sorption, and solubility) and pore size distribution using the lattice Boltzmann method and the cellular automaton particle dissolution model [43].
- A machine-learning-based approach and the database of basic and critical geometrical features for laser-based additive manufacturing from stainless steel for feature identification and part segmentation are developed to improve manufacturability, which can be extended to other technological processes and materials [44].

The authors of [31] identify the defects in additively manufactured machinery products made of stainless steel and various metallic alloys, such as cobalt, nickel, aluminum, and titanium alloys, which are earmarked for application in the airspace industry [45,46]. Mechanical machining, chemical etching, surface plastic deformation, ultrasonic cavitation abrasive finishing, laser ablation, and ion polishing in gas-discharge plasma are observed [47]. The method of gas-discharge plasma processing for finishing laser additive manufactured parts and subsequent wear-resistant coating is proposed for the first time. All the existing technologies are classified into three groups based on the possibility of volumetric processing and the nature of the exposure, which may be thermal, chemical, mechanical, or combined. The observed post-processing methods possess several disadvantages related to the nature of surface destruction, which influences the intensity of the wear on the working surfaces as a part of a unit or mechanism, along with the positive effect of plastic deformation and recrystallization of near-surface layers, erosion processes lead to stress states of the surfaces [48].

Innovative post-processing approaches, such as ion polishing in gas-discharge plasma, have no analogs in modern industry and may accelerate the transition to the next technological paradigm [49–51] by increasing the reliability and availability of additive manufacturing parts. Such approaches encompass the following stages:

- Granules of 40–100 µm in size are removed from the product's surface to achieve a roughness parameter R_a (Arithmetic Mean Deviation) of 30 µm. Negative-voltage microsecond pulses of up to 30 kV are applied to the product during its immersion in the plasma.
- The product is polished with concentrated ions or fast argon atoms under an angle greater than $60°$.
- A coating is deposited by evaporating liquid metal magnetron targets.

The condition of the surface layer and advanced double-layer (CrAlSi)N/DLC coating deposition [32] significantly influence the operational life of SiAlON ceramic cutting tools in machining Inconel 718-type chrome–nickel alloy. This type of alloy is often used to produce sophisticated aircraft turbine engine parts [52] and is machined by the most wear- and heat-resistant cutting tools that work under increased mechanical and thermal loads composed of oxide, nitride, and oxide-nitride cutting ceramics [53]. SiAlON is known for its excellent properties but is also very brittle, which can lead to machining difficulties. Industrially produced inserts feature numerous defects on their working surfaces; these serve as stress concentrators and provoke the destruction of the cutting edge by tearing the conglomerates of the material. This shortens the operational life of the cutting tool. Additional diamond grinding, lapping, and polishing and complex double-layered coating (consisted of an additional DLC coating and trinitride underlayer for better adhesion) deposition exhibit improved the properties of the cutting insert surface, resulting in a microhardness of 28 ± 2 GPa and an increase in the average friction coefficient at 800 °C, which was ~0.4. The average operational life of the insert under the increased loads improved, reaching 12.5 min, which was 1.8 and 1.25 times greater than that of the industrial cutting insert and double layer-coated insert without additional diamond grinding, lapping, and polishing.

The authors of [33] aim to resolve the issues with the electrical properties of the insulators that, under normal conditions, cannot be processed by electrophysical technologies [54]. This is especially relevant for insulating materials such as oxide and nitride ceramics [55], which are needed to produce cutting tools for machining titanium and nickel-based alloys used in the aircraft industry as was mentioned above, especially during the production of gas turbine engines [52,53]. During the cutting process, these alloys create intensive thermal and mechanical loads in the contact zone of the tools with a temperature up to 800–1200 °C and also tense state of up to 1.3×10^8–8.7×10^9 N/m². The chips tend to stick to the tool's cutting edge due to high adhesion, increasing the loads and promoting the snatching of conglomerates and faster tool wear. One of the most promising (though difficult-to-machine) cutting ceramics is Al_2O_3 (melting point of 2045–2345 °C, coefficient of linear thermal expansion of 7.0–8.0×10^6 K^{-1} (at 20–1000 °C), which possesses thermal conductivity of 20–25 W/(m·K) (at 20–100 °C), Mohs hardness of 9) in addition to its excellent mechanical and physical properties. The authors of this paper proposed an advanced electrical discharge machining method using assistive means such as conductive coatings and powder-mixed water-based medium. Hydrocarbon-based media, such as oils and kerosene, should be avoided when machining Al-based materials due to the risk of the formation of explosive Al_3C_4 products formed during its interactions with water, $H_2\uparrow$ and $O_2\uparrow$ gasses, which can be highly damaging to the machine and its filtration system. ZnO powder was chosen since it is a direct-gap semiconductor with a wide band gap E_g of 3.30–3.36 eV. ZnO powder-mixed deionized water-based medium with a ZnO concentration of 7–100 g/L and Cu-Ag and Cu mono- and multi-layer coatings, which were composed of 40μm-thickness copper tape and Ag adhesive, were used as assistive means. The total thickness of the assistive coatings was between 40 and 120 μm. A material removal rate of 0.0032–0.0053 mm³/s was achieved at a concentration of 14 g/L, a discharge pulse frequency of 2–7 kHz, and a pulse duration of 1 μs. The obtained data expand the scope of use of this sought-after material in the tool industry by highlighting new possibilities for its application.

The physical and mechanical properties of a Mg/2 wt.%CeO$_2$ nanocomposite at two cryogenic temperatures of -20 °C (253 K) and -196 °C (77 K) were researched to investigate this material's ability to operate in sub-zero conditions [34]. The nanocomposite was produced by the disintegrated melt deposition method followed by hot extrusion [56]. The shallow cryogen treatment at -20 °C reduced the porosity by 10.4%, and the deep cryogen treatment in liquid nitrogen (LN$_2$) [57] at -196 °C (boiling point) by 43.3%. The ignition temperature was reduced by 1 °C and amounted to 635 °C for the samples after the cold treatment, whereas it increased by 38 °C and amounted to 674 °C for deep cryogen-treated samples. The grain size of the treated samples was 2.9 ± 1.0 μm and 2.8 ± 0.6 μm, exceeding the sizes of the initial samples (2 ± 0.6 μm). Both samples had superior microhardness compared to the initial samples: 89 ± 5 HV and 92 ± 4 HV (correspondingly, 20% and 24% more than the initial samples). This can be related to their ability to increase the dislocation density, reduce porosity, and strain the lattice. The 0.2% compressive yield strength of the treated samples (186 ± 17 MPa for shallow cryogenic treatment, 203 ± 5 MPa for deep cryogenic treatment) was up to 14% greater than for untreated samples (178 ± 19 MPa), indicating that the sub-zero treatments improved the Mg nanocomposites' performance in strength-based constructions. The 0.2% compressive yield strength of the samples after deep cryogenic treatment was ~9% higher than that of the samples after the cold treatment. After shallow cryogenic treatment (-20 °C), the ultimate strength of the samples (441 ± 12 MPa) was reduced by 2.5% compared to that of the samples after deep cryogenic treatment (452 ± 15 MPa), and was 7% lower than for untreated samples (473 ± 16 MPa). The fracture strain was 16.5 ± 0.7% for untreated samples, 29.1 ± 1.0% for the samples after shallow cryogenic treatment, and 29.7 ± 1.2% for the samples after deep cryogenic treatment. The fracture surfaces of the samples showed no visible difference (45 ° shear fracture). Both types of treated samples demonstrated higher roughness than the initial ones. Overall, after treatment at -196 °C, the samples demonstrated an increase in the considered properties of the Mg nanocomposite, while the samples subjected to treatment at -20 °C exhibited similar values at lower costs.

Two innovative nanocrystalline soft magnetic nanoperm-type Fe-based alloys of the Fe-Zr-B-Cu system with the initial chemical composition of $Fe_{85}Zr_6B_8Cu_1$ (at.%) and $Fe_{80}Zr_5B_{13}Cu_1$ (at.%) were obtained by mechanical alloying [35]. Those alloys are sought after for applications such as magnetic sensors and actuators [58]. Those types of nanocrystalline soft magnetic alloys have recently been developed as a replacement for ferrites. Their key properties include the thermal stability of the magnetic phase and soft–hard behavior; they are also characterized by their low magnetic coercive force, high saturation magnetic flux density, and high magnetic permeability (response to the magnetic field). Reduced magnetic coercive force and increased magnetic permeability reduce core losses in devices under alternate-current magnetic fields and controlling those characteristics reduces energy consumption [59,60]. The high saturation magnetic flux density is favored for developing low-dimensional systems [61] such as those described below.

- For 0D systems, particles are confined to a single point (e.g., quantum dots) [36,37].
- For 1D systems, particles are confined to a line (e.g., carbon nanotubes).
- For 2D systems, interactions are confined to a plane (e.g., graphene) [38,39].

In other words, magnetic flux density is used to create consolidated systems composed of many particles. One of the critical characteristics of these alloys is the Curie point—the temperature above which alloys lose their permanent magnetic properties, which determines their range of operating temperatures. The magnetic properties also change depending on the alloy's phase (amorphous, crystalline), which is determined by powder production technology. Two main technologies are used to produce alloy powders: gas atomization and mechanical alloying. Both methods are implemented during powder metallurgy as the step before powder sintering and consolidation. Gas atomization allows us to obtain soft magnetic spherical particles of the mainly amorphous phase with a superior magnetic response, while mechanical alloying primarily produces smooth particles that are more specific and irregular. In gas atomization, with a high dispersion of particle sizes and a particle size growth, the particle's phase can also be both amorphous and nanocrystalline. Such powder production technology also prevents contamination of the cutting tools that happened during mechanical alloying. Mechanical alloying is mostly used to produce Fe-based nanocrystalline alloy powders. It should be noted that as the crystalline size increases, the alloy loses its soft behavior. The thermal stability of these alloys is also determined by crystalline growth, which depends on the apparent activation energy of crystallization. In this context, mechanical alloying is preferable for creating metastable alloys of nanocrystalline (supersaturated solid solutions or high-entropy alloys) and amorphous phases. The soft magnetic response can be improved by optimizing the milling modes in mechanical alloying and adding other elements. The main difference between the two developed alloys is in the Fe/B elements ratio of 85/8 and 80/13, where Fe determines the amount of magnetism. Correspondingly, the magnetization of $Fe_{85}Zr_6B_8Cu_1$ alloy is expected to exceed that of $Fe_{80}Zr_5B_{13}Cu_1$. At the same time, B is responsible for reducing the crystalline size, which increases the magnetic response by decreasing the coercive force. In such alloys, B usually determines the formation of the amorphous phase or more refined nanocrystalline phase, often leading to a loss of saturation magnetization. Zr and Cu were chosen equally in both alloys. Zr hinders crystalline growth due to the relatively large atom size, whereas Cu provokes a high density of nanocrystallization and hinders the growth of larger crystals. Heating these alloys can also influence the crystalline growth and reduce soft magnetic behavior. The thermal stability of the alloys was researched via thermal analysis of the apparent activation energy of the crystalline growth and the Curie point. Exothermic processes occurred in the 450–650 K temperature range and were associated with tension relaxation. The apparent activation energy of the main crystallization process was determined by Kissinger and isoconversion methods (differential calorimetry) and was compared with the previously obtained data. The lowest value of apparent activation energy (288 kJ/mol) was obtained for the $Fe_{85}Zr_6B_8Cu_1$ alloy. The transition temperatures, such as the peak crystallization and Curie points, were higher in the $Fe_{80}Zr_5B_{13}Cu_1$ alloy. The magnetic response was similar when the saturation magnetization was 5% higher in the $Fe_{85}Zr_6B_8Cu_1$ alloy than in the $Fe_{80}Zr_5B_{13}Cu_1$ alloy. The high-saturation magnetization and low coercive force are necessary for the feasibility of the applications in devices based on a soft magnetic behavior. Although the saturation magnetization is slightly lower in the $Fe_{80}Zr_5B_{13}Cu_1$ alloy, the decrease in the coercive force of $Fe_{80}Zr_5B_{13}Cu_1$ alloy is relatively significant due to the smaller size of the nanocrystals. The magnetic behavior, combined with a high thermal stability front crystalline growth associated with loss of soft magnetic behavior in $Fe_{80}Zr_5B_{13}Cu_1$ alloy, indicates that the reduction in Fe content does not significantly decrease its operating ability under the same conditions.

$Al_zGa_{1-z}As$ epitaxial layers ($0 \leq z \leq 0.3$) were produced by metal–organic chemical vapor deposition on a n-GaAs (100) substrate 2 inches in diameter with a SiO_2 mask of stripes of 100 µm in wide (100 µm wide SiO_2 mask/100 µm wide window) [36]. The monolithic integration of electro-optical elements is one of the issues of modern photonics that can be resolved by selective area epitaxy [62]. This technology facilitates the implementation of control and generation of optical radiation and electrical signals for the needs of optoelectronic devices, and other relevant operations in the production of single-mode lasers with mono-integrated modulators and couplers, multiwavelength single-mode laser systems, monolithic semiconductor sources of femtosecond laser pulses, and tunable semiconductor lasers with ultra-wide tuning ranges. The growth of nano-objects such as quantum dots [37] and nanowires is also induced by metal–organic chemical vapor deposition and molecular or chemical beam epitaxy. Epitaxial growth is produced using a passivating mask, which forms areas that suppress growth deposited on the substrate. Then, the epitaxial growth is produced in mask windows. The geometric dimensions of the mask and windows depend on the composition and properties of the grown epitaxial layers due to mass conservation during the growth process. For the first time, experimental and theoretical studies have been carried to evaluate the growth of layers of AlGaAs/GaAs solid solutions obtained using selective area epitaxy in ultrawide windows. The operating pressure in the reactor of the setup for epitaxial growth was 77 Torr. The samples were grown for 10 min on an n-GaAs (100) of 2 inches in diameter at a temperature of 750 °C and a rotation speed of 1000 rpm in hydrogen (H_2) (carrier gas). Trimethylgallium and trimethylaluminum were the atoms sources. Two sample types were grown epitaxially.

- One was grown without a mask (standard epitaxial growth).
- The second was grown with a mask of SiO_2 stripes of 1000 Å in thickness and 100 µm in width (selective area epitaxial growth).

Four $Al_{z_0}Ga_{1-z_0}As$ samples of each type were produced, with the following compositions (z_0) and growth rates:

- z_0 = 0, growth rate of 200 Å/min.
- z_0 = 0.11, growth rate of 225 Å/min.
- z_0 = 0.19, growth rate of 247 Å/min.
- z_0 = 0.3, growth rate of 286 Å/min.

During the initial production stage of the second type of samples, alternating stripes of the mask were produced by ion-plasma sputtering. The pattern was oriented in a [011] direction and produced by lithography and etching in a buffered oxide solution (buffered oxide etch 5:1). The patterned substrate was annealed at 750 °C for 20 min in the arsine flow (AsH_3) prior to the AlGaAs layer growth. The deposition of polycrystals with linear dimensions that did not exceed 100 nm was observed. The density of polycrystals slightly decreased towards the mask/window interface. A negligible number of polycrystals were detected in the mask area within 1 μm of the mask/window interface. A certain threshold concentration on the mask face, above which heterogeneous nucleation occurred, was observed for each reactant species. The threshold was higher for mask areas with improved roughness than those with high roughness. Polycrystals precipitate when one of the threshold concentrations is exceeded. The areas of the mask/window interface where nucleation did not occur were observed. The absence of polycrystals was observed on the surface of one of the samples. The microphotoluminescence spectra for samples of the second type were studied with a spatial resolution in the window region. The experimental data were compared with the simulation results obtained by the developed vapor-phase diffusion model. The mass diffusion constant and a surface reaction rate constant ratios were 85 μm for Ga and 50 μm for Al, which suggested that the proposed model could be used to predict the properties of the layers, such as the growth rate, layer thickness, and the layer composition distribution in the development of heterogeneous multilayer structures and optoelectronic devices.

Colloidal nanoparticles, such as $AgInS_2$ quantum dots [37], are a new material that improves the functionality of many sensory and electronic devices. The most relevant issue associated with the development of colloidal nanoparticle synthesis is adapting the developed technology to real manufacturing conditions. The new method proposed in this study is an additive printing of chips for microfluidic synthesis [63,64]. This method reduces costs, and it can be scaled and automated to increase productivity by up to 60% and improve optical properties, such as the position, shape, and width of the photoluminescent band and the photoluminescent quantum yield of quantum dots. An increase in the synthesis temperature of the $AgInS_2$ quantum dots led to a linear increase in photoluminescent quantum yield. The photoluminescent quantum yields of samples synthesized in the microfluidic chip at 40, 60, and 90 °C were 0.9, 1.8, and 3.6%. Since the samples produced by microfluidic synthesis of the $AgInS_2$ quantum dots at 90 °C exhibited the most significant photoluminescent quantum yield, they were synthesized at this temperature for 18 and 180 s to evaluate their optical properties and yield. The new method of flow hydrothermal synthesis for three-component $AgInS_2$ quantum dots by additive manufacturing, resulting in the formation of microfluidic chips, demonstrated the applicability of photopolymer resin-based chips without noticeable defects of the crystalline lattice and the degradation of mechanical properties that can negatively influence microfluidic chip channels. The microfluidic chip had a significantly greater mass and heat transfer coefficient than the conventional flask reactor. The photoluminescence quantum yield of samples synthesized by the developed technological method for 18 and 180 s was about 2.5 times higher than that of samples synthesized in laboratory conditions (in a flask).

Two-dimensional materials, particularly graphene [38], have recently become relevant in light of their specific carrier mobility in electronic device applications. Graphene layers are produced physically [65] and chemically (CVD and epitaxial) [66] on a metal, semiconducting, and insulating material basis. The chemical-based growing methods are considered to be the most promising for providing high-quality graphene to meet the high demands of the market. The most technologically promising graphene-growth method for a SiC substrate was achieved on Si- and C-faces. The type of face influences the properties of the graphene. The growth of the graphene on the Si-face leads to the formation of a C buffer underlayer between the graphene and Si-face. Charge transfer and substrate interaction of graphene grown on the Si-face affect its electronic properties compared to the C-face, which usually exhibits increased carrier mobility that depends on the layer's quantity. The most critical issue associated with graphene growth on the C-face is the thick layers and the inability to accurately monitor their number. Changes in the surrounding temperature can influence the properties of atomically thin graphene, making it more attractive for the sensory industry. A change in wetting properties is noted during surface functionalization, but the results of many studies on this topic are controversial. The primary sources of inconsistent graphene properties are surface charge, defects, and absorption of species (adsorbates). The adsorption of ambient molecular species on graphene surfaces leads to doping and changes in graphene's electrical and optical characteristics. However, when this process not properly controlled, it can lead to degradation, affecting the stability and reliability of the graphene-based device. The single-atom-thick graphene surface can absorb bulk aerosol and moisture contamination in the air. When exposed to moisture, graphene adsorbs contaminants on open surfaces at a temperature below the room's ambient temperature. It is important to keep such surfaces clear in order to maintain the properties of highly sensitive graphene. A change in the Raman spectrum of epitaxial graphene upon cooling due to moisture condensation can influence the evaluation of the material's quality. Raman spectroscopy is one of the most accurate research methods for detecting grown graphene contamination. A semi-insulating 6H-SiC substrate doped with vanadium (V) with a thickness of 369 μm was studied. The substrates were polished on both Si and C faces. The experimental graphene was produced using a furnace at a temperature of 1550 °C on the atomically flat 6H-SiC substrate for 25 min in a $<10^{-10}$ Torr vacuum. Si evaporates from the substrate at a high temperature when carbon atoms form graphene. Optical images of the graphene surface grown on the carbon face of the 6H-SiC substrate were obtained via an optical microscope fixed to a Raman spectrometer ranging from room temperature to −180 °C. A Raman spectrometer featuring a 532 nm excitation laser with a spot of 2 μm was utilized for the temperature-dependent spectra of the graphene. The moisture on the epitaxial graphene on the 6H-SiC substrate was evaluated by comparing Raman peaks with ice. Peaks in the 500–750 cm^{-1} frequency range and at ~1327 cm^{-1} were considered to be indicators of airborne contaminants. At the same time, a wide peak at ~1327 cm^{-1} was observed at room temperature due to water spots on the sample. This peak is of key importance, since it can be mistakenly considered as a D band of graphene when it is observed at lower than room temperature and is associated with graphene defects. The study emphasizes the importance of using Raman spectrometer to investigate graphene below room temperature and its moisture insulation.

The authors of [39] focus on the innovative technology of β-SiC film synthesis on a mono-silicon substrate by integrating porous silicon (por-Si) as an intermediate layer that increases adhesion between the film and substrate (β-SiC/por-Si/mono-Si nano-objects). SiC films deposited on the insulating ceramic substrate can also be used in graphene synthesis [67]. The results are relevant to the development of the semiconductor industry. The morphology showed that the produced SiC film possessed agglomerates of 2–6 μm, with 70–80 nm pores observed on those agglomerates. The synthesized β-SiC/por-Si/mono-Si heterostructure had crystallographic orientations (hkl) of (111) and (220) for Si and SiC corresponding to crystalline structures. X-ray diffraction analysis showed a shift toward lower angles (the peak at 2θ = 35.6 °), indicating quantum size effects corresponding to the nano-objects. The peak at an angular position of 2θ = 35.6 ° corresponds to β-SiC in a zinc-blend-type lattice, confirming the structural integrity and crystallinity of the SiC layer. A synthesized heterostructure is required for photodetectors, light-emitting diodes, and sensing technologies. The thermal conductivity of SiC and the insulating properties of por-Si can be used in advanced electronics. Due to the effects of quantum size, the proposed film-synthesizing approach can also be used in quantum computing. The proposed methodology substantially improves the lattice mismatch and adhesion that hinders conventional synthesis on the Si-substrates. Future development of the approach lies in the field of optimal porosity of the intermediate layer, mechanical properties in thermal and electrical conditions, and application of heterostructure in electronics and optoelectronics.

Another study included in this Special Issue proposes a new setup that can be used to remove toxic elements, such as arsenic (As), from water [40]. Many technologies have been developed to remove arsenic from drinking water [68,69], including oxidation (photochemical, photocatalytic, biological etc.) techniques, membrane-based technologies (micro-, ultra-

and nanofiltration, or reverse osmosis), coagulation/flocculation, ion exchange, adsorption onto solid media (activated Al_2O_3, Fe-based sorbents, zerovalent iron (Fe^0)), indigenous filters, miscellaneous sorbents, and metal–organic frameworks. One technique involves adsorption onto iron media, with regular replacement of the adsorbents. The formation of iron oxide/hydroxide (FeO(OH)) consists of a chemical reaction, a dehydration phase, and a grinding/granulating phase. Iron oxide/hydroxide can be obtained by combining a Fe^{+3} ion with an OH^- ion. The necessary amount of time is allotted for the $Fe(OH)_2$ to flocculate based on the reaction of $FeCl_3$ or $Fe_2(SO_4)_3$ reagents, and then it is passed through a filter press. The resulting sludge is dehydrated according to the following techniques: freezing and non-mechanical separation with translucent and granular sludge comprising 50% water, followed by thermal drying in a rotary drum or belt dryer until only 20% of the water remains. The next phase involves grinding/granulating the sludge, such as $Fe(OH)_2$. The resulting media is used to remove arsenic from water. The principle of adsorption on $Fe(OH)_2$ involves a reversible chemical exchange, which proves its effectiveness in removing arsenic and other substances from water; this is supported by many studies and has been accepted on an international basis. However, this method of replacement is quite costly. Adsorptive media technology requires replaceable Fe-based media that can only be used once. The replacement cost comprises approximately 80% of the overall service cost. Thus, a new portable setup based on the principles of iron media regenerating was developed and tested in Central Italy. In 2019–2023, the proposed system was used to regenerate iron media to restore the system's ability to adsorb arsenic in water. The legal threshold of As-content in water is 10 µg/L. When the level of arsenic concentration in water exceeds this threshold, iron media regeneration occurs, and the arsenic concentration in water is minimized. A system that can regenerate the media to make it more economically profitable is highly sought after by the industry. The advantages of this newly developed approach are the renewing of the filter bed with the restoration of its adsorption capacity, the absence of solid waste, the absence of disposal costs, the positive impact on the environment, the reduced service time, the fact that no equipment needs to be replaced, elimination of media production, and related material and transport costs.

Distilled water and groundwater were subjected to low-temperature plasma produced by barrier and bubble discharges [41]. Research on the effect of non-equilibrium low-temperature plasma of electric discharges in the air on water and aqueous solutions is of interest to the industry, particularly with regard to the development of installations for plasma-activated water (PAW) production [70,71]. PAW results from plasma action in water or aqueous solutions (e.g., phosphate-buffered saline, etc.) in the presence of oxygen O_2 or a mixture of O_2+N_2 at atmospheric pressure. PAW is used in biofilm removal, wound healing, bacterial inactivation, and to increase seed germination rates and subsequently accelerate plants' growth, inactivate pathogens, rescue fungus infections, and preserve crops due to the reactive oxygen and nitrogen species (RONS), relatively short-lived radicals (•OH, NO•), superoxide (O_2^-), peroxynitrate ($ONOO_2^-$), and peroxynitrite ($ONOO^-$). The stated effects on plant growth are attributed to the activity of water nitrates, nitrites, ammonium ions (NH_4^+) or $[NH_4]^+$), and hydrogen peroxide (H_2O_2). PAW is considered to be a sustainable and promising solution for biotechnological applications due to the transient nature of its biochemical activity and the potential economic and environmental benefits of using ambient air rather than rare or expensive chemicals. This approach can potentially reduce the technological costs of growing plants when used in tandem with other biotechnologies to improve germination and further plant growth [72,73]. There are many methods of discharge treatment of water solutions to obtain plasma-activated water. The spread approach in which the barrier discharge is used consists of the pulsed arc and corona discharge that occurs directly in a water solution when the high-voltage electrode is covered with a polyethylene or ceramic dielectric (barrier) layer (the heterophase method). The overloading leads to the decay of the useful chemicals in water solution when the formation of active oxygen and nitrogen species is limited by the low concentration of N_2 and O_2 dissolved in water. Thus, water should be constantly saturated with the indicated gasses or air to produce plasma-activated water in hydrodynamic setups. Another promising production method is based on the discharge that occurs directly in the volume and on the surface of air bubbles. Bubble discharges make initiating a discharge easier than electrohydraulic setups of heterophase methods. The pulse voltage is applied, and the type of plasma-forming gas determines the type of particles and their concentration. Devices that discharge treatment of water droplets or a thin water film to produce PAW have low productivity (measured in liters/hour). The types of electric discharges (barrier and in bubbles) considered in the study were chosen due to their prevalence in applied research. The discharges are constructively implemented and facilitate the development and production of technological installations. During the study, the ionic composition of two types (distilled and ground) of water treated with a low-temperature plasma formed by two pulsed discharges was revealed in atmospheric pressure air. After the exposure, the properties of both types of water, such as their magnesium and calcium ion concentrations, pH index, and electrical conductivity, were compared. The bubble discharge in groundwater showed maximum productivity for the NO_3^- anions. The barrier discharge in air, followed by water saturation with plasma products, is the most suitable for distilled water. The maximum energy input (thermalizing) into the stock solution is ensured in both treatments. From the point of view of energy consumption, both types of discharge treatment are suitable for obtaining approximately equal amounts of NO_3^- anions. This is a reasonably simple way to convert calcium carbonates ($CaCO_3$) from insoluble to soluble calcium nitrates ($Ca(NO_3)_2(H_2O)_x$). Insoluble carbonates pass into soluble nitrates when interacting with NO_3^- anions. Treatment with discharges did not significantly affect their concentration of potassium and sodium cations (K^+, Na^+) in water; the content of potassium and sodium cations did not change during 10 min of exposure and amounted to 1.065 and 9.395 mg/L. Carbonates K_2CO_3 and Na_2CO_3 are soluble salts with electrical conductivity of 280 µS/cm in water. As a result of the action of the discharge, additional NO_3^- and NO_2^- anions appear, leading to the formation of potassium and sodium nitrates (KNO_3, $NaNO_3$), which are also soluble salts. The complex compounds that affect the hardness of water, particularly Ca^{++} ions, are released into the solution. These features of the water treatment process using pulsed discharges should be considered when designing setups for industrial plasma-activated water production of groundwater for hydroponic plant growing technologies, in which a water solution enriched with NO_3^- anions is required.

The carbon dioxide-assisted polymer compression method is one of the methods used to produce porous polymer products with laminated sheets made of poly(ethylene terephthalate) fiber with a diameter of 8 µm [42]. Polymer fibers are placed in a specific direction along the sheet, and the intersections of the fibers are crimped in the presence of CO_2, forming a porous structure [74]. This orientation in a porous material is anisotropic [75]. The anisotropy of the permeation behavior in carbon dioxide-assisted polymer compression porous materials is of interest to the production of the drug-loaded tablet [76] and was assessed based on the aspect ratio of the dye solution permeation of the fiber-spread direction via the fabric-lamination direction. Quantitative evaluation of the anisotropy of permeation was performed, and the phenomenon was understood by linking it to the structure of the sample. Experiments were conducted using limited conditions of dye solution permeation with a slow injection rate to emphasize and examine the anisotropy of the structure. For the actual design of the component, the permeation rate and the amount of permeation are essential considerations. A dye solution was syringed into the 80-ply and 160-ply laminated porous polymer products. The aspect ratio decreased steadily with a decrease in porosity (0.63 for the 80-ply laminated product and 0.25 for the 160-ply laminated product) and was evaluated as 2.73 and 2.33, respectively. A 3D structural analysis showed that as the compression ratio increases, the fiber-to-fiber connection also increases due to an increased quantity of adhesion points, resulting in a decrease in the anisotropy of permeation. The hypothesis that an increase in the number of oriented fibers per unit volume could increase the anisotropy was disproved. Cross-sections of the obtained porous polymer with high porosity, which were subjected to X-ray computed tomography, showed less fiber-to-fiber bonding, and the number of fiber bonding points increased as the porosity decreased. Since permeation of the dye solution occurs along the fiber surface, more bonded fibers promote permeation between the upper and lower fiber surfaces, resulting in less anisotropy of permeation. Functional components, such as filters and tablets, are important industrial components. Therefore, structural anisotropy is essential when designing filters and drug-loading tablets using carbon dioxide-assisted polymer compression porous materials.

Information-analytical software has been developed in C# on the .NET framework by Mendeleev University of Chemical Technology of Russia (version 2.0) and is aimed to create digital models of structures of porous materials such as aerogels and new nanoporous materials to predict a set of properties (thermal and electrical conductivity, mechanical properties, sorption, and solubility) and pore size distribution [43]. Models facilitate the description of hydrodynamics of multicomponent systems, heat and mass transfer processes, dissolution, sorption, and desorption in processes in porous and nanoporous structures [77,78]. Digital models for different types of aerogels can be created. The pore size distribution was chosen as a criterion to compare the results obtained for each model with the experimental findings. The deviation of the resulting curves did not exceed 15%, showing a correlation between the digital and experimental results. The software includes both the existing and newly developed models. The existing models were used to model porous structures when the original models were developed for aerogels of silicon dioxide SiO_2, silica–resorcinol–formaldehyde, polyamide, carbon, polysaccharides (chitosan, cellulose), and protein and related processes (the dissolution of active pharmaceutical ingredients and mass transportation in porous media). The developed models have a wide range of input parameters for each type of aerogel, considering the features of the current sample. The software allows for modeling processes such as hydrodynamics inside digital porous structures using the lattice Boltzmann method and the cellular automaton particle dissolution model. The lattice Boltzmann method can be combined with cellular automata models, which calculate sorption, mass transfer, and dissolution inside porous structures. Software modules can be expanded with new cellular automata and other discrete models. Aerogels of silicon dioxide, silica–resorcinol–formaldehyde, polyamide, carbon, chitosan, cellulose, and protein were developed with the suggested original information-analytical software. Their thermal and electrical conductivity, mechanical properties, sorption, and solubility were predicted. These models establish a connection between structure geometry and properties, allowing for the development of materials with the required properties, such as new nanoporous materials. They also facilitate cellular automata models (original developments and independent implementation of existing models) with wide possibilities for varying their parameters and adding new modules. The software can potentially reduce the required number of full-scale experiments and, consequently, the costs of developing new porous materials.

Additive manufacturing technologies allow the production of products of complex shapes made from various types of materials [44]. Many of those technologies based on using laser source, namely laser powder bed fusion, are limited by the physical properties of the used materials, such as the thermal conductivity of the surrounding medium, the internal stresses, and the warpage or product weight [79]. One study aimed to solve the problem of creating machine learning algorithms for the needs of additive manufacturing [80] to identify the product's hard-to-manufacture geometrical features. Four features were considered:

- An overhanging surface with an angle in the range of 10–70° and a length of the overhanging plane in the range of 10–25 mm (critical of over 45°).
- Fine walls and slits with a thickness of 0.1–15 mm (critical of 0.1–5 mm).
- Horizontal and vertical holes with a diameter of 2–15 mm (critical of 6–15 mm).
- Helix tube (critical in the whole range of sizes and shapes).

The segmentation of these features permits the application of different manufacturing strategies to improve production ability. The algorithm is trained based on laser bed fusion of stainless steel. It identifies simple geometrical features which are hard to produce. The developed approach allows the treatment for the new product to be manufactured by laser powder bed fission with 88% efficiency. A database containing basic and hard-to-manufacture geometrical features was generated during the study. Every identified feature received its production limitations. Convolutional neural network architecture was trained to identify critical geometrical features and was introduced into the developed database. During testing, the developed algorithm produced segmentation of the feature and recognized untrained complex shapes, such as helix tubes. The approach confirmed its efficiency in complex geometry segmentation. This approach can be improved by the topology indexation and the definition of the algorithm input space. The output data of the developed classification are a collection of 3D geometries representing the uncritical basic volume part and the critical additive manufacturing features. The databases can be expanded and retrained for other technological processes and materials by defining of a new set of features. The study proposes the use of a new file format for additive manufacturing technologies that can be enriched with the necessary 3D feature data, such as .3mf. The open-source XML-based file format can include the features in a file which will be automatically processed by a build processor. Further development of build processors is necessary to adapt to the newly proposed 3D part processing.

3. Conclusions and Outlook

This Special Issue investigates various types of technologies, devices, and approaches used in the creation and manufacturing of innovative and progressive materials (including low-dimensional systems such as 0D–2D objects—quantum dots, quantum wells, graphene, etc.). The following critical aspects of the development of the new and improved industrial productions are also addressed.

- The surface quality of additive manufacturing parts of stainless steels and a wide range of alloys, their explosive ablation, ion polishing in gas-discharge plasma, and coating deposition [31].
- The surface quality of SiAlON after diamond grinding and diamond grinding–lapping–polishing and prior to double-layer trinitride and DLC coating deposition, and their effects on the durability of the cutting insert in tuning nickel superalloy [32].
- Advanced electrical discharge machining of alumina, which is achieved using assistive coating and powder suspension [33],
- Shallow (at -20 °C) and deep (at -196 °C) cryogenic treatment of magnesium nanocomposite produced through disintegrated melt deposition followed by hot extrusion [34].
- Mechanical alloying of two nanocrystalline soft magnetic Fe-based nanoperm-type alloys using Fe-Zr-B-Cu composition to produce low-dimensional systems (0D–2D objects) [35].
- The use of metal–organic chemical vapor deposition selective area epitaxy using an SiO_2 mask with ultrawide () windows (100 µm) on a GaAs substrate to produce strained quantum wells (0D objects) [36].
- A new manufacturing method for synthesizing $AgInS_2$ quantum dots (0D objects) in a 3D-printed microfluidic chip [37].
- Insulation of 2D material (a 2D object), such as epitaxial graphene chemically grown on 6H-SiC substrates, from moisture that adsorbs contaminants under cooling conditions (from 20 °C to -180 °C) and the influence of the moisture on its properties [38].
- Thermal synthesis of carbide ceramic (β-SiC) film on a silicon substrate using porous silicon as an intermediate layer for creating nano-objects [39].
- Water purification technology which removes arsenic via regenerating iron-based adsorptive media [40].
- Plasma-activated water generation of distilled water and groundwater by barrier and bubble-pulsed discharges for growing plants [41].

- Anisotropy of the permeation behavior in carbon dioxide-assisted polymer compressive porous materials [42].
- Analytical software for digital models of porous and nanoporous structures such as aerogels to create new materials based on SiO_2 [43].
- Machine learning algorithms that are used to identify the critical geometrical features produced by laser powder bed fusion of stainless steel [44].

Most of the proposals related to the creation of innovative materials and technologies for processing and production have huge industrial potential and are scalable or, in some cases, are already suitable for serial implementation. This was one of the key considerations of the Guest Editors when selecting articles for publication in this Special Issue. The developed technologies and approaches are expected to be introduced into modern production to accelerate the transition to the sixth technological paradigm.

Funding: This work was funded by the state assignment of the Ministry of Science and Higher Education of the Russian Federation, Project No. FSFS-2021-0006.

Acknowledgments: The Guest Editors appreciate the high requirements for the quality of presentation, the scientifically valuable content of the presented papers, and the kind efforts of the reviewers, editors, and assistants in contributing to this Special Issue.

Conflicts of Interest: The authors declare no conflicts of interest.

References

1. Myasoedova, T.N.; Kalusulingam, R.; Mikhailova, T.S. Sol-Gel Materials for Electrochemical Applications: Recent Advances. *Coatings* **2022**, *12*, 1625. [CrossRef]
2. Terekhov, I.V.; Chistyakov, E.M. Binders Used for the Manufacturing of Composite Materials by Liquid Composite Molding. *Polymers* **2022**, *14*, 87. [CrossRef] [PubMed]
3. Okunkova, A.A.; Volosova, M.A.; Kropotkina, E.Y.; Hamdy, K.; Grigoriev, S.N. Electrical Discharge Machining of Alumina Using Ni-Cr Coating and SnO Powder-Mixed Dielectric Medium. *Metals* **2022**, *12*, 1749. [CrossRef]
4. Kablov, E.N. The sixth technological order. *Sci. Life* **2010**, *4*, 16.
5. Golov, R.S.; Palamarchuk, A.G.; Anisimov, K.V.; Andrianov, A.M. Cluster Policy in a Digital Economy. *Russ. Eng. Res.* **2021**, *41*, 631–633. [CrossRef]
6. Simchenko, N.; Tsohla, S.; Chyvatkin, P. IoT & digital twins concept integration effects on supply chain strategy: Challenges and effect. *Int. J. Supply Chain. Manag.* **2019**, *8*, 803–808.
7. Barari, A.; Kishawy, H.A.; Kaji, F.; Elbestawi, M.A. On the surface quality of additive manufactured parts. *Int. J. Adv. Manuf. Technol.* **2017**, *89*, 1969–1974. [CrossRef]
8. Zakharov, O.V.; Brzhozovskii, B.M. Accuracy of centering during measurement by roundness gauges. *Meas. Tech.* **2006**, *49*, 1094–1097. [CrossRef]
9. Yadav, R.; Yadav, S.S.; Dhiman, R.; Patel, R. A Comprehensive Review on Failure Aspects of Additive Manufacturing Components under Different Loading Conditions. *J. Fail. Anal. Preven.* **2024**, *24*, 2341–2350. [CrossRef]
10. Sova, A.; Doubenskaia, M.; Grigoriev, S.; Okunkova, A.; Smurov, I. Parameters of the Gas-Powder Supersonic Jet in Cold Spraying Using a Mask. *J. Therm. Spray. Tech.* **2013**, *22*, 551–556. [CrossRef]
11. Monfared, V.; Ramakrishna, S.; Nasajpour-Esfahani, N.; Toghraie, D.; Hekmatifar, M.; Rahmati, S. Science and Technology of Additive Manufacturing Progress: Processes, Materials, and Applications. *Met. Mater. Int.* **2023**, *29*, 3442–3470. [CrossRef] [PubMed]
12. Orlova, E.; Riabtsev, M.A.; Varepo, L.G.; Trapeznikova, O.V. Implementation of additive technologies in the system of maintenance and repair of printing machines. *J. Phys. Conf. Ser.* **2021**, *1901*, 012015. [CrossRef]
13. Fu, X.; Lin, Y.; Yue, X.J.; Ma, X.; Hur, B.; Yue, X.Z. A Review of Additive Manufacturing (3D Printing) in Aerospace: Technology, Materials, Applications, and Challenges. In *Mobile Wireless Middleware, Operating Systems and Applications*; Tang, D., Zhong, J., Zhou, D., Eds.; EAI/Springer Innovations in Communication and Computing; Springer: Cham, Switzerland, 2022. [CrossRef]
14. Díaz, L.A.; Montes-Morán, M.A.; Peretyagin, P.Y.; Vladimirov, Y.G.; Okunkova, A.; Moya, J.S.; Torrecillas, R. Zirconia–alumina–nanodiamond composites with gemological properties. *J. Nanopartic. Res.* **2014**, *6*, 2257. [CrossRef]
15. Chen, C.; Xie, Y.; Yan, X.; Ahmed, M.; Lupoi, R.; Wang, J.; Ren, Z.; Liao, H.; Yin, S. Tribological properties of Al/diamond composites produced by cold spray additive manufacturing. *Addit. Manuf.* **2020**, *36*, 101434. [CrossRef]
16. Mazeeva, A.K.; Staritsyn, M.V.; Bobyr, V.V.; Manninen, S.A.; Kuznetsov, P.A.; Klimov, V.N. Magnetic properties of Fe–Ni permalloy produced by selective laser melting. *J. Alloys Compd.* **2020**, *814*, 152315. [CrossRef]
17. Nguyen, T.-N.-H.; Le, D.-B.; Nguyen, D.-T. Automation for feed motion of flat grinding machine improve accuracy and productivity machine. *Mater. Today. Proc.* **2023**, *81 Pt 2*, 427–433. [CrossRef]

18. Skeeba, V.Y.; Vakhrushev, N.V.; Titova, K.A.; Chernikov, A.D. Rationalization of modes of HFC hardening of working surfaces of a plug in the conditions of hybrid processing. *Obrab. Met.* **2023**, *25*, 63–86. [CrossRef]
19. Makarov, V.M. Well integrated technological systems: Prospects and problems of implementation. *Repair. Innov. Technol. Mod.* **2011**, *6*, 20–23.
20. Chen, C.; Lee, C.S.; Tang, Y. Fundamental Understanding and Optimization Strategies for Dual-Ion Batteries: A Review. *Nano-Micro Lett.* **2023**, *15*, 121. [CrossRef]
21. Malozyomov, B.V.; Martyushev, N.V.; Sorokova, S.N.; Efremenkov, E.A.; Qi, M. Mathematical Modeling of Mechanical Forces and Power Balance in Electromechanical Energy Converter. *Mathematics* **2023**, *11*, 2394. [CrossRef]
22. Mahal, R.K.; Taha, A.; Sabur, D.A.; Hachim, S.K.; Abdullaha, S.A.; Kadhim, M.M.; Rheima, A.M. A Density Functional Study on Adrucil Drug Sensing Based on the Rh-Decorated Gallium Nitride Nanotube. *J. Electron. Mater.* **2023**, *52*, 3156–3164. [CrossRef]
23. Lobiak, E.V.; Shlyakhova, E.V.; Bulusheva, L.G.; Plyusnin, P.E.; Shubin, Y.V.; Okotrub, A.V. Ni-Mo and Co-Mo alloy nanoparticles for catalytic chemical vapor deposition synthesis of carbon nanotubes. *J. Alloys Compd.* **2014**, *621*, 351–356. [CrossRef]
24. Pelevin, I.A.; Kaminskaya, T.P.; Chernyshikhin, S.V.; Larionov, K.B.; Dzidziguri, E.L. Atomic Force Microscopy's Application for Surface Structure Investigation of Materials Synthesized by Laser Powder Bed Fusion. *Compounds* **2024**, *4*, 562–570. [CrossRef]
25. Abrosimova, G.; Aksenov, O.; Volkov, N.; Matveev, D.; Pershina, E.; Aronin, A. Surface Morphology and Formation of Nanocrystals in an Amorphous $Zr_{55}Cu_{30}Al_{10}Ni_5$ Alloy under High-Pressure Torsion. *Metals* **2024**, *14*, 771. [CrossRef]
26. Pilania, G. Machine learning in materials science: From explainable predictions to autonomous design. *Comput. Mater. Sci.* **2021**, *193*, 110360. [CrossRef]
27. Malashin, I.; Tynchenko, V.; Gantimurov, A.; Nelyub, V.; Borodulin, A. Applications of Long Short-Term Memory (LSTM) Networks in Polymeric Sciences: A Review. *Polymers* **2024**, *16*, 2607. [CrossRef]
28. Mishin, Y. Machine-learning interatomic potentials for materials science. *Acta Mater.* **2021**, *214*, 116980. [CrossRef]
29. Chursin, A.; Boginsky, A.; Drogovoz, P.; Shiboldenkov, V.; Chupina, Z. Development of a Mechanism for Assessing Mutual Structural Relations for Import Substitution of High-Tech Transfer in Life Cycle Management of Fundamentally New Products. *Sustainability* **2024**, *16*, 1912. [CrossRef]
30. Gorlacheva, E.N.; Omelchenko, I.N.; Drogovoz, P.A.; Yusufova, O.M.; Shiboldenkov, V.A. Cognitive factors of production's utility assessment of knowledge-intensive organizations. *Nucleation Atmos. Aerosols* **2019**, *2171*, 090005.
31. Metel, A.S.; Grigoriev, S.N.; Tarasova, T.V.; Melnik, Y.A.; Volosova, M.A.; Okunkova, A.A.; Podrabinnik, P.A.; Mustafaev, E.S. Surface Quality of Metal Parts Produced by Laser Powder Bed Fusion: Ion Polishing in Gas-Discharge Plasma Proposal. *Technologies* **2021**, *9*, 27. [CrossRef]
32. Grigoriev, S.N.; Volosova, M.A.; Okunkova, A.A. Investigation of Surface Layer Condition of SiAlON Ceramic Inserts and Its Influence on Tool Durability When Turning Nickel-Based Superalloy. *Technologies* **2023**, *11*, 11. [CrossRef]
33. Okunkova, A.A.; Volosova, M.A.; Hamdy, K.; Gkhashim, K.I. Electrical Discharge Machining of Alumina Using Cu-Ag and Cu Mono- and Multi-Layer Coatings and ZnO Powder-Mixed Water Medium. *Technologies* **2023**, *11*, 6. [CrossRef]
34. Gupta, S.; Parande, G.; Gupta, M. Comparison of Shallow (−20 °C) and Deep Cryogenic Treatment (−196 °C) to Enhance the Properties of a Mg/2wt.%CeO_2 Nanocomposite. *Technologies* **2024**, *12*, 14. [CrossRef]
35. Daza, J.; Ben Mbarek, W.; Escoda, L.; Saurina, J.; Suñol, J.-J. Two Fe-Zr-B-Cu Nanocrystalline Magnetic Alloys Produced by Mechanical Alloying Technique. *Technologies* **2023**, *11*, 78. [CrossRef]
36. Shamakhov, V.; Slipchenko, S.; Nikolaev, D.; Soshnikov, I.; Smirnov, A.; Eliseyev, I.; Grishin, A.; Kondratov, M.; Rizaev, A.; Pikhtin, N.; et al. Features of Metalorganic Chemical Vapor Deposition Selective Area Epitaxy of $Al_zGa_{1-z}As$ ($0 \leq z \leq 0.3$) Layers in Arrays of Ultrawide Windows. *Technologies* **2023**, *11*, 89. [CrossRef]
37. Baranov, K.; Reznik, I.; Karamysheva, S.; Swart, J.W.; Moshkalev, S.; Orlova, A. Optical Properties of $AgInS_2$ Quantum Dots Synthesized in a 3D-Printed Microfluidic Chip. *Technologies* **2023**, *11*, 93. [CrossRef]
38. Saleem, M.F.; Khan, N.A.; Javid, M.; Ashraf, G.A.; Haleem, Y.A.; Iqbal, M.F.; Bilal, M.; Wang, P.; Ma, L. Moisture Condensation on Epitaxial Graphene upon Cooling. *Technologies* **2023**, *11*, 30. [CrossRef]
39. Suchikova, Y.; Kovachov, S.; Bohdanov, I.; Kozlovskiy, A.L.; Zdorovets, M.V.; Popov, A.I. Improvement of β-SiC Synthesis Technology on Silicon Substrate. *Technologies* **2023**, *11*, 152. [CrossRef]
40. Ceccarelli, I.; Filoni, L.; Poli, M.; Apollonio, C.; Petroselli, A. Regenerating Iron-Based Adsorptive Media Used for Removing Arsenic from Water. *Technologies* **2023**, *11*, 94. [CrossRef]
41. Panarin, V.; Sosnin, E.; Ryabov, A.; Skakun, V.; Kudryashov, S.; Sorokin, D. Comparative Effect of the Type of a Pulsed Discharge on the Ionic Speciation of Plasma-Activated Water. *Technologies* **2023**, *11*, 41. [CrossRef]
42. Aizawa, T. Anisotropy Analysis of the Permeation Behavior in Carbon Dioxide-Assisted Polymer Compression Porous Products. *Technologies* **2023**, *11*, 52. [CrossRef]
43. Lebedev, I.; Uvarova, A.; Menshutina, N. Information-Analytical Software for Developing Digital Models of Porous Structures' Materials Using a Cellular Automata Approach. *Technologies* **2024**, *12*, 1. [CrossRef]
44. Staub, A.; Brunner, L.; Spierings, A.B.; Wegener, K. A Machine-Learning-Based Approach to Critical Geometrical Feature Identification and Segmentation in Additive Manufacturing. *Technologies* **2022**, *10*, 102. [CrossRef]
45. Sarmah, P.; Gupta, K. A Review on the Machinability Enhancement of Metal Matrix Composites by Modern Machining Processes. *Micromachines* **2024**, *15*, 947. [CrossRef]

46. Qiao, J.; Yu, P.; Wu, Y.; Chen, T.; Du, Y.; Yang, J. A Compact Review of Laser Welding Technologies for Amorphous Alloys. *Metals* **2020**, *10*, 1690. [CrossRef]
47. Valikov, R.A.; Yashin, A.S.; Yakutkina, T.V.; Kalin, B.A.; Volkov, N.V.; Krivobokov, V.P.; Yanin, S.N.; Asainov, O.K.; Yurev, Y.N. Modification of the cylindrical products outer surface influenced by radial beam of argon ions at automatic mode. *J. Phys. Conf. Ser.* **2015**, *652*, 012068. [CrossRef]
48. Grigoriev, S.N.; Metel, A.S.; Tarasova, T.V.; Filatova, A.A.; Sundukov, S.K.; Volosova, M.A.; Okunkova, A.A.; Melnik, Y.A.; Podrabinnik, P.A. Effect of Cavitation Erosion Wear, Vibration Tumbling, and Heat Treatment on Additively Manufactured Surface Quality and Properties. *Metals* **2020**, *10*, 1540. [CrossRef]
49. Glaziev, S.Y. The discovery of regularities of change of technological orders in the central economics and mathematics institute of the soviet academy of sciences. *Econ. Math. Methods* **2018**, *54*, 17–30. [CrossRef]
50. Korotayev, A.V.; Tsirel, S.V. A spectral analysis of world GDP dynamics: Kondratiev waves, Kuznets swings, Juglar and Kitchin cycles in global economic development, and the 2008–2009 economic crisis. *Struct. Dyn.* **2010**, *4*, 3–57. [CrossRef]
51. Perez, C. Technological revolutions and techno-economic paradigms. *Camb. J. Econ.* **2010**, *34*, 185–202. [CrossRef]
52. Qi, H. Review of INCONEL 718 Alloy: Its History, Properties, Processing and Developing Substitutes. *J. Mater. Eng.* **2012**, *2*, 92–100.
53. Arunachalam, R.M.; Mannan, M.A.; Spowage, A.C. Residual stress and surface roughness when facing age hardened Inconel 718 with cBN and ceramic cutting tools. *Int. J. Mach. Tools Manuf.* **2004**, *44*, 879–887. [CrossRef]
54. Grigor'ev, S.N.; Kozochkin, M.P.; Fedorov, S.V.; Porvatov, A.N.; Okun'kova, A.A. Study of Electroerosion Processing by Vibroacoustic Diagnostic Methods. *Meas. Tech.* **2015**, *58*, 878–884. [CrossRef]
55. Wang, C.C.; Akbar, S.A.; Chen, W.; Patton, V.D. Electrical properties of high-temperature oxides, borides, carbides, and nitrides. *J. Mater. Sci.* **1995**, *30*, 1627–1641. [CrossRef]
56. Gupta, M.; Lai, M.; Saravanaranganathan, D. Synthesis, microstructure and properties characterization of disintegrated melt deposited Mg/SiC composites. *J. Mater. Sci.* **2000**, *35*, 2155–2165. [CrossRef]
57. Jovičević-Klug, P.; Podgornik, B. Review on the Effect of Deep Cryogenic Treatment of Metallic Materials in Automotive Applications. *Metals* **2020**, *10*, 434. [CrossRef]
58. Sheftel', E.; Bannykh, O. Films of Soft-Magnetic Fe-Based Nanocrystalline Alloys for High-Density Magnetic Storage Application. In *Nanostructured Thin Films and Nanodispersion Strengthened Coatings*; Voevodin, A.A., Shtansky, D.V., Levashov, E.A., Moore, J.J., Eds.; NATO Science Series II: Mathematics, Physics and Chemistry; Springer: Dordrecht, The Netherlands, 2004; Volume 155. [CrossRef]
59. Batista, T.D.; Luciano, B.A.; Freire, R.C.; Castro, W.B.; Araújo, E.M. Influence of magnetic permeability in phase error of current transformers with nanocrystalline alloys cores. *J. Alloys Compd.* **2014**, *615*, S228–S230. [CrossRef]
60. Dragoshanskii, Y.N.; Pudov, V.I.; Karenina, L.S. Optimizing the domains and reducing the magnetic losses of electrical steel via active coating and laser treatment. *Bull. Russ. Acad. Sci. Phys.* **2013**, *77*, 1286–1288. [CrossRef]
61. Li, X.; Sun, R.; Li, D.; Song, C.; Zhou, J.; Xue, Z.; Chang, C.; Sun, B.; Zhang, B.; Ke, H.; et al. A plastic iron-based nanocrystalline alloy with high saturation magnetic flux density and low coercivity via flexible-annealing. *J. Mater. Sci. Technol.* **2024**, *190*, 229–235. [CrossRef]
62. Liu, C.; Cai, Y.; Jiang, H.; Lau, K. Monolithic integration of III-nitride voltage-controlled light emitters with dual-wavelength photodiodes by selective-area epitaxy. *Opt. Lett.* **2018**, *43*, 3401–3404. [CrossRef]
63. Bezrukov, A.; Galeeva, A.; Krupin, A.; Galyametdinov, Y. Molecular Orientation Behavior of Lyotropic Liquid Crystal–Carbon Dot Hybrids in Microfluidic Confinement. *Int. J. Mol. Sci.* **2024**, *25*, 5520. [CrossRef] [PubMed]
64. Boken, J.; Kumar, D.; Dalela, S. Synthesis of Nanoparticles for Plasmonics Applications: A Microfluidic Approach. *Synth. React. Inorg. Met.-Org. Nano-Met. Chem.* **2015**, *45*, 1211–1223. [CrossRef]
65. Voznyakovskii, A.P.; Ilyushin, M.A.; Vozniakovskii, A.A.; Shugalei, I.V.; Savenkov, G.G. Safe Explosion Works Promoted by 2D Graphene Structures Produced under the Condition of Self-Propagation High-Temperature Synthesis. *Nanomanufacturing* **2024**, *4*, 45–57. [CrossRef]
66. Mostovoy, A.; Bekeshev, A.; Brudnik, S.; Yakovlev, A.; Shcherbakov, A.; Zhanturina, N.; Zhumabekova, A.; Yakovleva, E.; Tseluikin, V.; Lopukhova, M. Studying the Structure and Properties of Epoxy Composites Modified by Original and Functionalized with Hexamethylenediamine by Electrochemically Synthesized Graphene Oxide. *Nanomaterials* **2024**, *14*, 602. [CrossRef]
67. Galvão, N.; Vasconcelos, G.; Pessoa, R.; Machado, J.; Guerino, M.; Fraga, M.; Rodrigues, B.; Camus, J.; Djouadi, A.; Maciel, H. A Novel Method of Synthesizing Graphene for Electronic Device Applications. *Materials* **2018**, *11*, 1120. [CrossRef]
68. Meiramkulova, K.; Kydyrbekova, A.; Devrishov, D.; Nurbala, U.; Tuyakbayeva, A.; Zhangazin, S.; Ualiyeva, R.; Kolpakova, V.; Yeremeyeva, Y.; Mkilima, T. Comparative Analysis of Natural and Synthetic Zeolite Filter Performance in the Purification of Groundwater. *Water* **2023**, *15*, 588. [CrossRef]
69. Pervov, A.; Spitsov, D. Control of the Ionic Composition of Nanofiltration Membrane Permeate to Improve Product Water Quality in Drinking Water Supply Applications. *Water* **2023**, *15*, 2970. [CrossRef]
70. Xiao, A.; Liu, D.; Li, Y. Plasma-Activated Tap Water Production and Its Application in Atomization Disinfection. *Appl. Sci.* **2023**, *13*, 3015. [CrossRef]

71. Rathore, V.; Watanasit, K.; Kaewpawong, S.; Srinoumm, D.; Tamman, A.; Boonyawan, D.; Nisoa, M. Production of Alkaline Plasma Activated Tap Water Using Different Plasma Forming Gas at Sub-Atmospheric Pressure. *Plasma Chem. Plasma Process* **2024**, *44*, 1735–1752. [CrossRef]
72. Vasilieva, T.; Goñi, O.; Quille, P.; O'Connell, S.; Kosyakov, D.; Shestakov, S.; Ul'yanovskii, N.; Vasiliev, M. Chitosan Plasma Chemical Processing in Beam-Plasma Reactors as a Way of Environmentally Friendly Phytostimulants Production. *Processes* **2021**, *9*, 103. [CrossRef]
73. Rashid, M.; Rashid, M.; Alam, M.; Talukder, M. Stimulating effects of plasma activated water on growth, biochemical activity, nutritional composition and yield of Potato (*Solanum tuberosum* L.). *Plasma Chem. Plasma Process* **2022**, *1*, 15. [CrossRef]
74. Aizawa, T. Effect of Crystallinity on Young's Modulus of Porous Materials Composed of Polyethylene Terephthalate Fibers in the Presence of Carbon Dioxide. *Polymers* **2022**, *14*, 3524. [CrossRef] [PubMed]
75. Ren, S.; Xu, X.; Hu, K.; Tian, W.; Duan, X.; Yi, J.; Wang, S. Structure-oriented conversions of plastics to carbon nanomaterials. *Carbon. Res.* **2022**, *1*, 15. [CrossRef]
76. Machado, N.D.; Mosquera, J.E.; Martini, R.E.; Goñi, M.L.; Gañán, N.A. Supercritical CO_2-assisted Impregnation/Deposition of Polymeric Materials with Pharmaceutical, Nutraceutical, and Biomedical Applications: A Review (2015–2021). *J. Supercrit. Fluids* **2022**, *188*, 105671. [CrossRef]
77. Dosta, M.; Jarolin, K.; Gurikov, P. Modelling of mechanical behavior of biopolymer alginate aerogels using the bonded-particle model. *Molecules* **2019**, *24*, 2543. [CrossRef]
78. Menshutina, N.; Lebedev, I.; Lebedev, E.; Paraskevopoulou, P.; Chriti, D.; Mitrofanov, I. A Cellular Automata Approach for the Modeling of a Polyamide and Carbon Aerogel Structure and Its Properties. *Gels* **2020**, *6*, 35. [CrossRef]
79. Grigoriev, S.N.; Gusarov, A.V.; Metel, A.S.; Tarasova, T.V.; Volosova, M.A.; Okunkova, A.A.; Gusev, A.S. Beam Shaping in Laser Powder Bed Fusion: Péclet Number and Dynamic Simulation. *Metals* **2022**, *12*, 722. [CrossRef]
80. Kumar, S.; Gopi, T.; Harikeerthana, N.; Kumar Gupta, M.; Gaur, V.; Krolczyk, G.M.; Wu, C.S. Machine learning techniques in additive manufacturing: A state of the art review on design, processes and production control. *J. Intell. Manuf.* **2023**, *34*, 21–55. [CrossRef]

Disclaimer/Publisher's Note: The statements, opinions and data contained in all publications are solely those of the individual author(s) and contributor(s) and not of MDPI and/or the editor(s). MDPI and/or the editor(s) disclaim responsibility for any injury to people or property resulting from any ideas, methods, instructions or products referred to in the content.

Article

Investigation of Surface Layer Condition of SiAlON Ceramic Inserts and Its Influence on Tool Durability When Turning Nickel-Based Superalloy

Sergey N. Grigoriev, Marina A. Volosova * and Anna A. Okunkova

Department of High-Efficiency Processing Technologies, Moscow State University of Technology "STANKIN", Vadkovskiy per. 3A, 127055 Moscow, Russia
* Correspondence: m.volosova@stankin.ru; Tel.: +7-916-308-49-00

Abstract: SiAlON is one of the problematic and least previously studied but prospective cutting ceramics suitable for most responsible machining tasks, such as cutting sophisticated shapes of aircraft gas turbine engine parts made of chrome–nickel alloys (Inconel 718 type) with increased mechanical and thermal loads (semi-finishing). Industrially produced SiAlON cutting inserts are replete with numerous defects (stress concentrators). When external loads are applied, the wear pattern is difficult to predict. The destruction of the cutting edge, such as the tearing out of entire conglomerates, can occur at any time. The complex approach of additional diamond grinding, lapping, and polishing combined with an advanced double-layer (CrAlSi)N/DLC coating was proposed here for the first time to minimize it. The criterion of failure was chosen to be 0.4 mm. The developed tri-nitride coating sub-layer plays a role of improving the main DLC coating adhesion. The microhardness of the DLC coating was 28 ± 2 GPa, and the average coefficient of friction during high-temperature heating (up to 800 °C) was ~0.4. The average durability of the insert after additional diamond grinding, lapping, polishing, and coating was 12.5 min. That is superior to industrial cutting inserts and those subjected to (CrAlSi)N/DLC coating by 1.8 and 1.25 times, respectively.

Keywords: ceramic inserts; diamond grinding defects; DLC-coating; nickel-based superalloy; polishing; SiAlON; surface layer; tool durability; turning

1. Introduction

Sintered tool ceramic based on α/β modifications of SiAlON is a more efficient solution for high-performance machining of high-temperature nickel superalloys such as Inconel 718 type alloys compared to hard alloys widely used for cutting inserts [1–4]. Nickel superalloy is one of the primary structural materials for manufacturing components of power equipment, aircraft engines, and spacecraft due to improved mechanical, anti-corrosion properties and structural stability at elevated operating temperatures. The improved mechanical properties of nickel alloys predetermine the increased heat and power loads on the tools that accompany machining and contribute to the intensification of the physicochemical interaction in the contact zones of the flank surface of the cutting inserts with the workpiece and the face surface with descending chips and high-intensity tool wear [5–9].

As production experience shows, nickel alloys begin to soften at temperatures corresponding to cutting speeds of 280–300 m/min and above, after which their mechanical processing is accompanied by significantly lower heat and power loads on the tool. In the specified high-speed range, the carbide tool instantly loses its cutting properties, while SiAlON ceramics are effectively used in turning operations due to higher heat resistance [10–13]. However, even the most modern brands of tool ceramics tend to cause brittle fracture of the cutting part during the nickel alloy turning at cutting speeds of more than 250 m/min with an increase in the cross-section of the cut layer (feeds of more than

0.15 mm/rev) with all the apparent advantages and, as a result, simultaneous exposure to significant thermal and mechanical loads on the tool [14–16]. High wear rates of the cutting part are observed under these conditions, which is associated with intensive friction and adhesion on the flank surface of the insert in contact with the workpiece and on the face surface in contact with the descending chips, even when choosing ceramic cutting inserts (CCI) with reinforced geometry and increased cutting-edge strength. At the same time, the worn section of the flank surface of the insert is characterized by the presence of craters with traces of adhesion, while relatively large recesses in the shape of holes are formed on the face surface, which often leads to micro-splitting and chipping of the cutting edge [17–21]. Therefore, with all the apparent advantages of CCI and the great potential of its use for critical product mechanical engineering, the actual share of its industrial use in the total global market volume of cutting tools is at a low level of about 11–12% [21,22].

Various researchers associate the low efficiency of CCI operation with the structural feature of tool ceramics, as well as the different volume and surface defects that form at the stages of the tool life cycle [23–25]. Among various defects, the critical role is played by defects of the surface layer since this layer primarily perceives operational loads, and the CCI efficiency depends on its condition [26–28]. The authors of this work propose the SiAlON CCI microstructural model presented in Figure 1. It explains the possible mechanisms of ceramic destruction under the influence of mechanical (P) and thermal (T) loads during cutting. The development of destruction of the surface layer is possible by one of three mechanisms:

1. intragranular destruction with gradual separation (abrasion) of microparticles of the surface layer;
2. grain-boundary destruction with separation of individual elements of the microstructure;
3. mixed destruction, in which there is a separation of grain conglomerates occurring inside the elements of the microstructure.

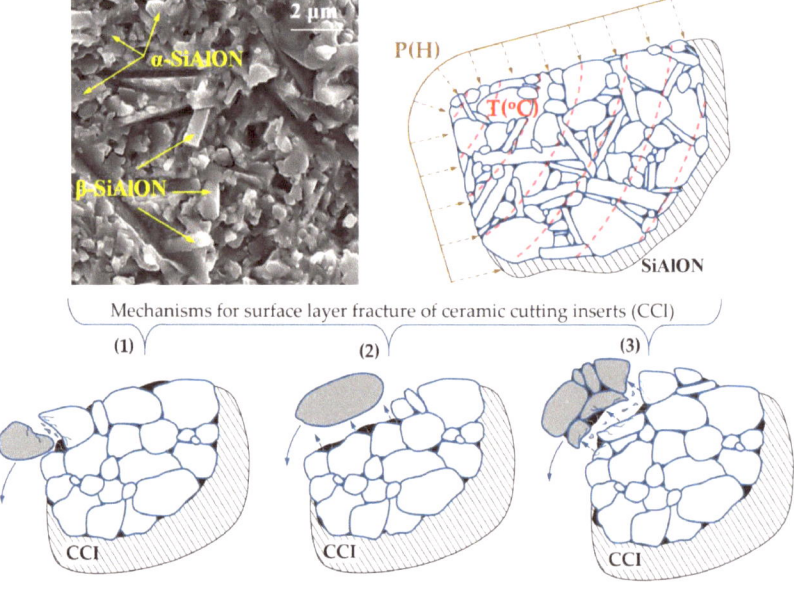

Figure 1. Variants of the destruction of the surface layer of SiAlON tool ceramics under external mechanical (P) and thermal (T) loads: intragranular (1), grain-boundary (2), and mixed (3).

The presence of numerous defects has a highly unfavorable effect on the resistance to the destruction of the surface layer under intense thermal and mechanical loads. Defects being stress concentrators contribute to the accelerated destruction of CCI contact surfaces and largely affect the stability of tool performance indicators. In particular, many authors have noted enormous variations in the durability (low operational stability) of cutting tools equipped with ceramic inserts, which hinders the industrial use of CCI in mechanical engineering.

The deposition of standard wear-resistant tool coatings cannot solve these problems since the increased defectiveness of the substrate significantly reduces the quality of the formed coatings and limits their functionality [29–33].

The authors of this work comprehensively investigated the influence of the condition of the surface layer of ceramic inserts made of Al_2O_3 + TiC, including the role of the formed wear-resistant coatings, on the performance of the tool when turning hardened bearing steels [28]. It was experimentally proven that the defects formed during diamond sharpening (grinding) of the ceramic inserts reduce the efficiency of the tool operation and contribute to premature destruction of the cutting part, which can occur at various stages of tool operation, in particular during the run-in period and at the stage of regular wear. It is possible to significantly minimize surface layer defects, the presence of which significantly affects the tool performance, by acting on the surface layer using various diamond abrasive machining methods [28,34–37].

The issues of the influence of the use of various diamond abrasive machining methods on the condition of the surface layer of SiAlON cutting tools and on the functioning of subsequently formed wear-resistant coatings on the performance of the tool when turning heat-resistant nickel alloys have so far remained out of the focus of attention of research groups. At the same time, these issues are highly significant from the point of view of a deeper understanding of the wear characteristics of SiAlON CCI and the scientifically justified use of various technological approaches to increase their performance, as well as more comprehensive industrial applications for machining heat-resistant nickel superalloys of the Inconel 718 type.

The purpose of the presented work was to study the influence of the surface layer condition of SiAlON ceramic cutting inserts processed by diamond grinding and polishing and subsequent (CrAlSi)N/DLC coating deposition on the tool operability in the high-speed turning of a heat-resistant nickel alloy with increased cross-sections of the cut layer.

The object of the research was the least previously studied SiAlON cutting insert. The focus of researchers until now has been mainly on square and rhombic plates made of Al_2O_3 + TiC ceramics [9,13,26,28], and less often Si_3N_4 [30,31]. At the same time, ceramic round plates made of SiAlON are designed for processing parts of gas turbine engines from heat-resistant chromium-nickel alloys. Along with titanium alloys, they are the primary structural materials for manufacturing responsible parts of aircraft gas turbine engines.

The novelty of this work lies in evaluating the surface layer condition influence on the operational ability of the ceramic cutting insert made of SiAlON cutting ceramic with the deposed double-layer coating of (CrAlSi)N/DLC structure, where the tri-nitride (CrAlSi)N sub-layer of the coating is responsible for better DLC coating adhesion knowing the problematic behavior of SiAlON to coatings, in conditions of extreme mechanical and thermal loads.

The practical significance of the work lies in evaluating the effect of diamond grinding and polishing in combination with the advanced double-layer (CrAlSi)N/DLC coating effect on the performance of industrially produced SiAlON ceramic cutting inserts in chrome–nickel alloy (Inconel 718 type) turning with increased cross-sections of the cut layer.

2. Materials and Methods

2.1. Cutting Tools, Material to Be Processed, and Laboratory Testing Methods

CRSNR 3232P 19 turning holders (Sandvik AB, Sandviken, Sweden) for external machining on universal lathes and CNC machines were used as a cutting tool for the research, in which RNGN-190800 round ceramic inserts of the AS500 brand manufactured by TaeguTec (Daegu, Republic of Korea) with a diameter of 19.05 mm and a thickness of

7.9 mm were mechanically fastened (Figure 2). When the insert is placed in the holder, the clearance angle is 5°, the rake angle is −5°, and the lead angle is 46°. It should be noted that the geometry of the cutting inserts was chosen from the tasks of providing curved surfaces for manufacturing gas turbine engine parts. SiAlON can be considered one of the most inconvenient materials for coating and machining that exhibits superior wear resistance compared to carbide tools, especially under extreme load conditions (semi-finishing in the temperature range up to 800 °C with increased depth of cut and feed). The tool material was SiAlON ceramic of the following composition: 79 vol% Si_5AlON_7, 17 vol% Si_3N_4, 4 vol% Yb_2O_3. The content of the main phases was revealed by X-ray diffraction analysis and processing of the results using the PANalytical HighScore Plus software by PANalytical B.V. (version 3.0) and the ICCD PDF-2 database (version 2023).

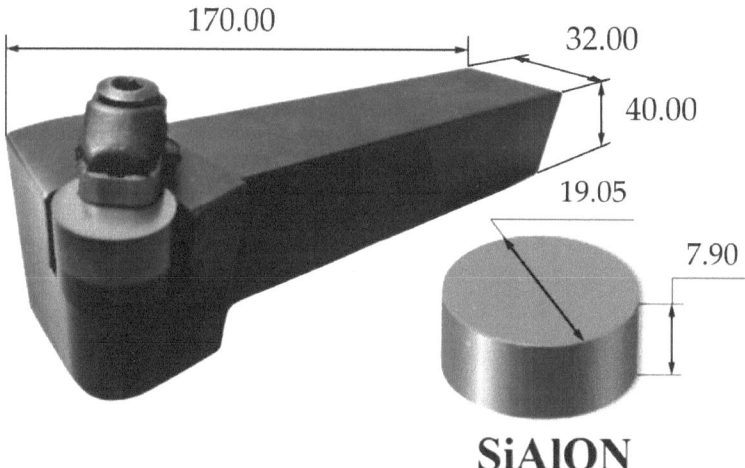

Figure 2. Design of a prefabricated turning cutter with mechanically fastened round ceramic inserts used in experiments.

XH45MBTJuBP nickel-based heat-resistant alloy, according to the national standard of the Russian Federation 5632-2014 (the closest analog of Inconel 718), was used as the material to be processed. Table 1 shows the composition of the XH45MBTJuBP alloy. The basis of this alloy is an austenitic solid solution of the nickel–chromium–iron system. This alloy is used for highly loaded elements of load-bearing structures and other parts of gas turbine engines operating in various climatic conditions at temperatures up to 800 °C. A hot-rolled bar with a diameter of 100 mm was used as a workpiece for the study. The nickel alloy had a hardness of 322 (HB) and a strength of 1130 MPa.

Table 1. Chemical composition of XH45MBTJuBP alloy used in the research.

Element	Ni	Fe	Cr	Mo	Nb	W	Ti	Al	C, Si, Mn, S, P
Content (%)	45.3	27.9	15.0	4.0	1.3	2.5	2.0	1.0	1.0

Workpieces were processed on a ZMM CU500MRD lathe (ZMM, Nova Zagora, Bulgaria) under cutting conditions providing intense heat–power loads on CCI: cutting speed V = 300 m/min, feed S = 0.2 mm/rev, and cutting depth t = 0.8 mm. Increased values were selected from the practice of semi-finishing difficult-to-machine alloys based on nickel and titanium. Increased feed at a simultaneously high cutting speed provides more intense heat and power loads on the tool when the probability of premature failure of the ceramic insert increases sharply. It is under these conditions that the role of the surface layer increases. It should be noted, within this study, that one value of the cutting speed was chosen so as not to overload the article with data and to conduct scientific groundwork for further research.

Ten CCI faces were tested to identify the nature of wear development over the cutting time. In order to minimize the error of experiments, all experiments were duplicated 10 times. The size of the wear area along the flank surface was measured every 2 min of turning on a Stereo Discovery V12 Zeiss optical microscope (Carl Zeiss AG, Oberkochen, Germany). The wear chamfer value of 400 μm was taken as the failure criterion.

When choosing the criterion of the flank wear, it should be noted that finishing of 0.3 mm is an optimum value, but only for tools that are subjected to regrinding. The cutting ceramic insert is not related to this group of cutting tools. They are much more expensive than hard alloy tools and cannot be reused after achieving the criterion. In most cases, this type of cutting tool is used carefully with a relatively small depth of cut. In this study, we used a relatively large depth of cut and feed to achieve the forementioned purposes to maximize tool life that is not subject to regrinding. Thus, the chosen cutting modes are not suitable for finishing but for semi-finishing, after which finishing is foreseen. In this condition, the tool is expected to develop its resource to the maximum. Therefore, flank wear of 400 μm is chosen.

2.2. Preparation of Cutting Ceramic Inserts with Different Condition of the Surface Layer

Industrially produced round-shaped SiAlON inserts were subjected to additional diamond abrasive machining operations such as lapping and polishing to form the experimental groups of ceramic inserts with different surface layer conditions. Additional processing of ceramic inserts was carried out on the Lapmaster Wolters lapping and polishing machine (Mt Prospect, IL, USA) with unique lapping and polishing wheels using various diamond suspensions (with a grain size of 50/40, 40/28 during lapping, and 10/7, 5/3 during polishing) with a cutting speed of 3 m/s. Finishing took 28 min, and polishing 16 min. Grinding and finishing modes were chosen based on literature data [35,37] and the authors' experience [9,16]. There were two groups of ceramic inserts made of SiAlON: industrial samples subjected to diamond grinding at production (I), experimental samples prepared in laboratory conditions of MSUT "STANKIN", subjected to additional lapping and polishing (II). Lapping and polishing as additional operations of abrasive machining of ceramic inserts were chosen because these processes can significantly minimize the degree of imperfection of the surface layer formed during diamond grinding, as noted by various researchers [34–37].

2.3. Coating of Ceramic Inserts

The choice was made in favor of a diamond-like (DLC) coating, which is deposited on a pre-formed nitride (CrAlSi)N sublayer based on the experimental data previously obtained by the authors of this study, when coating cutting tools designed for processing heat-resistant nickel alloys [38–43]. DLC coatings demonstrate the lowest coefficient of friction and depth of worn track when contacting the ball under high-temperature heating conditions during tribological tests and reduce the intensity of friction and adhesive interaction of the contact pads of the cutting tool when cutting nickel alloys [44,45].

DLC coatings, along with the described advantages, have specific features that should be taken into account when choosing the method of their deposition. First, there is a limitation of the maximum thickness of no more than 2.0 μm. Exceeding it leads to an increase in the level of internal stresses and an increase in the probability of their delamination when exposed to external loads [46,47]. Therefore, DLC coatings should be deposited on a preformed sublayer to increase the thickness of the formed coating on the cutting tool. In addition, the use of DLC coatings for high-speed cutting conditions of heat-resistant alloys is limited by their relatively low thermal stability. Formation of a thermally stable sublayer, for example, (CrAlSi)N, as well as doping of the DLC coating with various elements, for example, Si, allows the noted disadvantage to be minimized and ensures high efficiency of the DLC coatings [48–50].

The total thickness of the (CrAlSi)N/DLC coating that was formed on SiAlON CCI in the present study was 3.9 µm, including a 1.9 µm (CrAlSi)N sublayer and a 2.0 µm DLC coating.

It should be noted that the role of the (CrAlSi)N sublayer in cutting is in providing more favorable conditions for the functioning of the external DLC, increasing its adhesive bond strength to the substrate [10,41].

The technological process of (CrAlSi)N/DLC coating deposition was implemented on a multifunctional STANKIN unit (MSUT Stankin, Moscow, Russia), equipped with a set of systems and devices that provides purification of the processed samples and coating deposition by vacuum-arc evaporation of cathodes as well as chemical vapor deposition [51–55]. A schematic diagram and a general view of the unit are shown in Figure 3. The inner (CrAlSi)N layer was deposited by vacuum-arc evaporation of cathode materials, and the outer DLC layer was formed by the PACVD method by decomposition of hydrocarbon-containing gases in a gas discharge plasma. The DLC coating technology was optimized for ceramic tools, and the approaches developed by Platit AG (Selzach, Switzerland) were used as the scientific basis.

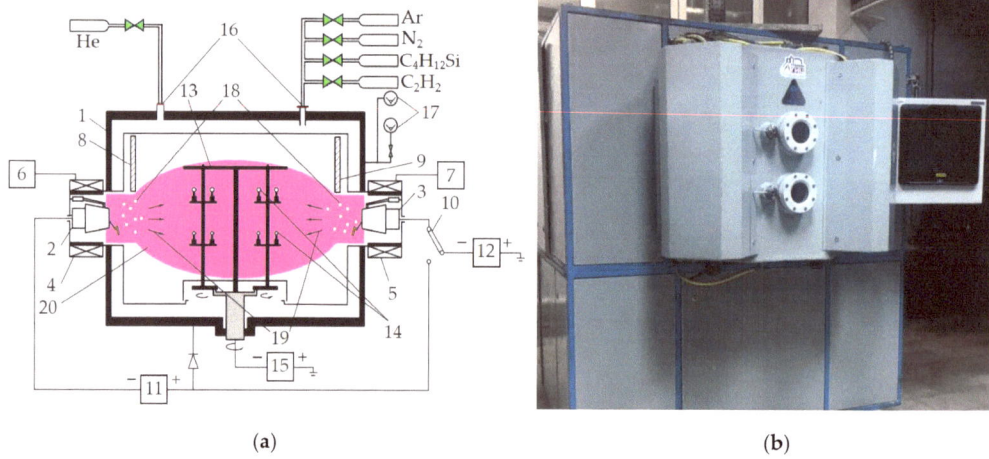

Figure 3. (**a**) Schematic diagram of the STANKIN technological unit for coating ceramic inserts, where 1 is vacuum chamber; 2,3 are cathodes (Cr and AlSi); 4,5 are cathode coils; 6,7 are power sources of cathode coils; 8,9 are cathode shutters; 10 is switchboard; 11,12 are cathode current sources; 13 is equipment for placing samples with planetary rotation; 14 is ceramic inserts; 15 is reference voltage power supply; 16 is gas supply systems; 17 is vacuum gauges; 18 is ions; 19 is ion movement directions; 20 is plasma; (**b**) a general view.

The complete technological cycle of coating deposition included six stages: sample heating, gas discharge purification, ion purification, a (CrAlSi)N sublayer deposition, a gradient DLC-Si layer deposition, and an external DLC layer deposition. Table 2 shows data on the modes that were assigned at each of the stages of coating formation. The modes mentioned were worked out in previous years and were chosen to ensure maximum adhesion of coatings [10,38,41].

2.4. Investigation of the Properties of the Surface Layer of Ceramic Inserts

A scanning electron microscopy method was used on Tescan VEGA3 LMH equipment (Brno, Czech Republic) for microstructural analysis of the surface of ceramic inserts with different surface layer conditions.

Table 2. Range of factors of technological cycle of coating deposition.

Stage of the Process	Technological Modes	Measuring Units	Values
Sample heating	Chamber pressure	Pa	0.03
	Rotation speed of the tooling with samples (constant at all stages)	rpm	5
	Heating temperature	°C	500
	Heating time	min	60
Purification in a gas discharge	Composition of the working gas	-	Ar
	Chamber pressure	Pa	1.2
	Bias voltage	V	−650
	Chamber temperature	°C	500
	Current at the cathode AlSi (open shutter)	A	90
	Current at the cathode Cr (closed shutter)	A	105
	Purification time	min	20
Ion purification	Composition of the working gas	-	Ar
	Chamber pressure	Pa	2.2
	Bias voltage	V	−800
	Chamber temperature	°C	500
	Current at the cathode Cr	A	90
	Purification time	min	20
(CrAlSi)N sublayer deposition	Composition of the working gas	-	95% N_2/5% Ar
	Chamber pressure	Pa	0.9
	Bias voltage	V	−40
	Chamber temperature	°C	500
	Current at the cathode AlSi	A	100
	Current at the cathode Cr	A	100
	Deposition time	min	90
Gradient DLC-Si layer deposition	Composition of the working gas	-	72% N_2/20% Ar/8% $C_4H_{12}Si$
	Chamber pressure	Pa	1.5
	Bias voltage	V	−500
	Chamber temperature	°C	180
	Deposition time	min	20
External DLC-layer deposition	Composition of the working gas	-	55% Ar/45% C_2H_2
	Chamber pressure	Pa	0.8
	Bias voltage	V	−500
	Chamber temperature	°C	180
	Deposition time	min	100

A Dektak XT stylus profilometer (Bruker AXS GmbH, Karlsruhe, Germany) was used to construct profilograms of the ceramic insert surface layer conditions. The specified device performs electromechanical measurements by contact scanning of the required surface area with a highly sensitive diamond tip at a given speed. Specialized software based on the analysis of measurement results processes information and visualizes the results in the form of 3D profilograms. Two parameters of the condition of the surface layer were evaluated, provided by the ISO 4287-2014 standard: R_t is the sum of the most significant height of the profile peak and the most significant depth of the profile cavity (the total height of the profile by which the depth of the defective layer can be judged); R_a is the arithmetic mean of the absolute values of the profile. According to the manufacturer's data, the measurement error is ±10% of the measured value when measuring from 100 μm up to 500 Å.

Crack resistance and microhardness of ceramic inserts with different surface layer conditions were determined on a universal QnessQ10A microhardness tester (Qness GmbH, Mammelzen, Germany) by the Vickers pyramid indentation method. The load on the indenter was 2 kg when assessing microhardness and 5 kg when assessing crack resistance. The indentation diagonals and the length of cracks propagating from the corners of the indentations were measured. Crack resistance (K_c) and microhardness (HV) were determined according to the known dependencies based on it [56].

An original method proposed by the authors of this study was used to assess the effect of surface defects on the ability of ceramic cutting edges to resist chipping when exposed to an external load. The approach described in [57,58] was taken as a basis. The possibility of obtaining new information about the mechanical behavior of ceramics when their edges are chipped by an indenter was substantiated by Griffiths' theory of brittle fracture. The authors of this study developed an improved method, the schematic diagram shown in Figure 4, taking into account the operational loads acting on ceramic inserts. The tests were performed on a Revetest scratch tester (Anton Paar, Corcelles-Cormondrèche, Switzerland) equipped with an acoustic emission (AE) signal registration sensor when exposed to a surface layer with a diamond indenter with a smoothly increasing load on the indenter (P) from 20 to 40 N. The diamond pyramid as an indenter was chosen purposefully to create the maximum stress concentration near the cutting edge. The identity of the location of the indenter application point relative to the cutting edge was strictly controlled. The chip had a shape close to a tetrahedron as a result of the force acting directly near the cutting edge. The experiments showed that the chip's length B and width L are not informative parameters for assessing the ability of the cutting edges to resist chipping (the chip size in all samples had similar values). The force at which the chip occurred turned out to be an indicator sensitive to changes in the condition of the surface layer of the ceramic samples. The moment of chipping and corresponding load were identified by the spectrum of the AE signal, which sharply increased at the moment of destruction. The high information content of the AE signal in the destruction of various materials has been noted in several works [59–63].

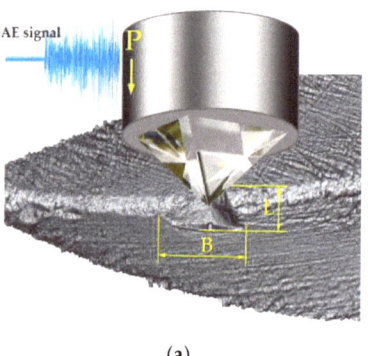
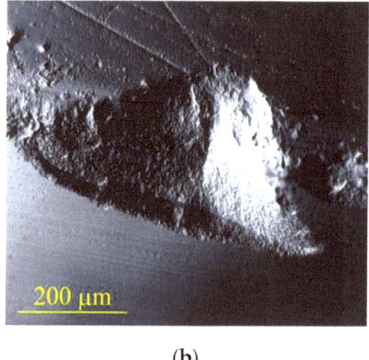

(a) (b)

Figure 4. Schematic diagram of the method used to evaluate the resistance of the CCI cutting edges to chipping: (a) the impact of the destructive load created by the diamond indenter near the cutting edge; (b) SEM-image of the chipping area of the cutting edge.

Tests were carried out on a Calowear CSM Instruments device (Peseux, Switzerland) under pressure on samples with a force of 0.2 N to study the ability of the surface layer of ceramic inserts to resist abrasion under abrasive conditions. The samples were affected by a rotating sphere of hardened steel when an abrasive suspension was supplied into the contact zone. Optical analysis of the wear holes' geometric dimensions and their measurement on a stylus profiler made it possible to quantify and qualitatively assess the volume wear of samples [64]. The optical measurement error was calculated by the formula [65]:

$$\delta_l = \pm 3 + \frac{L}{30} + \frac{g \cdot L}{4000}, \quad (1)$$

$$\delta_t = \pm 3 + \frac{\Delta}{50} + \frac{g \cdot \Delta}{2500}, \quad (2)$$

where δ_l is the longitudinal measurement error, µm; δ_t is the transversal measurement error, µm; Δ is the measured length, mm; g is the product height above microscope table glass (taken equal to zero), mm. The evaluation of the change in the coefficient of friction of ceramic inserts with different surface layer conditions was performed on a TNT-S-AH0000 tribometer (Anton Paar, Corcelles-Cormondrèche, Switzerland) when the ceramic insert rotates relative to a fixed ceramic ball with a diameter of 6 mm at a load of 1 N, a sliding speed of 10 cm/s and a test temperature of 800 °C [66].

The adhesion strength of the formed coatings with ceramic substrates by scratch testing with fixation of the spectrum of acoustic emission signals was evaluated on the NANOVEA M1 device (Irvine, CA, USA). The tests were performed with a linearly increasing load up to 50 N and a loading speed of 5 N/min. The acoustic emission spectra and corresponding forces were recorded during the test. The results of three measurements identified the normal load, which corresponded to the coating delamination moment, by [67–69].

3. Results and Discussion

3.1. Influence of Various Diamond Abrasive Machining Methods on the Condition and Characteristics of the Surface Layer of SiAlON Ceramic Inserts

The results of microstructural SEM-analysis of the surface layer of industrially manufactured CCI from SiAlON, subjected to diamond grinding at the finishing, show that the samples contain numerous surface defects (Figure 5), and the surface has a sophisticated morphological pattern. The surface layer of SiAlON tool ceramics has a specific relief, including a set of defects as a result of the impact of the diamond grains and the friction of the wheel binder on the surface of the high-density ceramics as well as local plastic deformation occurring during high-speed heating of the surface areas of the ceramics together with their rapid cooling. The observed defects (Figure 5) can be classified into the following types: (1) micro ridges, which are the result of plastic deformation, (2) ripped-out single grains and (3) ripped-out conglomerate grains, under force loads, (4) deep grooves, shaped by diamond grains of the grinding wheel, (5) micro-cracks, arising as a result of the intense thermomechanical impact, and (6) grooves, formed because of slipping of "passive" diamond grains over the ceramic surface.

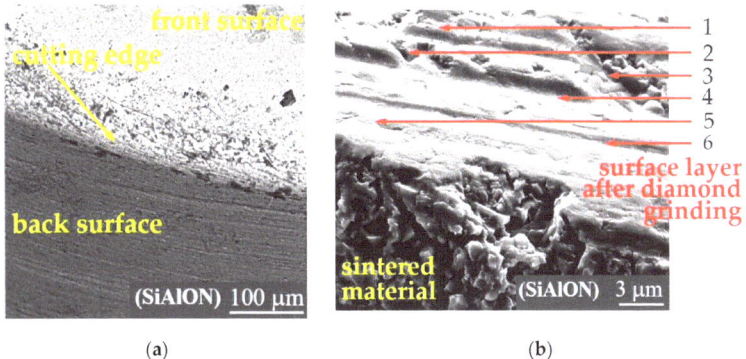

Figure 5. SEM-images of the general view of the cutting part of a round SiAlON ceramic insert (**a**) and the microstructure of the fracture boundary of sintered ceramics and the surface layer after diamond grinding (**b**): 1, micro ridges; 2, ripped-out single grains; 3, ripped-out conglomerate grains; 4, deep grooves; 5, micro cracks; 6, grooves.

The obtained 3D profilograms and SEM images of the microstructure of the surface layer of SiAlON ceramic inserts of the two studied groups are shown in Figure 6. The characteristic defects of the surface layer of industrially produced inserts (group I) with the typical topography [70] are noted in Figure 5 and visible on the profilograms (Figure 6a). The experimental data shown in Figure 6b demonstrate the pronounced positive changes occurring in the surface layer of industrially produced CCI due to additional diamond lapping and polishing (group II). Their use minimizes the defects formed during diamond grinding of the ceramic inserts (the maximum R_t value was 0.37 μm compared to 3.6 μm for group I inserts) and improves the surface quality (the R_a value was 0.014–0.026 μm compared to 0.28–0.3 μm for group I inserts).

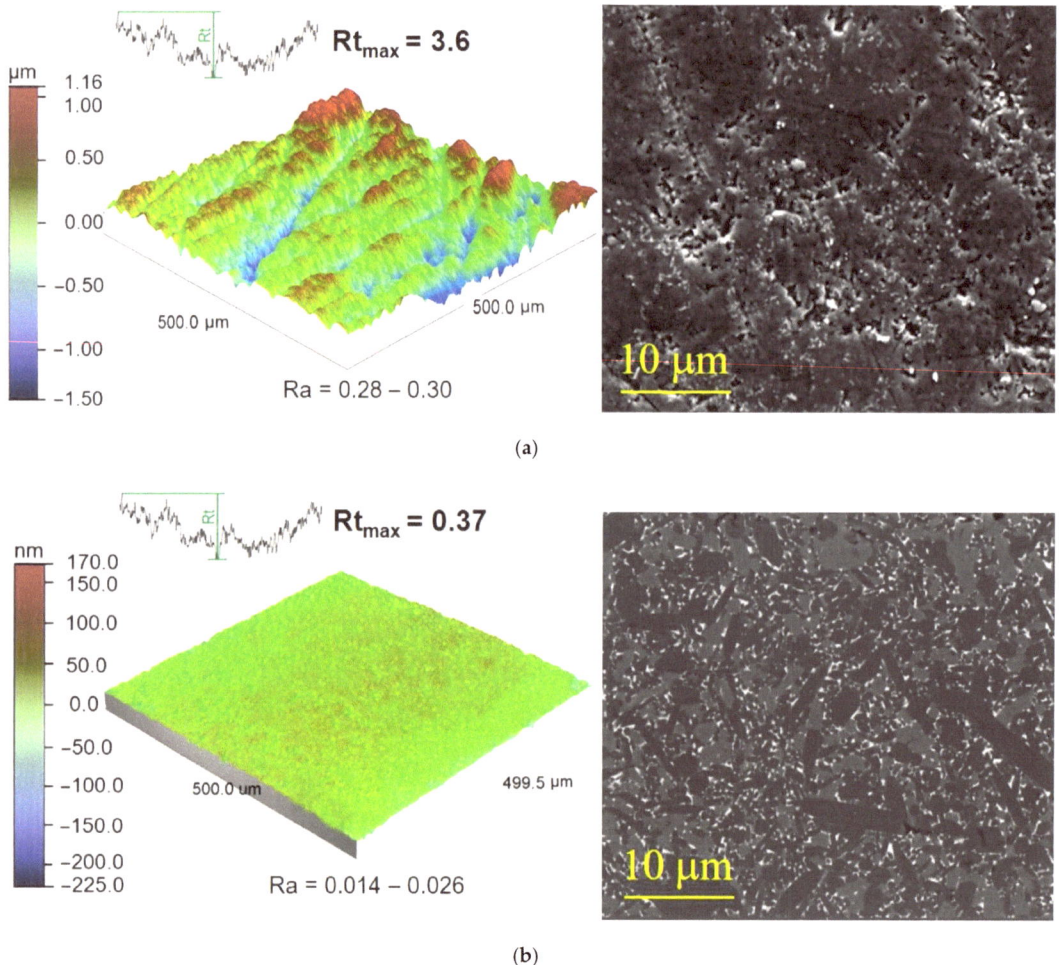

Figure 6. 3D profilograms (left) and SEM images (right) of the microstructure of the surface layer of SiAlON ceramic inserts after diamond grinding (**a**), and after diamond grinding, lapping, and polishing (**b**).

Table 3 summarizes data on the average values of the SiAlON CCI surface layer characteristics after various diamond abrasive machining methods. It can be seen that the average value of R_t after diamond grinding, lapping, and polishing was reduced many times—from 3.1 to 0.29 μm—while the parameter R_a was reduced from 0.29 to 0.019 μm.

Table 3. Characteristics of the SiAlON ceramic inserts' surface layer after various diamond abrasive machining methods.

No.	Diamond Abrasive Machining	Average Values of the Surface Layer Characteristics (According to the Measurement Result of 10 Samples)			
		Crack Resistance K_c, MPa·m$^{1/2}$	Microhardness HV, GPa	Roughness R_a, μm	Defect Layer Depth R_t, μm
1	Diamond grinding (group I)	4.91 ± 0.35	15.99 ± 0.05	0.29 ± 0.025	3.1 ± 0.031
2	Diamond grinding, lapping, and polishing (group II)	5.26 ± 0.27	16.09 ± 0.05	0.019 ± 0.002	0.29 ± 0.032

The measurement results obtained by indentation crack resistance (K_c) of the SiAlON ceramic inserts' surface layer after various diamond abrasive machining methods (Table 3) revealed that the estimated indicator demonstrates a certain dependence on the defectiveness of the surface layer. The average K_c value was increased by 7%—from 4.91 to 5.26 MPa·m$^{1/2}$—due to additional diamond lapping and polishing of the inserts (group II) compared to industrially produced CCI (group I). There was no noticeable effect on the microhardness of the condition of the CCI surface layer (Table 3).

Table 4 shows the experimentally obtained data of measured loads on the indenter corresponding to the chipping of the SiAlON ceramic inserts' cutting edges after various diamond abrasive machining methods (chipping and corresponding load were estimated based on the AE signal spectra). It has been found that additional lapping operations combined with polishing have a significant effect on the average value of the destructive load and the value of the spread of this indicator. For SiAlON CCI, an increase in the destructive load by ~30% was recorded.

Table 4. Loads on the indenter corresponding to the chipping of the SiAlON ceramic inserts' cutting edges after various diamond abrasive machining methods (measuring error of ±1.5 N).

		Diamond Abrasive Machining Methods	
No.	Destructive Load, N	Diamond Grinding (Group I)	Diamond Grinding, Lapping, and Polishing (Group II)
1	Average value	28.9	37.6
2	Max value	33	40
3	Min value	26	36

Figure 7 shows the results of a study of the abrasion resistance of a SiAlON CCI surface layer formed by various diamond machining methods under the abrasive action of a rotating sphere. The presented dependences of the volume of worn material on the test time show that the presence of diamond grinding defects in the SiAlON ceramic inserts' surface layer (I group) significantly reduces the ability of the ceramics to resist abrasive wear. The ceramic inserts' lapping and polishing (group II) provide the lowest defectiveness of the surface layer and a decrease in the volume wear of the ceramic samples by ~1.8 times. It can be assumed that the so-called "edge effect" contributes to the decrease in the intensity of abrasion [71], considering that the microhardness of the SiAlON CCI surface layer of the two studied groups differs insignificantly (Table 3). The contact pads of the samples with a minimum number of defects restrain the development of a wear hole formed from the mechanical and abrasive effects of a rotating sphere. The boundaries of the surface layer containing multiple defects, shown in Figure 5, have a reduced ability to resist microfracture when exposed to an external load, and, as a result, the wear hole grows faster.

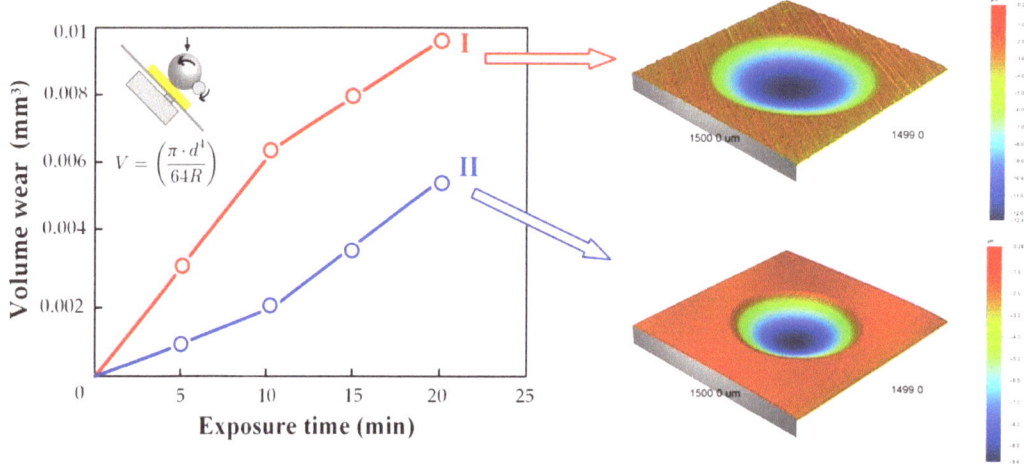

Figure 7. Dependences of the volume wear of SiAlON ceramic inserts with the different conditions of the surface layer after various diamond abrasive machining methods on the time of exposure to a rotating sphere and abrasive particles during testing, where (I) is diamond grinding, (II) is diamond grinding, lapping, and polishing.

Figure 8 shows the dependences of the coefficient of friction (COF) during high-temperature heating of SiAlON CCI after various abrasive machining methods such as diamond grinding (I) and additional lapping and polishing (II). The characteristic curves of the coefficient of friction changing over time demonstrate that minimizing the level of defects in the surface layer of samples (group II) provides an average of ~20% reduction in COF relative to samples with numerous defects (group I). Another feature that attracts attention is that at the stage of the run-in of contact surfaces in group I samples with high defectiveness, COF changes nonmonotonically, which may indicate the intensive adhesive setting of the SiAlON CCI surface layer with a counter body. For the group II samples, the change in COF is uniform throughout the entire test distance, which indicates more favorable conditions for frictional interaction. The observed patterns can be explained by a significant difference in the surface layer roughness achieved by the selected diamond abrasive machining methods.

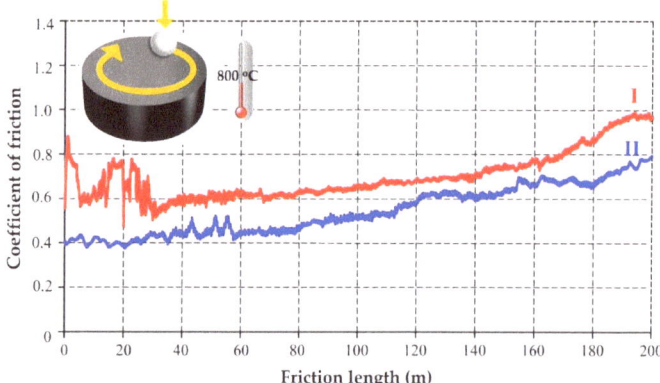

Figure 8. Dependences of the coefficient of friction during high-temperature heating of SiAlON ceramic inserts with different surface layer conditions on the length of the friction path, where (I) is diamond grinding, (II) is diamond grinding, lapping, and polishing.

3.2. The Influence of Various Diamond Abrasive Machining Methods on the Durability of SiAlON Ceramic Inserts in Nickel-Based Superalloy Turning

Figure 9a,b shows groups of "backface wear–cutting time" curves, constructed according to the results of laboratory tests of 10 faces of industrially produced round cutting inserts made of SiAlON when turning ХН45МВТЮБР alloy at the cutting mode mentioned above. Experimental data illustrate (Figure 9a) the disadvantages associated with the CCI abovementioned low operational stability, which limits their industrial use. It can be seen that the curves of wear development over the cutting time for cutting inserts after diamond grinding (with the maximum number of surface defects) have a pronounced fan-shaped character with different wear rates of the cutting faces at the stage of run-in and regular wear. As a result, there is a considerable variation in the resistance until the failure criterion is reached. The authors of this work observed a similar pattern of wear development of ceramic inserts after diamond grinding when studying the turning of hardened steel 100CrMn6 using CCI made of Al_2O_3 + TiC ceramics [28].

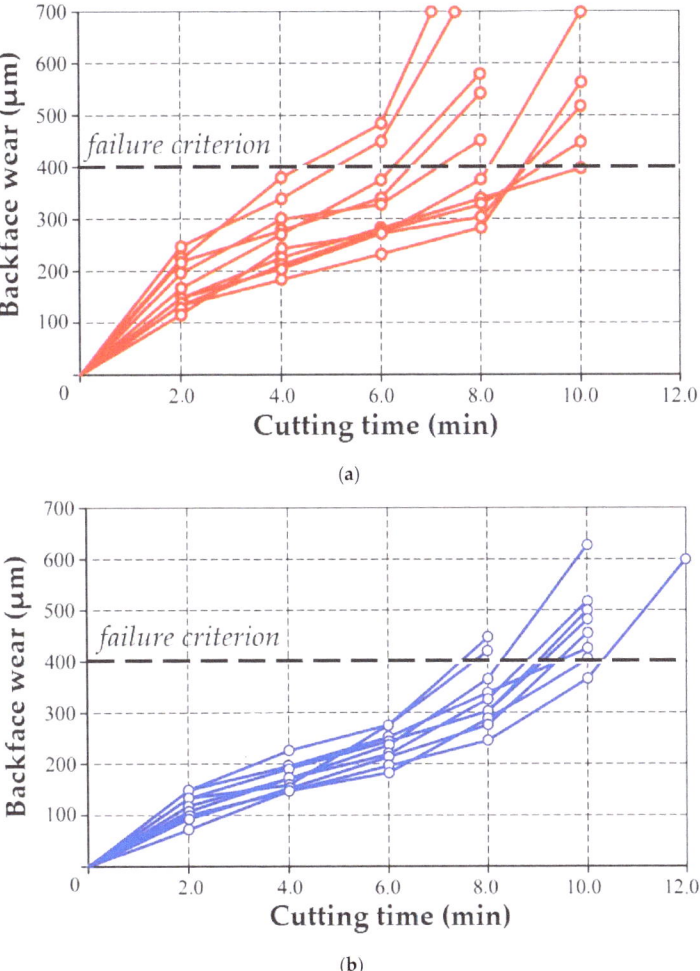

Figure 9. Group of "backface wear–cutting time" curves of the cutting faces of SiAlON ceramic inserts after diamond grinding (**a**), and diamond grinding, lapping, and polishing (**b**) when turning ХН45МВТЮБР heat-resistant nickel alloy (V = 300 m/min, S = 0.2 mm/rev, t = 0.8 mm).

It should be noted once again that the surface layer of industrially produced inserts is replete with numerous defects that are stress concentrators, and when external loads are applied, the wear pattern is difficult to predict. Thus, the destruction of the cutting part can occur at any time (Figure 9a), while it can resemble the tearing out of entire conglomerates (Figure 1, option 3) When minimizing defects, i.e., repeated reduction of stress concentrators, the probability of destruction of the cutting part occurs mainly as a result of the gradual abrasion of the grains of the ceramic (Figure 1, option 1).

The additional lapping and polishing (with a minimum number of surface defects) have a significant impact on the nature of the development of cutting-edge wear over time (Figure 9b): the "backface wear–cutting time" curves have the form of an intertwining bunch of curves with significantly less variation in resistance values, while the ceramic inserts show higher operational stability.

It should be noted that the proposed research approach is unique (there are no other works related to the improving surface layer of SiAlON inserts for deposition of the unique coating structure for turning nickel-based alloys). The standard operating characteristic of a tool, the average tool life used by researchers, is an uninformative indicator for cutting ceramics. It can be seen (Figure 9a) that for ceramic inserts with a defective surface layer, the curves of development of wear over time along the back surface have a pronounced fan-shaped nature of the realizations of a random variable. Such a pattern of wear development over time is difficult to predict, and the operating time, until the accepted failure criterion is reached (in our case, 400 μm), has a large spread of values, which does not provide high operational reliability. Moreover, the approach proposed by the authors to reduce the defectiveness of the surface layer demonstrates a favorable effect on tool reliability, which is a key problem that hinders the use of ceramics in industry.

At the same time, if we compare the average values of wear on the flank surface of SiAlON CCI after diamond grinding (group I) and diamond grinding, lapping, and polishing (group II) on the turning time of the ХН45МВТЮБР heat-resistant nickel alloy (Figure 10), the final difference in the average resistance of inserts with minimum and maximum defectiveness indices (according to the results of tests of 10 faces) is no more than 30% (9 min and 7 min, respectively). Therefore, the main effect of minimizing surface layer defects should be considered to be the reduction in the resistance spread, which provides more stable (predictable) operating conditions.

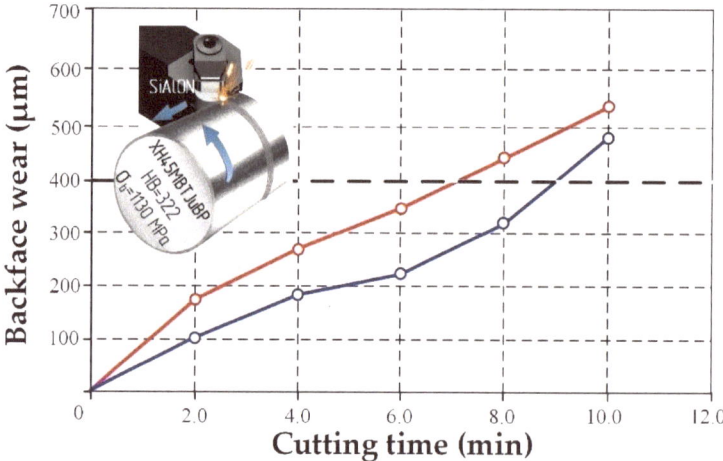

Figure 10. Dependences of the average backface wear value of SiAlON ceramic inserts after diamond grinding (red curve) and diamond grinding, lapping, and polishing (blue curve) on the cutting time when turning ХН45МВТЮБР heat-resistant nickel alloy (V = 300 m/min, S = 0.2 mm/rev, t = 0.8 mm).

3.3. Influence of the Condition of the Surface Layer of SiAlON Ceramic Inserts on the Quality of Formed (CrAlSi)N/DLC Coatings and Their Wear Resistance in Nickel-Based Superalloy Turning

Figure 11 shows an image of the microstructure of a two-layer (CrAlSi)N/DLC coating formed on a SiAlON ceramic sample, the sublayer of which (CrAlSi)N has a columnar structure traditional for vacuum-arc nitride coatings. The outer DLC layer is characterized by an amorphous structure. The measurements showed that the microhardness of the formed outer DLC layer is 28 ± 2 GPa, and the average coefficient of friction during high-temperature heating (up to 800 °C) is ~0.4.

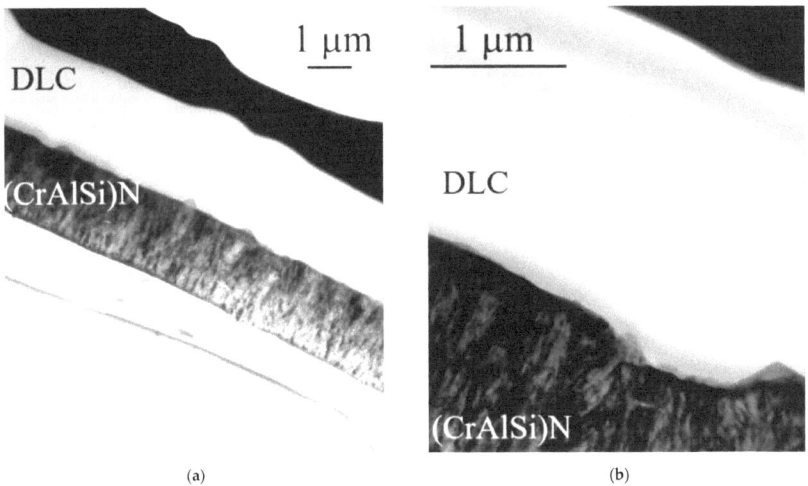

Figure 11. Microstructure of (CrAlSi)N/DLC coating formed on SiAlON ceramic inserts: (**a**) general view; (**b**) layer and sublayer.

Figure 12 shows images of the surface microstructure and profilograms of surface samples with (CrAlSi)N/DLC coatings deposited on SiAlON ceramic inserts with different states of the surface layer (groups I and II). It can be seen that the morphology of the deposited coatings significantly depends on the degree of defectiveness of the SiAlON CCI surface layer [72]. A thin-film coating cannot "heal" numerous defects that abound in industrially produced SiAlON inserts subjected to diamond grinding (the defects are listed in Figure 5) [73,74]. The coating only partly fills in rough surface defects—an experimental assessment of the surface layer of the samples by the R_t parameter made it possible to establish that (CrAlSi)N/DLC coatings can reduce the depth of the defect layer by no more than 35–40%. Comparison of the data in Figure 12b allows us to conclude that the presence of various defects on the ceramic substrate during the deposition of thin-film coatings contributes to the formation of defects in their growth in the form of porosity, discontinuities, and deformation of crystallites (Figure 12a). With a minimum number of defects in the surface layer of ceramic samples, the deposited coatings are characterized only by morphological features characteristic of the PACVD process of DLC layer deposition (Figure 12b). The coatings do not contain visible pores and discontinuities, and the microstructure is represented by rounded crystallites [75,76].

The studies also found significant differences in the strength of the adhesive bond of (CrAlSi)N/DLC coatings depending on the condition (defectiveness) of CCI from SiAlON. It was found that coatings formed on defective substrates subjected to diamond grinding began to peel off at loads of ~33 N, and coatings deposited on samples after lapping and polishing began to peel off at loads of 42–44 N. This can be explained by the high density of micro defects in the SiAlON CCI surface layer leading to the formation of a large amount of porosity near the "coating-substrate" interface and an increase in internal stresses in the coatings [28].

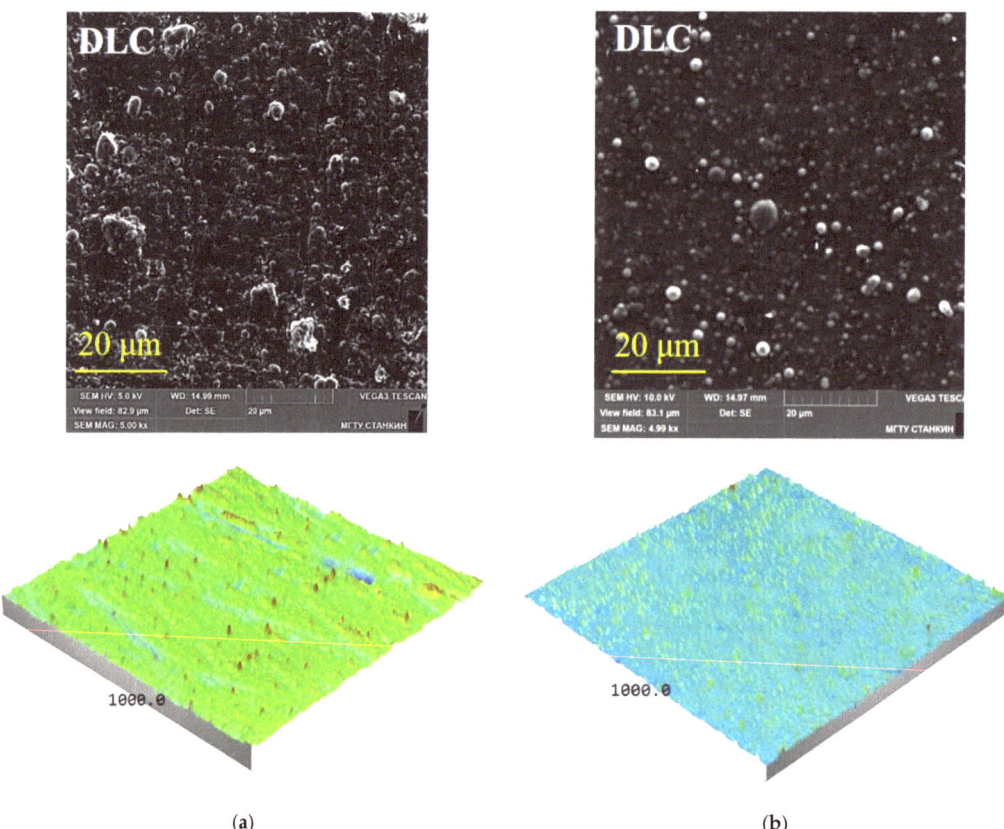

Figure 12. SEM images of the microstructure (top) and 3D profilograms (bottom) of (CrAlSi)N/DLC coatings deposited on SiAlON ceramic inserts with different surface layer conditions: (**a**) after diamond grinding; (**b**) after diamond grinding, lapping, and polishing.

The dependences of the volume of worn material on the test time presented in Figure 13, when compared with the data in Figure 7, allow us to conclude that the (CrAlSi)N/DLC coatings formed on ceramic inserts with different surface layer conditions are highly resistant to abrasion [77]. Compared with industrially produced SiAlON CCI, the formation of (CrAlSi)N/DLC coatings significantly reduces the volume wear of SiAlON ceramics when the surface layer is exposed to a rotating sphere and abrasive particles. A significant effect was found in the case of deposition of (CrAlSi)N/DLC coatings on samples of group I with a large number of defects (reduction of abrasive wear by 2.1 times) and their formation on samples of group II with a minimum number of defects (reduction of abrasive wear by 4.7 times). It can be assumed that the effective resistance to abrasive wear is provided by the high microhardness and low coefficient of friction of the SiAlON CCI contact surfaces achieved after (CrAlSi)N/DLC coating. The maximum effect found for SiAlON samples after diamond grinding, lapping, polishing, and (CrAlSi)N/DLC coating is achieved through a combination of high microhardness, low coefficient of friction, and high adhesion strength of the coatings.

It is necessary to perform cutting resistance tests for a more profound and comprehensive assessment of the influence of the surface layer condition of SiAlON CCI with deposed (CrAlSi)N/DLC coatings on the wear resistance of tool contact pads when interacting with high-temperature nickel superalloys under conditions of increased mechanical and thermal loads.

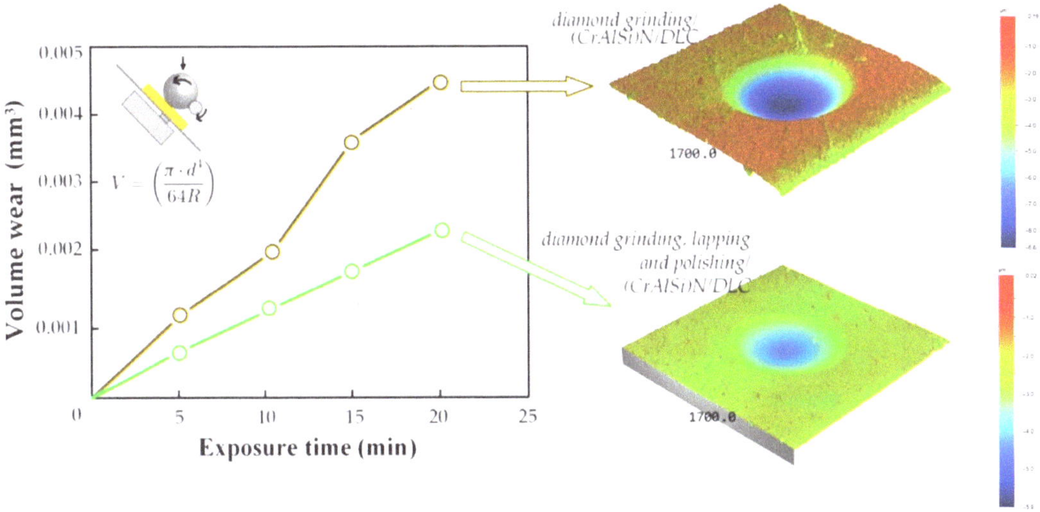

Figure 13. Dependences of the volume wear of SiAlON ceramic inserts with (CrAlSi)N/DLC coatings deposited on samples after various diamond abrasive machining methods on the abrasive exposure time.

Figure 14 shows experimentally obtained dependences of wear on the back surface of SiAlON ceramic inserts during turning of XH45MBTJuBP high-temperature nickel alloy, according to the test results of round CCI of three types:

- SiAlON after diamond grinding (industrially produced CCI), red curve;
- SiAlON after diamond grinding and (CrAlSi)N/DLC coating, brown curve;
- SiAlON after diamond grinding, lapping, polishing, and (CrAlSi)N/DLC coating, green curve.

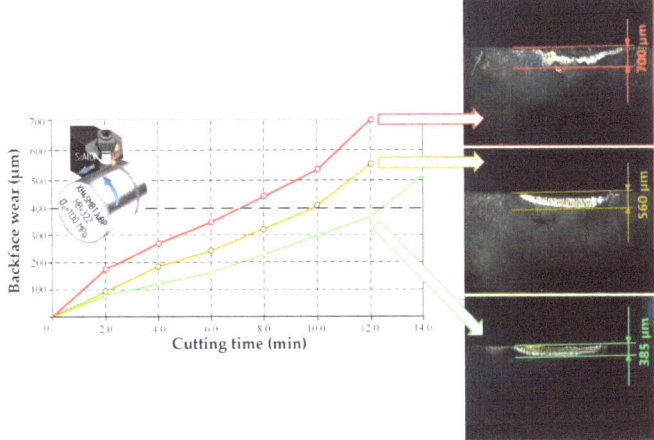

Figure 14. Dependences of the average backface wear of SiAlON ceramic inserts after diamond grinding (red curve), diamond grinding and (CrAlSi)N/DLC coating (brown curve), and diamond grinding, lapping, polishing, and (CrAlSi)N/DLC coating (green curve) on the cutting time when turning XH45MBTJuBP heat-resistant nickel alloy (V = 300 m/min, S = 0.2 mm/rpm, t = 0.8 mm).

The experimental results showed that the samples of SiAlON after diamond grinding, lapping, and polishing with (CrAlSi)N/DLC coating demonstrated the highest resistance to the accepted failure criterion (400 μm). The average durability was 12.5 min. At the same time, it was 9.9 min for the samples with (CrAlSi)N/DLC coatings formed on defective ceramic inserts and only 7 min for uncoated samples.

4. Conclusions

The study aimed to solve the actual scientific and technical problem of high accident brittleness and relatively short operational life of SiAlON cutting ceramics by changing the surface layer condition and double-layer coating of the (CrAlSi)N/DLC structure in the condition of extreme mechanical and thermal loads. The unique complex approach includes improving the surface of the ceramic inserts by additional diamond riding, lapping, and polishing when the tri-nitride sublayer improves the DLC coating adhesion in the conditions of chrome–nickel (Inconel 718 type) alloy machining.

The practical effect of diamond grinding and polishing in combination with the advanced double-layer (CrAlSi)N/DLC coating effect on the performance of industrially produced SiAlON ceramic cutting inserts in a chrome–nickel alloy (Inconel 718 type) turning was proven, and can be recommended for industrial application in production of responsible aircraft gas turbine engine parts.

The following conclusions were drawn:

1. The surface layer of industrially produced SiAlON ceramic inserts after diamond grinding combines numerous defects such as deep grooves, micro-cracks, ripped-out single grains, and conglomerates of grains of the sintered ceramic. The presence of a defective layer significantly reduces the resistance of the ceramic inserts' edges to chipping under external mechanical loads and also reduces the contact pads' ability to resist abrasive wear.
2. The conducted studies allowed us to obtain data proving the strong influence of the condition of the surface layer (presence of defects) of SiAlON ceramic inserts on their operational stability (resistance spread) when turning heat-resistant nickel superalloy under conditions of increased cutting speeds and cross-section of the cut layer. At the same time, using various diamond abrasive machining methods, in particular, additional lapping and polishing, allows for minimization of the defective layer formed during diamond grinding and reduction of the resistance spread.
3. Deposition of thin-film two-layer (CrAlSi)N/DLC coatings on the surface of industrially produced SiAlON ceramic inserts significantly improves the characteristics of tool ceramics, such as the microhardness of the surface layer increasing, and the coefficient of friction decreasing during high-temperature heating while the abrasion resistance also increases. However, (CrAlSi)N/DLC coatings are not able to "heal" numerous defects of the surface layer but can only reduce the depth of the defective layer. The defects in the surface layer of SiAlON ceramic inserts contribute to forming porous and discontinuous coatings with low adhesive bond strength.
4. The maximum effect when turning heat-resistant nickel superalloys under increased cutting speed and cross-section of the cut layer is achieved by using a combined surface treatment method, where lapping and polishing minimize the diamond grinding defects and the subsequent (CrAlSi)N/DLC coating provides an increase in the microhardness of the surface layer and a decrease in the coefficient of friction on the contact surfaces. The developed approach makes it possible to increase the resistance of SiAlON ceramic inserts by a factor of 1.78 compared to industrially produced inserts.

Author Contributions: Conceptualization, S.N.G.; methodology, M.A.V.; software, M.A.V. and A.A.O.; validation, M.A.V.; formal analysis M.A.V.; investigation, M.A.V.; resources, S.N.G.; data curation, A.A.O.; writing—original draft preparation, M.A.V.; writing—review and editing, M.A.V.; visualization, M.A.V. and A.A.O.; supervision, S.N.G.; project administration, M.A.V.; funding acquisition, S.N.G. All authors have read and agreed to the published version of the manuscript.

Funding: This work was supported financially by the Ministry of Science and Higher Education of the Russian Federation (project No. FSFS-2021-0006).

Data Availability Statement: Data are available in a publicly accessible repository.

Acknowledgments: The study was carried out on the equipment of the Center of collective use of MSUT "STANKIN" supported by the Ministry of Higher Education of the Russian Federation (project No. 075-15-2021-695 from 26 July 2021, unique identifier RF 2296.61321X0013).

Conflicts of Interest: The authors declare no conflict of interest.

References

1. Molaiekiya, F.; Stolf, P.; Paiva, J.M.; Bose, B.; Goldsmith, J.; Gey, C.; Engin, S.; Fox-Rabinovich, G.; Veldhuis, S.C. Influence of process parameters on the cutting performance of SiAlON ceramic tools during high-speed dry face milling of hardened Inconel 718. *Int. J. Adv. Manuf. Technol.* **2019**, *105*, 1083–1098. [CrossRef]
2. Zheng, G.; Zhao, J.; Zhou, Y.; Li, A.; Cui, X.; Tian, X. Performance of graded nano-composite ceramic tools in ultra-high-speed milling of Inconel 718. *Int. J. Adv. Manuf. Technol.* **2013**, *67*, 2799–2810. [CrossRef]
3. Grguras, D.; Kern, M.; Pusavec, F. Suitability of the full body ceramic end milling tools for high speed machining of nickel based alloy Inconel 718. *Procedia CIRP* **2018**, *77*, 630–633. [CrossRef]
4. Vereschaka, A.A.; Vereschaka, A.S.; Grigoriev, S.N.; Kirillov, A.K.; Khaustova, O.U. Development and research of environmentally friendly dry technological machining system with compensation of physical function of cutting fluids. *Procedia CIRP* **2013**, *7*, 311–316. [CrossRef]
5. Uhlmann, E.; Hübert, C. Tool grinding of end mill cutting tools made from high performance ceramics and cemented carbides. *CIRP Ann.* **2011**, *60*, 359–362. [CrossRef]
6. Fadok, J. Advanced Gas Turbine Materials, Design and Technology. In *Advanced Power Plant Materials, Design and Technology*; Roddy, D., Ed.; Woodhead Publishing Series in Energy: Cambridge, UK, 2010; pp. 3–31.
7. Bitterlich, B.; Bitsch, S.; Friederich, K. SiAlON based ceramic cutting tools. *J. Eur. Ceram. Soc.* **2008**, *28*, 989–994. [CrossRef]
8. Zheng, G.; Zhao, J.; Gao, Z.; Cao, Q. Cutting performance and wear mechanisms of Sialon–Si3N4 graded nano-composite ceramic cutting tools. *Int. J. Adv. Manuf. Technol.* **2011**, *58*, 19–28. [CrossRef]
9. Seleznev, A.; Pinargote, N.W.S.; Smirnov, A. Ceramic Cutting Materials and Tools Suitable for Machining High-Temperature Nickel-Based Alloys: A Review. *Metals* **2021**, *11*, 1385. [CrossRef]
10. Grigoriev, S.N.; Volosova, M.A.; Fedorov, S.V.; Okunkova, A.A.; Pivkin, P.M.; Peretyagin, P.Y.; Ershov, A. Development of DLC-Coated Solid SiAlON/TiN Ceramic End Mills for Nickel Alloy Machining: Problems and Prospects. *Coatings* **2021**, *11*, 532. [CrossRef]
11. Grigoriev, S.N.; Vereschaka, A.A.; Fyodorov, S.V.; Sitnikov, N.N.; Batako, A.D. Comparative analysis of cutting properties and nature of wear of carbide cutting tools with multi-layered nano-structured and gradient coatings produced by using of various deposition methods. *Int. J. Adv. Manuf. Technol.* **2017**, *90*, 3421–3435. [CrossRef]
12. Smirnov, K.L. Sintering of SiAlON ceramics under high-speed thermal treatment. *Powder Metall. Met. Ceram.* **2012**, *51*, 76–82. [CrossRef]
13. Vereschaka, A.S.; Grigoriev, S.N.; Sotova, E.S.; Vereschaka, A.A. Improving the efficiency of the cutting tools made of mixed ceramics by applying modifying nano-scale multilayered coatings. *Adv. Mat. Res.* **2013**, *712*, 391–394.
14. Sharman, A.R.C.; Hughes, J.J.; Ridgway, K. Workpiece surface integrity and tool life issues when turning Inconel 718 nickel-based superalloy. *Mach. Sci. Technol.* **2004**, *8*, 399–414. [CrossRef]
15. Tian, X.; Zhao, J.; Zhao, J.; Gong, Z.; Dong, Y. Effect of cutting speed on cutting forces and wear in high-speed face milling of Inconel 718 with Sialon ceramic tools. *Int. J. Adv. Manuf. Technol.* **2013**, *69*, 2669–2678. [CrossRef]
16. Kuzin, V.V.; Grigoriev, S.N. Method of Investigation of the Stress-Strain State of Surface Layer of Machine Elements from a Sintered Nonuniform Material. *Appl. Mech. Mater.* **2013**, *486*, 32–35. [CrossRef]
17. Rizzo, A.; Goel, S.; Luisa Grilli, M.; Iglesias, R.; Jaworska, L.; Lapkovskis, V.; Novak, P.; Postolnyi, B.O.; Valerini, D. The Critical Raw Materials in Cutting Tools for Machining Applications: A Review. *Materials* **2020**, *13*, 1377. [CrossRef] [PubMed]
18. Yap, T.C. Roles of Cryogenic Cooling in Turning of Superalloys, Ferrous Metals, and Viscoelastic Polymers. *Technologies* **2019**, *7*, 63. [CrossRef]
19. Vereschaka, A.; Tabakov, V.; Grigoriev, S.; Sitnikov, N.; Milovich, F.; Andreev, N.; Bublikov, J. Investigation of wear mechanisms for the rake face of a cutting tool with a multilayer composite nanostructured Cr–CrN-(Ti,Cr,Al,Si)N coating in high-speed steel turning. *Wear* **2019**, *438*, 203069. [CrossRef]
20. Jiang, C.P.; Wu, X.F.; Li, J.; Song, F.; Shao, Y.F.; Xu, X.H.; Yan, P. A study of the mechanism of formation and numerical simulations of crack patterns in ceramics subjected to thermal shock. *Acta Mater.* **2012**, *60*, 4540–4550. [CrossRef]
21. Vereschaka, A.A.; Grigoriev, S.N.; Volosova, M.A.; Batako, A.; Vereschaka, A.S.; Sitnikov, N.N.; Seleznev, A.E. Nano-scale multi-layered coatings for improved efficiency of ceramic cutting tools. *Int. J. Adv. Manuf. Technol.* **2017**, *90*, 27–43. [CrossRef]

22. Kuzin, V.V.; Grigor'ev, S.N.; Volosova, M.A. Microstructural Model of the Surface Layer of Ceramics After Diamond Grinding Taking into Account Its Real Structure and the Conditions of Contact Interaction with Elastic Body. *Refract. Ind. Ceram.* **2020**, *61*, 303–308. [CrossRef]
23. Vigneau, J.; Bordel, P.; Geslot, R. Reliability of eramic cutting tools. *CIRP Ann.* **1988**, *37*, 101–104. [CrossRef]
24. Wachtman, J.B.; Cannon, W.R.; Matthewson, M.J. *Mechanical Properties of Ceramics*, 2nd ed.; Wiley: Hoboken, NJ, USA, 2009; p. 479. ISBN 978-0-471-73581-6.
25. Bensouilah, H.; Aouici, H.; Meddour, I.; Yallese, M.A.; Mabrouki, T.; Girardin, F. Performance of coated and uncoated mixed ceramic tools in hard turning process. *Measurement* **2016**, *82*, 1–18. [CrossRef]
26. Aslantas, K.; Ucun, İ.; Çicek, A. Tool life and wear mechanism of coated and uncoated Al2O3/TiCN mixed ceramic tools in turning hardened alloy steel. *Wear* **2012**, *274*, 442–451. [CrossRef]
27. Matthew, B.; Sam, T.; Keith, R. Correlation between tool life and cutting force coefficient as the basis for a novel method in accelerated MWF performance assessment. *Procedia CIRP* **2021**, *101*, 366–369. [CrossRef]
28. Volosova, M.A.; Stebulyanin, M.M.; Gurin, V.D.; Melnik, Y.A. Influence of Surface Layer Condition of Al_2O_3+TiC Ceramic Inserts on Quality of Deposited Coatings and Reliability during Hardened Steel Milling. *Coatings* **2022**, *12*, 1801. [CrossRef]
29. Vereschaka, A.; Tabakov, V.; Grigoriev, S.; Sitnikov, N.; Milovich, F.; Andreev, N.; Sotova, C.; Kutina, N. Investigation of the influence of the thickness of nanolayers in wear-resistant layers of Ti-TiN-(Ti,Cr,Al)N coating on destruction in the cutting and wear of carbide cutting tools. *Surf. Coat. Technol.* **2020**, *385*, 125402. [CrossRef]
30. Liu, W.; Chu, Q.; Zeng, J.; He, R.; Wu, H.; Wu, Z.; Wu, S. PVD-CrAlN and TiAlN coated Si_3N_4 ceramic cutting tools—1. Microstructure, turning performance and wear mechanism. *Ceram. Int.* **2017**, *43*, 8999–9004. [CrossRef]
31. Long, Y.; Zeng, J.; Wu, S. Cutting performance and wear mechanism of Ti–Al–N/Al–Cr–O coated silicon nitride ceramic cutting inserts. *Ceram. Int.* **2014**, *40*, 9615–9620. [CrossRef]
32. Vereschaka, A.; Grigoriev, S.; Tabakov, V.; Migranov, M.; Sitnikov, N.; Milovich, F.; Andreev, N. Influence of the nanostructure of Ti-TiN-(Ti,Al,Cr)N multilayer composite coating on tribological properties and cutting tool life. *Tribol. Int.* **2020**, *150*, 106388. [CrossRef]
33. Grigoriev, S.; Vereschaka, A.; Milovich, F.; Tabakov, V.; Sitnikov, N.; Andreev, N.; Sviridova, T.; Bublikov, J. Investigation of multicomponent nanolayer coatings based on nitrides of Cr, Mo, Zr, Nb, and Al. *Surf. Coat. Technol.* **2020**, *401*, 126258. [CrossRef]
34. Arai, S.A.; Wilson, S.A.; Corbett, J.; Whatmore, R.W. Ultra-precision grinding of PZT ceramics—Surface integrity control and tooling design. *Int. J. Mach. Tools Manuf.* **2009**, *49*, 998–1007. [CrossRef]
35. Canneto, J.J.; Cattani-Lorente, M.; Durual, S.; Wiskott, A.H.W.; Scherrer, S.S. Grinding damage assessment on four high-strength ceramics. *Dent. Mater.* **2016**, *32*, 171–182. [CrossRef] [PubMed]
36. Zhang, C.; Liu, H.; Zhao, Q.; Guo, B.; Wang, J.; Zhang, J. Mechanisms of ductile mode machining for AlON ceramics. *Ceram. Int.* **2020**, *46*, 1844–1853. [CrossRef]
37. Available online: https://www.ctemag.com/news/articles/grinding-ceramic-medical-parts-requires-diamond-grit-patience (accessed on 11 January 2023).
38. Grigoriev, S.N.; Volosova, M.A.; Vereschaka, A.A.; Sitnikov, N.N.; Milovich, F.; Bublikov, J.I.; Fyodorov, S.V.; Seleznev, A.E. Properties of (Cr,Al,Si)N-(DLC-Si) composite coatings deposited on a cutting ceramic substrate. *Ceram. Int.* **2020**, *46*, 18241–18255. [CrossRef]
39. Vopát, T.; Sahul, M.; Haršáni, M.; Vortel, O.; Zlámal, T. The Tool Life and Coating-Substrate Adhesion of AlCrSiN-Coated Carbide Cutting Tools Prepared by LARC with Respect to the Edge Preparation and Surface Finishing. *Micromachines* **2020**, *11*, 166. [CrossRef]
40. Bobzin, K.; Brögelmann, T.; Kruppe, N.C.; Carlet, M. Nanocomposite (Ti,Al,Cr,Si)N HPPMS coatings for high performance cutting tools. *Surf. Coat. Technol.* **2019**, *378*, 124857. [CrossRef]
41. Grigoriev, S.; Volosova, M.; Fedorov, S.; Mosyanov, M. Influence of DLC Coatings Deposited by PECVD Technology on the Wear Resistance of Carbide End Mills and Surface Roughness of AlCuMg2 and 41Cr4 Workpieces. *Coatings* **2020**, *10*, 1038. [CrossRef]
42. Wei, C.; Yang, J.-F. A finite element analysis of the effects of residual stress, substrate roughness and non-uniform stress distribution on the mechanical properties of diamond-like carbon films. *Diam. Relat. Mater.* **2011**, *20*, 839–844. [CrossRef]
43. Lubwama, M.; Corcoran, B.; McDonnell, K.A.; Dowling, D. Flexibility and frictional behaviour of DLC and Si-DLC films deposited on nitrile rubber. *Surf. Coat. Technol.* **2014**, *239*, 84–94. [CrossRef]
44. Hainsworth, S.V.; Uhure, N.J. Diamond like carbon coatings for tribology: Production techniques, characterisation methods and applications. *Int. Mat. Rev.* **2007**, *52*, 153–174. [CrossRef]
45. Nakazawa, H.; Kamata, R.; Miura, S.; Okuno, S. Effects of frequency of pulsed substrate bias on structure and properties of silicon-doped diamond-like carbon films by plasma deposition. *Thin Solid Film.* **2015**, *574*, 93–98. [CrossRef]
46. Martinez-Martinez, D.; De Hosson, J.T.M. On the deposition and properties of DLC protective coatings on elastomers: A critical review. *Surf. Coat. Technol.* **2014**, *258*, 677–690. [CrossRef]
47. Liu, X.Q.; Yang, J.; Hao, J.Y.; Zheng, J.Y.; Gong, Q.Y.; Liu, W.M. A near-frictionless and extremely elastic hydrogenated amorphous carbon film with self-assembled dual nanostructure. *Adv. Mater.* **2012**, *24*, 4614–4617. [CrossRef] [PubMed]
48. Zou, C.W.; Wang, H.J.; Feng, L.; Xue, S.W. Effects of Cr concentrations on the microstructure, hardness, and temperature-dependent tribological properties of Cr-DLC coatings. *Appl. Surf. Sci.* **2013**, *286*, 137–141. [CrossRef]

49. Wu, Y.; Li, H.; Ji, L.; Ye, Y.; Chen, J.; Zhou, H. Vacuum tribological properties of a-C:H film in relation to internal stress and applied load. *Tribol. Int.* **2014**, *71*, 82–87. [CrossRef]
50. Zhang, T.F.; Pu, J.J.; Xia, Q.X.; Son, M.J.; Kim, K.H. Microstructure and nano-wear property of Si-doped diamond like carbon films deposited by a hybrid sputtering system. *Mater. Today Proc.* **2016**, *3*, S190–S196. [CrossRef]
51. Grigoriev, S.N.; Melnik, Y.A.; Metel, A.S.; Panin, V.V.; Prudnikov, V.V. A compact vapor source of conductive target material sputtered by 3-keV ions at 0.05-Pa pressure. *Instrum. Exp. Technol.* **2009**, *52*, 731–737. [CrossRef]
52. Metel, A.; Bolbukov, V.; Volosova, M.; Grigoriev, S.; Melnik, Y. Equipment for deposition of thin metallic films bombarded by fast argon atoms. *Instrum. Exp. Technol.* **2014**, *57*, 345–351. [CrossRef]
53. Sobol, O.V.; Andreev, A.A.; Grigoriev, S.N.; Volosova, M.A.; Gorban, V.F. Vacuum-arc multilayer nanostructured TiN/Ti coatings: Structure, stress state, properties. *Met. Sci. Heat Treat.* **2012**, *54*, 28–33. [CrossRef]
54. Sobol, O.V.; Andreev, A.A.; Grigoriev, S.N.; Gorban, V.F.; Volosova, M.A.; Aleshin, S.V.; Stolbovoi, V.A. Effect of high-voltage pulses on the structure and properties of titanium nitride vacuum-arc coatings. *Met. Sci. Heat Treat.* **2012**, *54*, 195–203. [CrossRef]
55. Metel, A.S.; Grigoriev, S.N.; Melnik, Y.A.; Bolbukov, V.P. Characteristics of a fast neutral atom source with electrons injected into the source through its emissive grid from the vacuum chamber. *Instrum. Exp. Technol.* **2012**, *55*, 288–293. [CrossRef]
56. Mei, Z.; Lu, Y.; Lou, Y.; Yu, P.; Sun, M.; Tan, X.; Zhang, J.; Yue, L.; Yu, H. Determination of Hardness and Fracture Toughness of Y-TZP Manufactured by Digital Light Processing through the Indentation Technique. *Biomed Res. Int.* **2021**, *2021*, 11.
57. Gogotsi, G.A. Edge chipping resistance of ceramics: Problems of test method. *J. Adv. Ceram.* **2013**, *2*, 370–377. [CrossRef]
58. Gogotsi, G.A.; Galenko, V.Y. Sensitivity of Brittle Materials to Local Stress Concentrations on Their Fracture. *Strength Mater.* **2022**, *54*, 250–255. [CrossRef]
59. Feng, P.; Borghesani, P.; Smith, W.A.; Randall, R.B.; Peng, Z. A Review on the Relationships Between Acoustic Emission, Friction and Wear in Mechanical Systems. *Appl. Mech. Rev.* **2020**, *72*, 020801. [CrossRef]
60. Babici, L.M.; Tudor, A.; Romeu, J. Stick-Slip Phenomena and Acoustic Emission in the Hertzian Linear Contact. *Appl. Sci.* **2022**, *12*, 9527. [CrossRef]
61. Hase, A. Early Detection and Identification of Fatigue Damage in Thrust Ball Bearings by an Acoustic Emission Technique. *Lubricants* **2020**, *8*, 37. [CrossRef]
62. Hase, A.; Mishina, H.; Wada, M. Correlation between Features of Acoustic Emission Signals and Mechanical Wear Mechanisms. *Wear* **2012**, *292*, 144–150. [CrossRef]
63. Xu, C.; Li, B.; Wu, T. Wear Characterization under Sliding–Rolling Contact Using Friction-Induced Vibration Features. *Proc. Inst. Mech. Eng. Part J. Eng. Tribol.* **2022**, *236*, 634–647. [CrossRef]
64. Steier, V.F.; Ashiuchi, E.S.; Reißig, L.; Araújo, J.A. Effect of a Deep Cryogenic Treatment on Wear and Microstructure of a 6101 Aluminum Alloy. *Adv. Mater. Sci. Eng.* **2016**, *2016*, 12. [CrossRef]
65. Shulepov, I.A.; Kashkarov, E.B.; Stepanov, I.B.; Syrtanov, M.S.; Sutygina, A.N.; Shanenkov, I.; Obrosov, A.; Weiß, S. The Formation of Composite Ti-Al-N Coatings Using Filtered Vacuum Arc Deposition with Separate Cathodes. *Metals* **2017**, *7*, 497. [CrossRef]
66. Grigoriev, S.N.; Teleshevskii, V.I. Measurement Problems in Technological Shaping Processes. *Meas. Tech.* **2011**, *54*, 744–749. [CrossRef]
67. Choudhary, R.K.; Mishra, P. Use of Acoustic Emission During Scratch Testing for Understanding Adhesion Behavior of Aluminum Nitride Coatings. *J. Mater. Eng Perform* **2016**, *25*, 2454–2461. [CrossRef]
68. Vereschaka, A.A.; Volosova, M.A.; Grigoriev, S.N.; Vereschaka, A.S. Development of wear-resistant complex for high-speed steel tool when using process of combined cathodic vacuum arc deposition. *Procedia CIRP* **2013**, *9*, 8–12. [CrossRef]
69. Kazlauskas, D.; Jankauskas, V.; Tučkutė, S. Research on Tribological Characteristics of Hard Metal WC-Co Tools with TiAlN and CrN PVD Coatings for Processing Solid Oak Wood. *Coatings* **2020**, *10*, 632. [CrossRef]
70. Shao, L.; Zhou, Y.; Fang, W.; Wang, J.; Wang, X.; Deng, Q.; Lyu, B. Preparation of Cemented Carbide Insert Cutting Edge by Flexible Fiber-Assisted Shear Thickening Polishing Method. *Micromachines* **2022**, *13*, 1631. [CrossRef]
71. Volosova, M.; Grigoriev, S.; Metel, A.; Shein, A. The Role of Thin-Film Vacuum-Plasma Coatings and Their Influence on the Efficiency of Ceramic Cutting Inserts. *Coatings* **2018**, *8*, 287. [CrossRef]
72. Kiryukhantsev-Korneev, P.; Sytchenko, A.; Sheveyko, A.; Moskovskikh, D.; Vorotylo, S. Two-Layer Nanocomposite TiC-Based Coatings Produced by a Combination of Pulsed Cathodic Arc Evaporation and Vacuum Electro-Spark Alloying. *Materials* **2020**, *13*, 547. [CrossRef]
73. Mpilitos, C.; Amanatiadis, S.; Apostolidis, G.; Zygiridis, T.; Kantartzis, N.; Karagiannis, G. Development of a Transmission Line Model for the Thickness Prediction of Thin Films via the Infrared Interference Method. *Technologies* **2018**, *6*, 122. [CrossRef]
74. Volosova, M.A.; Fyodorov, S.V.; Opleshin, S.; Mosyanov, M. Wear Resistance and Titanium Adhesion of Cathodic Arc Deposited Multi-Component Coatings for Carbide End Mills at the Trochoidal Milling of Titanium Alloy. *Technologies* **2020**, *8*, 38. [CrossRef]
75. Grigoriev, S.; Vereschaka, A.; Zelenkov, V.; Sitnikov, N.; Bublikov, J.; Milovich, F.; Andreev, N.; Mustafaev, E. Specific features of the structure and properties of arc-PVD coatings depending on the spatial arrangement of the sample in the chamber. *Vacuum* **2022**, *200*, 111047. [CrossRef]

76. Grigoriev, S.; Vereschaka, A.; Zelenkov, V.; Sitnikov, N.; Bublikov, J.; Milovich, F.; Andreev, N.; Sotova, C. Investigation of the influence of the features of the deposition process on the structural features of microparticles in PVD coatings. *Vacuum* **2022**, *202*, 111144. [CrossRef]
77. Grigoriev, S.; Volosova, M.; Fyodorov, S.; Lyakhovetskiy, M.; Seleznev, A. DLC-coating Application to Improve the Durability of Ceramic Tools. *J. Mater. Eng Perform* **2019**, *28*, 4415–4426. [CrossRef]

Disclaimer/Publisher's Note: The statements, opinions and data contained in all publications are solely those of the individual author(s) and contributor(s) and not of MDPI and/or the editor(s). MDPI and/or the editor(s) disclaim responsibility for any injury to people or property resulting from any ideas, methods, instructions or products referred to in the content.

Communication

Comparison of Shallow (−20 °C) and Deep Cryogenic Treatment (−196 °C) to Enhance the Properties of a Mg/2wt.%CeO₂ Nanocomposite

Shwetabh Gupta [†], Gururaj Parande and Manoj Gupta *

Department of Mechanical Engineering, National University of Singapore, 9 Engineering Drive 1, Singapore 117575, Singapore
* Correspondence: mpegm@nus.edu.sg; Tel.: +65-6516-6358
† Current address: Blackett Laboratory, Imperial College, London SW7 2AZ, UK.

Abstract: Magnesium and its composites have been used in various applications owing to their high specific strength properties and low density. However, the application is limited to room-temperature conditions owing to the lack of research available on the ability of magnesium alloys to perform in sub-zero conditions. The present study attempted, for the first time, the effects of two cryogenic temperatures (−20 °C/253 K and −196 °C/77 K) on the physical, thermal, and mechanical properties of a Mg/2wt.%CeO₂ nanocomposite. The materials were synthesized using the disintegrated melt deposition method followed by hot extrusion. The results revealed that the shallow cryogenically treated (refrigerated at −20 °C) samples display a reduction in porosity, lower ignition resistance, similar microhardness, compressive yield, and ultimate strength and failure strain when compared to deep cryogenically treated samples in liquid nitrogen at −196 °C. Although deep cryogenically treated samples showed an overall edge, the extent of the increase in properties may not be justified, as samples exposed at −20 °C display very similar mechanical properties, thus reducing the overall cost of the cryogenic process. The results were compared with the data available in the open literature, and the mechanisms behind the improvement of the properties were evaluated.

Keywords: magnesium; nanocomposite; cryogenic treatment; mechanical properties; grain size

1. Introduction

Cryogenic treatment for metals has been established for almost 300 years primarily for enhancing the resistance to wear and localized indentation in steels [1–5]. In most of the studies, liquid nitrogen is used as a cryogenic medium corresponding to a temperature of −196 °C. Another commonly used medium is dry ice (−84 °C/189 K) [5]. There has been no systematic research attempt made, and there is no such attempt available in the open literature that uses comparatively higher temperatures in the cryogenic domain to investigate the properties of materials.

In the context of the cryogenic treatment of metallic materials, researchers have investigated materials based on iron, aluminum, and magnesium [1–3]. Among these materials, magnesium-based materials are gaining prominence due to their lightweight (~33% lighter than aluminum-based materials) and nontoxic and nutritional characteristics [6–10]. These properties are currently sought after to mitigate global warming and the toxification of land and water bodies. Among magnesium-based materials, magnesium nanocomposites have emerged as highly promising materials. Numerous studies have demonstrated that incorporating nanoparticles into magnesium and its alloys can significantly enhance its thermal properties, static and dynamic responses, and machining and wear properties [2,8,11–18]. The literature review indicates that no prior work has so far been conducted to explore the effects of cryogenic treatment on the response of magnesium nanocomposites. Furthermore, no work has been completed to compare the effects of a shallow cryogenic treatment

(−20 °C in freezer) and a deep cryogenic treatment (−196 °C in liquid nitrogen) on the physical, thermal, and mechanical responses of magnesium nanocomposites.

Accordingly, the present study aims to address these research gaps by investigating the above for an Mg/2wt.%CeO$_2$ nanocomposite, highlighting the capability of shallow cryogenic treatment (−20 °C) as a cost-effective way to improve multiple properties of the magnesium nanocomposite.

2. Materials and Methods

2.1. Materials Processing

Raw materials such as monolithic magnesium turnings (99.9% purity; supplier: ACROS Organics, Waltham, MA, USA) and cerium oxide (CeO$_2$) nanoparticles (size 15–30 nm; supplier: Alfa Aesar, Ward Hill, MA, USA) were utilized to produce the composite through the disintegrated melt deposition (DMD) method [19]. The raw materials were arranged in a multilayer arrangement and heated to a superheating temperature of 750 °C. The materials were arranged in a graphite crucible, and argon was used as the protective gas. To ensure the homogenization of the temperature and a uniform distribution, the molten magnesium composite melt was stirred for 150 s using a mild steel impeller. Post stirring, the melt was bottom poured into a steel mold to obtain a solid ingot. The ingots were machined, soaked at 400 °C for 1 h, and hot extruded at 350 °C using an extrusion ratio of 20.25:1 to obtain cylindrical rods of 8 mm diameter. The sub-zero immersion time was chosen based on the recommendations made in the open literature [1,3]. The temperature used for the shallow cryogenic treatment (RF) was −20 °C, and the temperature for the deep cryogenic treatment (LN) was −196 °C on the extruded samples for 24 h, respectively.

2.2. Characterization

2.2.1. Density and Porosity

The experimental densities were calculated using Archimedes' principle. Five samples were measured using an A&D GH-252 electronic balance with a standard deviation of ±1 mg. The theoretical densities of Mg (1.738 g·cm^{-3}) and CeO$_2$ (7.132 g·cm^{-3}), respectively, were used for the theoretical density calculation of the nanocomposite using the rule of mixtures. Porosity values were computed by comparing the experimental and theoretical densities.

2.2.2. Microstructure

The grain size of the samples was analyzed using a JEOL JSM-6010 Scanning Electron Microscope (SEM) as per ASTM E112-13. The samples were prepared by grinding, polishing, and etching before SEM observation. Oxalic acid was used as the etchant.

X-ray diffraction (XRD) studies along the longitudinal direction were performed using the Shimadzu LAB-XRD-6000 automated spectrometer with Cu K$_α$ radiation of 1.54 Å wavelength and a scan speed of 2 °min^{-1}.

2.2.3. Thermal Properties

A Shimadzu DSC-60 differential scanning calorimeter (DSC) was used to analyze the effect of the cryogenic treatments on the thermal response of the samples. An argon gas flow rate of 25 mL·min^{-1} with a heating rate of 5 °C min^{-1} and a temperature range of 30–600 °C was used.

Thermogravimetric analysis (TGA) was performed to ascertain the ignition temperatures of the samples. Purified air with a flow rate of 50 mL·min^{-1} and a heating rate of 10 °C min^{-1} within a temperature range of 30–1000 °C were used.

2.2.4. Mechanical Properties

The microhardness was measured on the cryogenically treated samples using a Shimadzu-HMV automatic digital microhardness tester with a Vickers indenter as per

ASTM standard E384-08. An indentation load of 245.2 mN with a dwell time of 15 s was used, and a minimum of 20 readings per sample were taken.

Quasi-static room-temperature compression testing was carried out using an MTS E44 fully automated servohydraulic mechanical testing machine at a strain rate set at 5×10^{-3} min^{-1}. A minimum of three samples with a length-to-diameter (L/D) ratio of 1 were tested.

A JEOL JSM-6010 Scanning Electron Microscope was used to analyze the post-fracture behavior.

3. Results and Discussion

3.1. Density and Porosity Measurements

The results of the density measurements and porosity are summarized in Table 1. The density increased and porosity reduced after both cryogenic treatments. The percentage reduction was ~10.4% after the cold treatment (RF samples) and ~44% in the case of the deep cryogenic treatment (LN samples).

Table 1. The density and porosity measurements of Mg/2wt.%CeO$_2$ nanocomposite before and after cryogenic treatments.

Material	Theoretical Density (g·cm^{-3})	Before CT		After CT		Change in Porosity (%)
		Experimental Density (g·cm^{-3})	Porosity (%)	Experimental Density (g·cm^{-3})	Porosity (%)	
Pure Mg [a]	1.7380	1.732 ± 0.0005	0.3190	–	–	
Mg-2CeO$_2$ (AE)	1.7648	1.745 ± 0.002	1.099	–	–	
Mg-2CeO$_2$ (RF)	1.7648	1.7454 ± 0.008	1.102	1.7474 ± 0.001	0.9875	↓10.4
Mg-2CeO$_2$ (LN)	1.7648	1.7476 ± 0.0009	0.9764	1.755 ± 0.002	0.5445	↓43.3

[a] Values generated in the laboratory using similar raw materials and processing methods [13]. Note: AE—As Extruded; AE + RF—As Extruded + Shallow Cryogenic Treatment; AE + LN—As Extruded + Liquid Nitrogen Treatment. Note: '↓' represents the decrease in porosity before and after cryogenic treatment.

The reduction in porosity can be attributed to the compressive stresses generated during the cryogenic treatments with lower temperatures, yielding more remarkable effects [1–3]. The reduction in porosity in both cases also suggests that, under both cryogenic treatments, the material is capable of deforming inward into free space provided by pores. Furthermore, the reduction in porosity can also be attributed to the ability of the pores to serve as sinks for the dislocations generated during the cryogenic treatments [1,20].

3.2. Microstructure

Two aspects of the microstructure were characterized: the grain size and texture. The results of the grain size measurements are shown in Table 2. The grain size analysis of the samples is shown in Figure 1. The average grain size increased (up to 45%) for both types of cryogenic treatments. The increase in grain size, while unexpected, can be attributed to the simultaneous effects of (a) the capabilities of grains to orient themselves during cryogenic treatments [21], (b) the influence of compressive stresses to create order at the grain boundary region, leading to the merger of small grains with big grains (Figure 2), and (c) the capability of the defects to migrate to the grain boundaries [1]. The results also revealed that the average aspect ratio decreased with a decrease in the cryogenic temperature from $-20\,°C$ to $-196\,°C$.

Table 2. The grain size measurements of the samples.

Composition	Grain Size (µm)	Aspect Ratio
Pure Mg	21 ± 0.8	1.4 ± 0.2
Mg-2CeO$_2$ (AE)	2 ± 0.6	1.4 ± 0.3
Mg-2CeO$_2$ (RF)	2.9 ± 1.0	1.3 ± 0.2
Mg-2CeO$_2$ (LN)	2.8 ± 0.6	1.2 ± 0.3

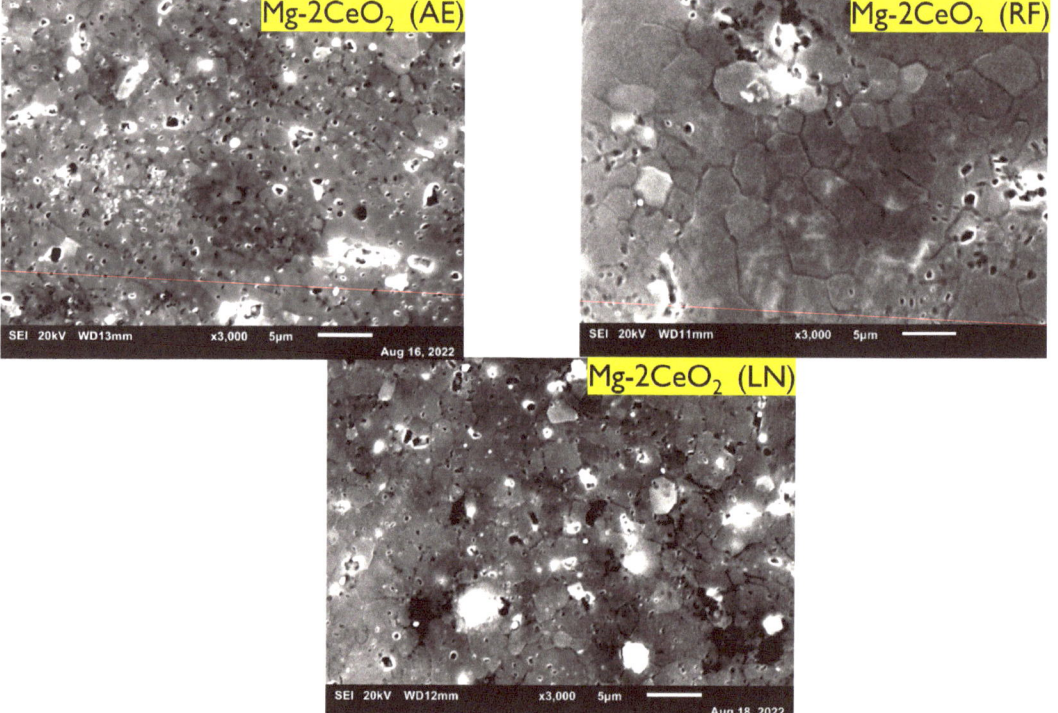

Figure 1. Microstructural characterization of the samples.

The results of the XRD studies indicated the dominance of the basal texture in the as-extruded (AE), cold-treated (RF), and DCT (LN) samples (Figure 3). However, the relative intensities (I/I_{max}) of the LN samples were higher compared to the RF samples, indicating that the RF samples have a stronger fiber texture when compared to the LN samples. The results thus suggest that variation in the cryogenic temperature leads to a variation in the degree of the microstructural evolution, which will have varying effects on the properties, provided that such a difference is substantial.

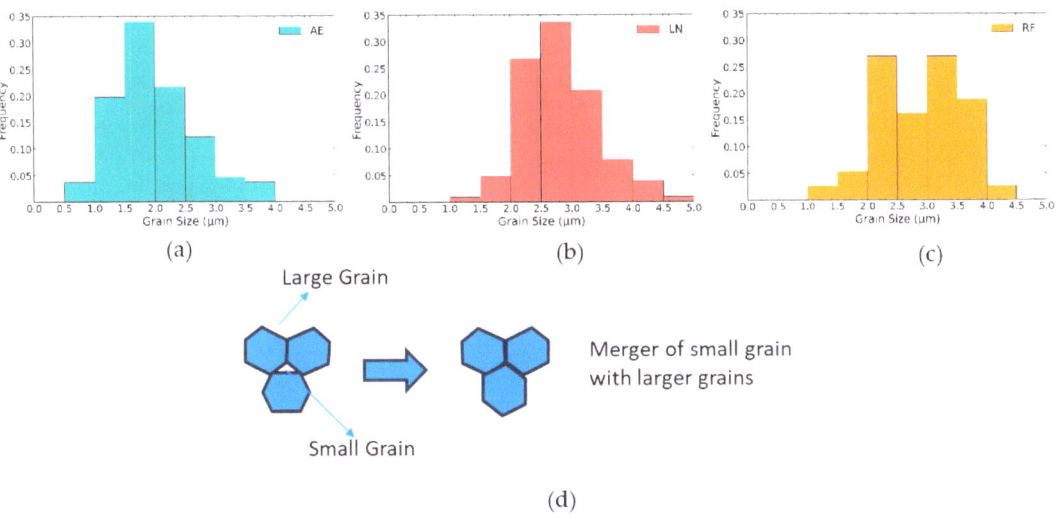

Figure 2. (**a**–**c**) Histograms of the frequency distributions of the various grain sizes within the Mg/2%wt. CeO$_2$ samples. It can be observed that the cryogenic treatment results in the frequency of grains smaller than 2 μm being significantly decreased along with an increase in the frequency of larger grains, indicating the merger of smaller grains and larger grains during the cryogenic treatments. (**d**) The proposed mechanism of the grain merger is shown.

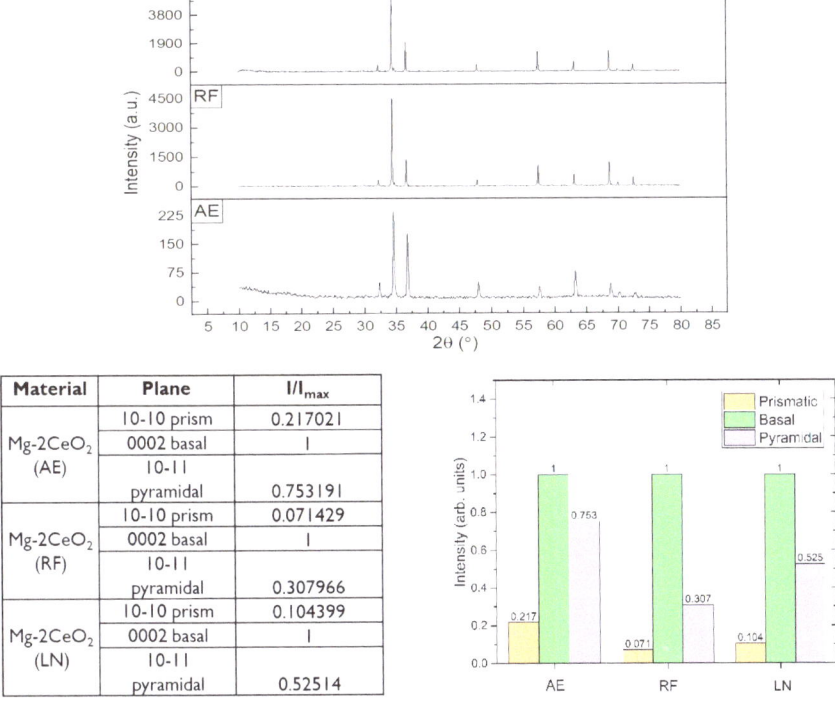

Material	Plane	I/I$_{max}$
Mg-2CeO$_2$ (AE)	10-10 prism	0.217021
	0002 basal	1
	10-11 pyramidal	0.753191
Mg-2CeO$_2$ (RF)	10-10 prism	0.071429
	0002 basal	1
	10-11 pyramidal	0.307966
Mg-2CeO$_2$ (LN)	10-10 prism	0.104399
	0002 basal	1
	10-11 pyramidal	0.52514

Figure 3. The results of the X-ray diffraction studies.

3.3. Thermal Response

The thermal response of the samples was evaluated in terms of the DSC studies (Figure 4) and the determination of the ignition temperature (Table 3). Visible peaks were observed at a temperature of ~470 °C and ~450 °C for the LN and RF samples, respectively. This observation was made in the author's past work on LN samples [22]. These findings suggest the release of stresses accumulated during the CT treatment in the case of the LN and RF samples. Further work is required in this area to understand this mechanism. It is worth noting that as the matrix is pure Mg, peaks due to dissolution or precipitation of other elements can be ruled out.

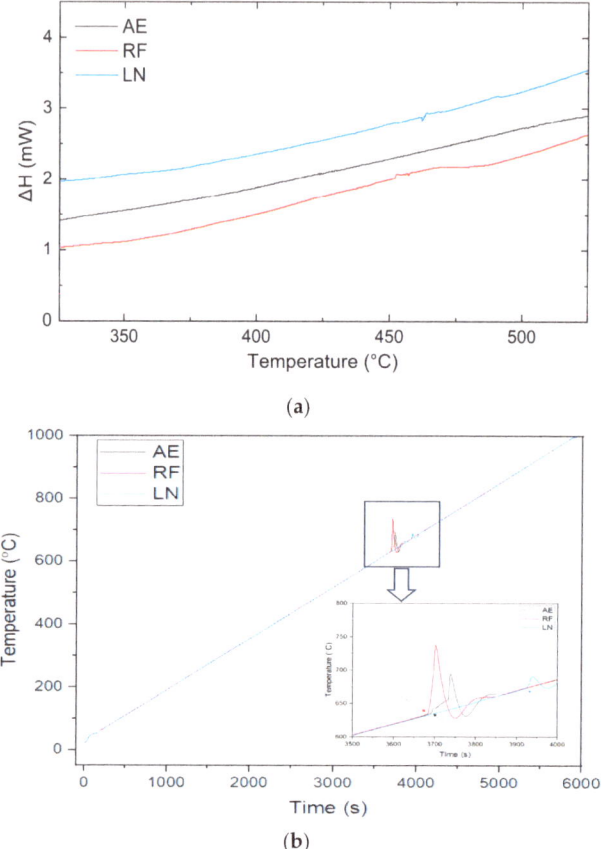

Figure 4. The results of the thermal analysis studies; (**a**) the DSC and (**b**) the ignition temperature of the samples.

The ignition temperature studies (Figure 4) indicated that the RF samples exhibited a similar ignition temperature, while the LN samples displayed a 38 °C increase compared to the as-extruded samples (Table 3). These results suggest that an increase in the ignition temperature in the LN samples can be attributed to an increase in the dislocation density during the LN treatment [22,23]. In contrast, the RF samples did not generate sufficient dislocations, thereby displaying no change in ignition characteristics.

It is noteworthy that the ignition temperatures of all the nanocomposite samples (AE, RF, and LN) remained superior to those of the commonly used commercial magnesium alloys (AZ and ZK series and WE 43 alloy), with the LN samples delivering the most favorable results.

Table 3. The ignition temperature measurements of the samples.

Composition	Ignition Temperature (°C)
Pure Mg	580
Mg-2CeO$_2$ (AE)	636
Mg-2CeO$_2$ (RF)	635 (↓1)
Mg-2CeO$_2$ (LN)	674 (↑38)
AZ31 [a]	628
AZ61 [a]	559
WE43 [a]	644
AZ91 [a]	600
ZK40A [a]	500
ZK60A [a]	499
AM50 [a]	585
AZ81A [a]	543

[a]—[24]. Note: (↑↓) changes are with respect to Mg-2CeO$_2$ (AE).

3.4. Mechanical Response

The mechanical response of the samples (AE, RF, and LN) was assessed in terms of the hardness (Table 4) and bulk compressive properties (Table 5).

Table 4. The microhardness measurements of the samples.

Composition/Treatment	Microhardness (HV)
Pure Mg	55 ± 3
Mg-2CeO$_2$ (AE)	74 ± 3
Mg-2CeO$_2$ (RF)	89 ± 5 (↑20%)
Mg-2CeO$_2$ (LN)	92 ± 4 (↑24%)

Note: (↑%) changes are with respect to Mg-2CeO$_2$ (AE).

Both the RF and LN samples exhibited superior hardness when compared to the AE samples. This can be attributed to the capability of the sub-zero temperature exposure's ability to (a) increase the dislocation density [1,20–22], (b) reduce porosity (Table 1), and (c) strain the lattice due to the induced compressive stresses [1,20,21]. All these factors increase the resistance to local deformation as experienced by the samples during hardness testing. Note that while the average microhardness of the RF samples remained marginally lower when compared to the LN samples, the difference is statistically insignificant, as their standard deviations overlap with each other.

The bulk compressive response of the AE, RF, and LN samples is interpreted each for 0.2 CYS, UCS, and failure strain. The 0.2 CYS of the RF and LN samples remained notably higher (up to 14% for LN samples) compared to the AE samples, indicating the capability of the sub-zero treatments to enhance the applicability of Mg nanocomposites for strength-based designs, which are typically based on yield strength. Between the RF and LN samples, the 0.2 CYS of the LN samples remained ~9% higher when compared to the RF samples. This increase in 0.2 CYS for the RF and LN samples indicates an increase in the stress required to initiate the motion of the unlocked dislocations [25]. This can primarily be attributed to the increase in the compressive stresses in the matrix rather than the grain size, as the grain size of both the RF and LN samples remained higher than the AE samples (Table 2). The results underscore the dominant role played by the induced compressive stresses due to sub-zero treatments, mitigating the Hall–Petch softening in the case of the RF and LN samples.

The ultimate compressive strength (UCS) of both the RF and LN samples remained lower than the AE samples by a maximum of ~7% (RF samples). Between the RF and LN samples, the difference in the UCS can be considered insignificant given the overlap in standard deviations. The average UCS, however, remains higher for the LN samples by ~2.5% over the RF samples. These findings suggest the reduced work-hardening capability of the RF and LN samples. When computing Ds (difference in UCS and 0.2 CYS), the values were 299 MPa for AE, 255 MPa for RF, and 249 MPa for the LN samples. Both the RF and LN samples clearly exhibited lower work-hardening capabilities when compared to the AE samples, while the difference in work-hardening capabilities remained negligible and in favor of the RF samples.

Table 5. The compressive property measurements of the samples.

Composition/ Treatment	0.2 CYS (MPa)	UCS (MPa)	Fracture Strain (%)	Energy Absorbed (MJ/mm^3)
Pure Mg	63 ± 4	278 ± 5	24 ± 1	45
Mg-2CeO$_2$ (AE)	178 ± 19	473 ± 16	16.5 ± 0.7	44 ± 2
Mg-2CeO$_2$ (RF)	186 ± 17(↑5%)	441 ± 12(↓7%)	29.1 ± 1.0 (↑76%)	73 ± 4 (↑65%)
Mg-2CeO$_2$ (LN)	203 ± 5 (↑14%)	452 ± 15 (↓4%)	29.7 ± 1.2 (↑80%)	76 ± 6 (↑72%)
Mg-2Nd-4Zn [a]	242	502	8	
AM50	110	312	11.5	
AZ91D	130	300	12.4	
AZ31	NR	250	28	
Mg-5Zn/5BG	NR	112.8	NR	
WE43	261 ± 16	420 ± 13	16.3 ± 1.0	NA
WE43 + Apatite	229 ± 6	380.1 ± 9.0	11.7 ± 0.5	
ME21	87	260	25	
WE54	210	325	27	
ZK60	159	472	12.4	
Mg4Zn3Gd1Ca	260 ± 3	585 ± 18	12.6 ± 0.3	
Mg4Zn3Gd1Ca-ZnO	355 ± 5	703 ± 40	10.6 ± 0.3	

[a]—Cryogenic treatment in liquid nitrogen (−196 °C) for 1 day [26]. Note: (↑↓%) changes are with respect to Mg-2CeO$_2$ (AE).

The fracture strain of both the RF and LN samples showed a remarkable improvement by a maximum of ~80% (LN samples). Between the RF and LN samples, the difference in the fracture strain remained negligible (29.1 and 29.7, respectively). The results underscore the unique capability of sub-zero treatments in enhancing the fracture strain, demonstrating that exposure to −20 °C provides similar benefits as exposure to liquid nitrogen (−196 °C). The increase in the fracture strain of the sub-zero-treated samples (RF and LN samples) can be attributed to (a) the reduction in porosity (reduced crack initiation sites), (b) the enhanced matrix-reinforcement bonding (reduced crack initiation sites) [22], and (c) the reduced work hardening, leading to an increase in the uniform plastic deformation zone before the cracks initiated and catastrophically propagated (Figure 5). These results indicate that, following cryogenic treatment, dislocations can move easily and over long distances when compared to the non-treated samples.

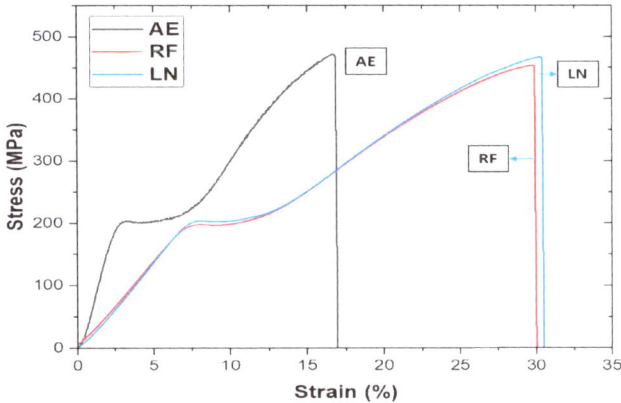

Figure 5. The compressive stress–strain diagrams for the AE, RF, and LN samples.

Fractographic studies conducted on the AE, RF, and LN samples are presented in Figures 6 and 7. Macrographs of all three samples show an approximately 45° fracture in relation to the compression axis. There was no remarkable difference between them visually.

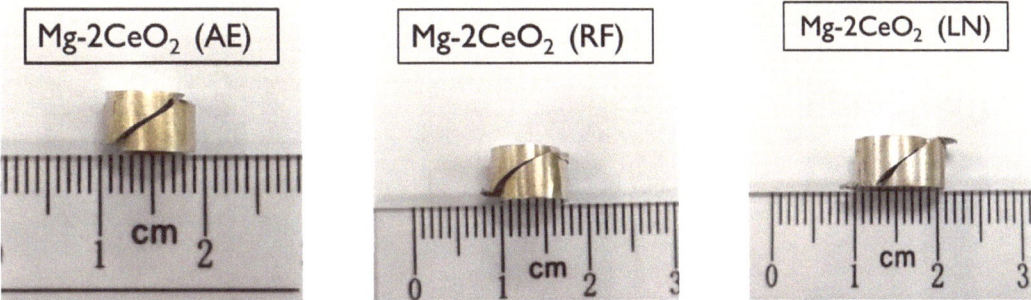

Figure 6. The macroscopic images of the compressively fractured samples.

Figure 7 presents the SEM micrographs of the fractured samples. A notable distinction is observed: the AE samples exhibit a relatively higher degree of flatness, indicative of a relatively brittle fracture, in contrast to the RF and LN samples, which exhibit a higher degree of roughness, signifying comparatively more plastic deformation.

The influence of deep cryogenic treatment on the elastic properties of magnesium materials depends on a complex interplay between dislocations and grain structure [27,28]. Dislocation pinning is promoted during cryogenic treatment, leading to an improvement in the elastic modulus. Furthermore, cryogenic treatment promotes the development of a preferred grain orientation (texture), resulting in the elastic properties becoming directionally dependent. Hence, specific textures might increase the elastic modulus in one direction but decrease it in another. Hence, while the overall trend often points toward an increased elastic modulus due to the grain refinement and dislocation pinning, the specific effects can vary greatly depending on the magnesium's material composition, microstructure, and treatment parameters. It is noteworthy that these effects will be different for nanocomposites, that no information in the open literature is available at this stage, and that systematic studies are required to isolate these effects for nanocomposites.

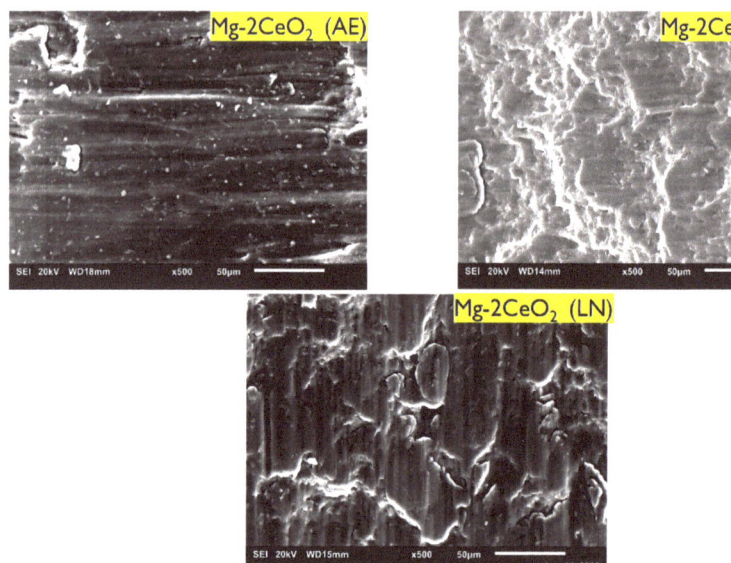

Figure 7. SEM images of the compressively fractured samples.

4. Conclusions

a. The porosity reduction of ~10.4% and ~43.3% was observed when compared with the AE samples when the samples were exposed to −20 °C (RF) and −196 °C (LN), respectively.
b. The DSC studies revealed the release of residual stresses in the case of the LN samples but not for the AE and RF samples.
c. The ignition temperature of the LN samples improved by 38 °C but decreased by only 1 °C for the RF samples when compared to the AE samples.
d. When exposed to shallow cryogenic treatment (−20 °C), the Mg-2CeO$_2$ nanocomposite showed a ~5%, ~76%, and ~65% increment in the 0.2 CYS, fracture strain, and energy absorption values, respectively, as compared to the untreated samples. By comparison, when exposed to deep cryogenic treatment (−196 °C), the Mg-2CeO$_2$ nanocomposite showed a ~14%, ~80%, and ~72% increment in the 0.2 CYS, fracture strain, and energy absorption values, respectively, as compared to the untreated samples. Overall, the UCS values for both conditions were slightly lower than the untreated conditions.
e. The fracture surfaces of the AE, RF, and LN samples did not reveal any noticeable difference at the visual level (45 ° shear fracture). The RF and LN samples showed a higher degree of surface roughness, indicating a higher fracture strain when compared to the AE samples.
f. The future outlook for the expansion of this field of research will be to focus in depth on the mechanism behind the improvement of the properties in a cryogenic setting and to identify suitable lightweight magnesium materials that can be suitable for such applications and to engineer them for high cryogenic performance.

Author Contributions: S.G.: Conceptualization, Methodology, Software, Formal analysis, Investigation, Data Curation, Writing—Original Draft, Writing Review and Editing, Visualization. G.P.: Formal analysis, Investigation, Data Curation, Writing—Review and Editing. M.G.: Conceptualization, Methodology, Writing—Review and Editing, Supervision, Project administration, Funding acquisition. All authors have read and agreed to the published version of the manuscript.

Funding: This research received no external funding.

Institutional Review Board Statement: Not applicable.

Informed Consent Statement: Not applicable.

Data Availability Statement: The data presented in this study are available on request from the corresponding author. The data are not publicly available due to confidentiality of research in some aspects.

Acknowledgments: The authors would like to thank the experimental support extended by Khin Sandar Tun and Hong Wei from the Department of Mechanical Engineering, National University of Singapore.

Conflicts of Interest: The authors declare that they have no known competing financial interests or personal relationships that could appear to have influenced the work reported in this paper.

References

1. Sonar, T.; Lomte, S.; Gogte, C. Cryogenic Treatment of Metal—A Review. *Mater. Today Proc.* **2018**, *5 Pt 3*, 25219–25228. [CrossRef]
2. Dieringa, H. Influence of Cryogenic Temperatures on the Microstructure and Mechanical Properties of Magnesium Alloys: A Review. *Metals* **2017**, *7*, 38. [CrossRef]
3. Baldissera, P.; Delprete, C. Deep cryogenic treatment: A bibliographic review. *Open Mech. Eng. J.* **2008**, *2*, 1–11. [CrossRef]
4. Das, D.; Ray, K.K.; Dutta, A.K. Influence of temperature of sub-zero treatments on the wear behaviour of die steel. *Wear* **2009**, *267*, 1361–1370. [CrossRef]
5. Barron, R.F. Cryogenic treatment of metals to improve wear resistance. *Cryogenics* **1982**, *22*, 409–413. [CrossRef]
6. Prasad, S.V.S.; Prasad, S.B.; Verma, K.; Mishra, R.K.; Kumar, V.; Singh, S. The role and significance of Magnesium in modern day research-A review. *J. Magnes. Alloys* **2022**, *10*, 1–61. [CrossRef]
7. Liu, L.; Chen, X.; Pan, F. A review on electromagnetic shielding magnesium alloys. *J. Magnes. Alloys* **2021**, *9*, 1906–1921. [CrossRef]
8. Shang, Y.; Pistidda, C.; Gizer, G.; Klassen, T.; Dornheim, M. Mg-based materials for hydrogen storage. *J. Magnes. Alloys* **2021**, *9*, 1837–1860. [CrossRef]
9. Staiger, M.P.; Pietak, A.M.; Huadmai, J.; Dias, G. Magnesium and its alloys as orthopedic biomaterials: A review. *Biomaterials* **2006**, *27*, 1728–1734. [CrossRef]
10. Kujur, M.S.; Manakari, V.; Parande, G.; Prasadh, S.; Wong, R.; Mallick, A.; Gupta, M. Effect of samarium oxide nanoparticles on degradation and invitro biocompatibility of magnesium. *Mater. Today Commun.* **2021**, *26*, 102171. [CrossRef]
11. Teo, Z.M.B.; Parande, G.; Manakari, V.; Gupta, M. Using low-temperature sinterless powder method to develop exceptionally high amount of zinc containing Mg–Zn–Ca alloy and Mg–Zn–Ca/SiO$_2$ nanocomposite. *J. Alloys Compd.* **2021**, *853*, 156957. [CrossRef]
12. Bupesh Raja, V.K.; Parande, G.; Kannan, S.; Sonawwanay, P.D.; Selvarani, V.; Ramasubramanian, S.; Ramachandran, D.; Jeremiah, A.; Akash Sundaraeswar, K.; Satheeshwaran, S.; et al. Influence of Laser Treatment Medium on the Surface Topography Characteristics of Laser Surface-Modified Resorbable Mg3Zn Alloy and Mg3Zn1HA Nanocomposite. *Metals* **2023**, *13*, 850. [CrossRef]
13. Parande, G.; Tun, K.S.; Neo, H.J.N.; Gupta, M. An Investigation into the Effect of Length Scale (Nano to Micron) of Cerium Oxide Particles on the Mechanical and Flammability Response of Magnesium. *J. Mater. Eng. Perform.* **2022**, *32*, 2710–2722. [CrossRef]
14. Wei, J.; He, C.; Qie, M.; Li, Y.; Tian, N.; Qin, G.; Zuo, L. Achieving high performance of wire arc additive manufactured Mg–Y–Nd alloy assisted by interlayer friction stir processing. *J. Mater. Process. Technol.* **2023**, *311*, 117809. [CrossRef]
15. Lopes, V.; Puga, H.; Gomes, I.V.; Peixinho, N.; Teixeira, J.C.; Barbosa, J. Magnesium stents manufacturing: Experimental application of a novel hybrid thin-walled investment casting approach. *J. Mater. Process. Technol.* **2022**, *299*, 117339. [CrossRef]
16. Yu, Z.; Chen, J.; Yan, H.; Xia, W.; Su, B.; Gong, X.; Guo, H. Degradation, stress corrosion cracking behavior and cytocompatibility of high strain rate rolled Mg-Zn-Sr alloys. *Mater. Lett.* **2020**, *260*, 126920. [CrossRef]
17. Maier, P.; Hort, N. *Magnesium Alloys for Biomedical Applications*; MDPI: Basel, Switzerland, 2020; Volume 10, p. 1328.
18. Joost, W.J.; Krajewski, P.E. Towards magnesium alloys for high-volume automotive applications. *Scr. Mater.* **2017**, *128*, 107–112. [CrossRef]
19. Prasadh, S.; Manakari, V.; Parande, G.; Wong, R.C.W.; Gupta, M. Hollow silica reinforced magnesium nanocomposites with enhanced mechanical and biological properties with computational modeling analysis for mandibular reconstruction. *Int. J. Oral Sci.* **2020**, *12*, 31. [CrossRef]
20. Huang, H.; Zhang, J. Microstructure and mechanical properties of AZ31 magnesium alloy processed by multi-directional forging at different temperatures. *Mater. Sci. Eng. A* **2016**, *674*, 52–58. [CrossRef]
21. Jiang, Y.; Chen, D.; Chen, Z.; Liu, J. Effect of Cryogenic Treatment on the Microstructure and Mechanical Properties of AZ31 Magnesium Alloy. *Mater. Manuf. Process.* **2010**, *25*, 837–841. [CrossRef]
22. Gupta, S.; Parande, G.; Tun, K.S.; Gupta, M. Enhancing the Physical, Thermal, and Mechanical Responses of a Mg/2wt.% CeO$_2$ Nanocomposite Using Deep Cryogenic Treatment. *Metals* **2023**, *13*, 660. [CrossRef]
23. Kogure, Y.; Hiki, Y. Effect of dislocations on low-temperature thermal conductivity and specific heat of copper-aluminum alloy crystals. *J. Phys. Soc. Jpn.* **1975**, *39*, 698–707. [CrossRef]

24. Tekumalla, S.; Gupta, M. An insight into ignition factors and mechanisms of magnesium based materials: A review. *Mater. Des.* **2017**, *113*, 84–98. [CrossRef]
25. Dieter, G.E.; Bacon, D. *Mechanical Metallurgy*; McGraw-Hill New York: New York, NY, USA, 1976; Volume 3.
26. Dong, N.; Sun, L.; Ma, H.; Jin, P. Effects of cryogenic treatment on microstructures and mechanical properties of Mg-2Nd-4Zn alloy. *Mater. Lett.* **2021**, *305*, 130699. [CrossRef]
27. Wang, J.; Xie, J.; Ma, D.; Mao, Z.; Liang, T.; Ying, P.; Wang, A.; Wang, W. Effect of deep cryogenic treatment on the microstructure and mechanical properties of Al–Cu–Mg–Ag alloy. *J. Mater. Res. Technol.* **2023**, *25*, 6880–6885. [CrossRef]
28. Jia, J.; Meng, M.; Zhang, Z.; Yang, X.; Lei, G.; Zhang, H. Effect of deep cryogenic treatment on the microstructure and tensile property of Mg-9Gd-4Y–2Zn-0.5Zr alloy. *J. Mater. Res. Technol.* **2022**, *16*, 74–87. [CrossRef]

Disclaimer/Publisher's Note: The statements, opinions and data contained in all publications are solely those of the individual author(s) and contributor(s) and not of MDPI and/or the editor(s). MDPI and/or the editor(s) disclaim responsibility for any injury to people or property resulting from any ideas, methods, instructions or products referred to in the content.

Article

Two Fe-Zr-B-Cu Nanocrystalline Magnetic Alloys Produced by Mechanical Alloying Technique

Jason Daza, Wael Ben Mbarek, Lluisa Escoda, Joan Saurina and Joan-Josep Suñol *

Department of Physics, University of Girona, 17003 Girona, Spain; jason.daza@cadscrits.udg.edu (J.D.); u1930157@campus.udg.edu (W.B.M.); joan.saurina@udg.edu (J.S.)
* Correspondence: joanjosep.sunyol@udg.edu

Abstract: Fe-rich soft magnetic alloys are candidates for applications as magnetic sensors and actuators. Spring magnets can be obtained when these alloys are added to hard magnetic compounds. In this work, two nanocrystalline Fe-Zr-B-Cu alloys are produced by mechanical alloying, MA. The increase in boron content favours the reduction of the crystalline size. Thermal analysis (by differential scanning calorimetry) shows that, in the temperature range compressed between 450 and 650 K, wide exothermic processes take place, which are associated with the relaxation of the tensions of the alloys produced by MA. At high temperatures, a main crystallisation peak is found. A Kissinger and an isoconversional method were used to determine the apparent activation of the exothermic processes. The values are compared with those found in the scientific literature. Likewise, adapted thermogravimetry allowed for the determination of the Curie temperature. The functional response has been analysed by hysteresis loop cycles. According to the composition, the decrease of the Fe/B ratio diminishes the soft magnetic behaviour.

Keywords: mechanical alloying; thermal analysis; soft magnetic; Fe based

1. Introduction

In magnetic alloys, it is important to check the thermal stability of the magnetic phase and their soft-hard behaviour. It is also known that magnetic properties change if the alloy is amorphous or crystalline. Mechanical alloying is a production technique that is applied in the development and manufacturing of powdered Fe-rich nanocrystalline alloys [1,2]. Mechanical alloying is a powder metallurgy technology applied before sintering or consolidation [3,4]. The thermal stability of these alloys' front crystalline growth is determinant, due to the loss of soft behaviour as the crystalline size increases. A key parameter to determine this thermal stability is the apparent activation energy of crystallisation [5,6]. The MA method is one of the most preferred because of its high potential to produce metastable alloys, such as nanocrystalline and amorphous alloys.

Soft ferromagnetic alloys are characterised by low coercive field, H_c, high saturation magnetic flux density, B_S, and high permeability, μ [7]. Low coercivity and high permeability favour low core losses in applications under alternate current magnetic fields. Thus, the control of these parameters is associated with the optimisation of energy savings [8]. Regarding the saturation magnetic flux density, higher values favour application in low dimensional systems as the consequent miniaturisation [9]. Likewise, the thermal behaviour of these alloys allows for the establishment of working temperature limits; the Curie temperature marks the transition from ferromagnetic to paramagnetic and the loss of soft magnetism. Magnetic thermogravimetry has been applied to determine this limiting temperature [10]. Regarding MA, this technique favours the formation of nanocrystalline (including super saturated solid solutions or high entropy alloys) and amorphous soft magnetic alloys. The optimised selection of the milling parameters of MA improve the soft magnetic response [11]. One of the pathways to modifying the soft response is the controlled addition of a small percentage of other elements. It has been found that the addition of non-magnetic elements, such as Cr and Nb, reduces both the magnetisation of saturation and the coercivity [12].

There are other powder techniques, such as atomisation [13]. Atomisation is also used to obtain soft magnetic particles with superior magnetic response [14–16]. In addition, studies analyse the influence of the atomisation production parameters (pressure, gas temperature and thermal conductivity, cooling rate) in the final product [17]. The atomisation process facilitates the formation of spherical particles, whereas MA particles are usually smoothed (but are not spherical). Thus, MA powders have more specific surfaces and irregular shaped particles than atomised particles of a similar radius. Gas atomisation favours the formation of the amorphous phase, however, if a high dispersion of particle sizes is produced, depending on the particle size, the structure can be amorphous or a mixture of amorphous and nanocrystalline [18]. With this technique, contamination from the MA milling tools is avoided.

In recent decades, several families of nanocrystalline soft magnetic alloys have been developed as alternatives to traditional ferrites such as Finemet [19], Nanoperm [20], Hitperm [21], or Nanomet [22]. Figure 1 shows the typical values of these nanocrystalline alloys, representing the initial permeability as a function of the saturation of the magnetic flux density. For comparison, information about traditional ferrites, sendust, permalloy, Si steels, Fe-based amorphous, and Co-based amorphous is also provided in this graph. High values of saturation flux density favour the use of these materials in technological applications in which miniaturisation is desirable. Contrastingly, a high magnetic permeability usually favours energy savings.

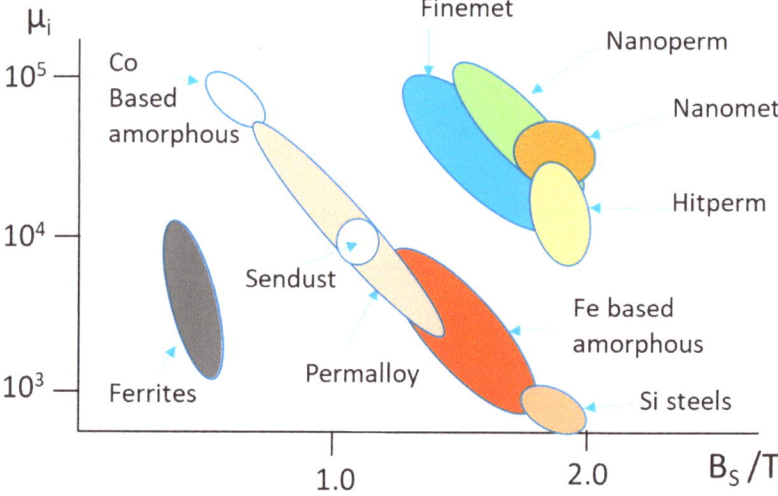

Figure 1. Permeability versus magnetic flux density of several families of soft magnetic alloys.

In this work, two nanocrystalline Fe-Zr-B-Cu Nanoperm-type alloys have been produced by mechanical alloying. The thermal stability has been determined through thermal analysis by checking the apparent activation energy of the main crystalline growth process and the Curie temperature.

The main difference between the two Fe-Zr-B-Cu alloys produced is the relative content of Fe and B. In one the ratio is 85/8 and in the other it is 80/13. As iron is the element that contributes magnetism to the alloy, it is expected that the magnetisation of the sample with a ratio of 85/8 is greater than that of the ratio of 80/13. However, it is known that the reduction of the nanocrystalline size favours the magnetic response (usually by decreasing coercivity) and that the addition of boron in the Fe base alloys favours either the formation of an amorphous phase or the formation of a nanocrystalline phase with smaller nanocrystals. Likewise, the amorphous induced phase formation is usually associated with a loss of the magnetisation of saturation [23], which is an undesired effect for soft magnetic behaviour. It is therefore a question of ascertaining in two specific compositions whether one effect can counteract, at least partially, the other or not. Furthermore, it is a question of verifying the thermal stability of both alloys, since the crystalline growth (provoked on heating the samples) usually entails the loss of soft magnetic behaviour. The percentage of copper and zirconium is the same in both alloys, therefore its effect on the different thermal and thermomagnetic responses, or on the microstructure, can be considered lesser due to the relative content of Fe and Nb. Looking at its role, the Nb being an element of high atomic size, it will normally be located on the neighbouring crystalline grains, hindering the crystalline growth, whereas the well-dispersed copper atoms favour the formation of multiple nanocrystals and a higher density of nanocrystals, making the formation of large-sized crystals difficult.

2. Materials and Methods

In this work, two Fe-rich soft magnetic alloys were produced by mechanical alloying (MA). The initial compositions were $Fe_{85}Zr_6B_8Cu_1$ (at.%) and $Fe_{80}Zr_5B_{13}Cu_1$ (at.%); both Nanoperm-type. These alloys are labelled as A and B, respectively. The precursors were elemental Fe (6–8 µm, 99.9% purity), Zr (5 µm, 98% purity), B (20 µm, 99.7% purity), and Cu (45 µm, 99% purity) powders. The MA was performed in a planetary ball mill device (Fritsch P7 model). The MA process was achieved up to 50 h, under argon atmosphere, with hardened steel vials and balls as MA media, a ball-to-powder weight ratio (BPR) of about 10:1, and a rotation speed of 700 r.p.m. The Ar atmosphere was undertaken in a cycling process (first vacuum near 10^{-5} atm., second Ar addition to 1.1 atm.) performed three times. To prevent an excess in the internal vial temperature, the MA was performed in cycles (10 min on followed by 5 min off). Thus, a period of 75 h corresponds to 50 h of MA. The extractions were performed after waiting for the vials to cool down to prevent high surface oxidation of the metallic particles when vials were opened. The experimental conditions were chosen to optimise the alloys' production by the mechanical alloying technique. In the preparation of soft magnetic samples, a key parameter is to prevent excessive oxidation of the metallic particles because the formation of oxides usually reduces the magnetic response, and the oxide layer hinders the interaction between the magnetic domains and the associated exchange coupling when a bulk specimen is built.

The particles' powder morphology and distribution size were checked by scanning electron microscopy (SEM) in a DSM960A Zeiss apparatus (Zeiss, Jena, Germany). The final composition of the alloys was checked by inductive coupled plasma (ICP) in a Liberty-RL ICP Varian equipment. The nanocrystalline state (bcc Fe-rich phase) was confirmed by X-Ray diffraction (XRD) patterns, collected using D-500 Siemens (Bruker, Billerica, MA, USA) equipment with CuKα radiation (λ = 0.15406 nm). The thermal stability of the mechanically alloyed powders was studied by differential scanning calorimetry (DSC) using a LabSys Evo 1600 °C apparatus (Setaram, Caliure-et-Cuire, France). The DSC curves were measured in the temperature range of 350–950 K at different heating rates: 5, 10, 20, 30, and 40 K/min under argon flow (20 mL/min). The thermogravimetry measurements were performed in a TGA Star^e Mettler Toledo model under Ar atmosphere at a heating rate of 10 K/min. The magnetic parameters, such as intrinsic coercivity, Hc, saturation magnetisation, Ms, and saturation to remanence ratio, Mr/Ms, were determined by analysing the magnetic hysteresis loops collected in a Lakeshore 7404 vibrating sample magnetometer (VSM) at room temperature, under an applied magnetic field of 15 kOe.

3. Results

3.1. Morphology and Structure

The alloys were produced in powder form. Figure 2 shows two micrographs of alloys A and B milled for 50 h. The rounded shape of the particles, with smooth contours and a micrometric size is verified. Moreover, a relatively wide distribution in particle sizes was found. In order to check the particles' size and distribution, the particle size of five micrographs of each sample were measured. The results are shown in Figure 3. The distribution does not correspond to a typical distribution function, either symmetric or asymmetric. Therefore, to compare both distributions, it was decided that the calculation of the median particle size would be performed, as well as calculating the values of the particle size of the first and last 10%. The calculated values were as follows: the first 10% of particles' sizes were 8.1 µm and 4.8 µm, the median particle sizes were 16.9 µm and 21.6 µm, and the last 10% of particles' sizes were 30.7 µm and 40.6 µm, respectively. As is observable, the corresponding values for the distribution of sample B were higher than those for sample A.

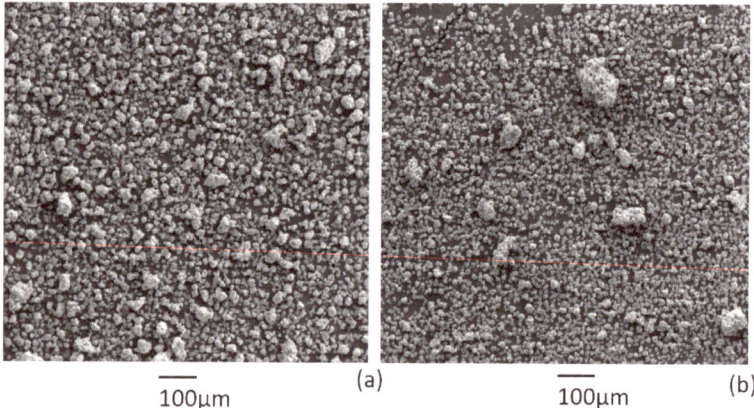

Figure 2. SEM micrographs of alloys A (**a**) and B (**b**) milled for 50 h.

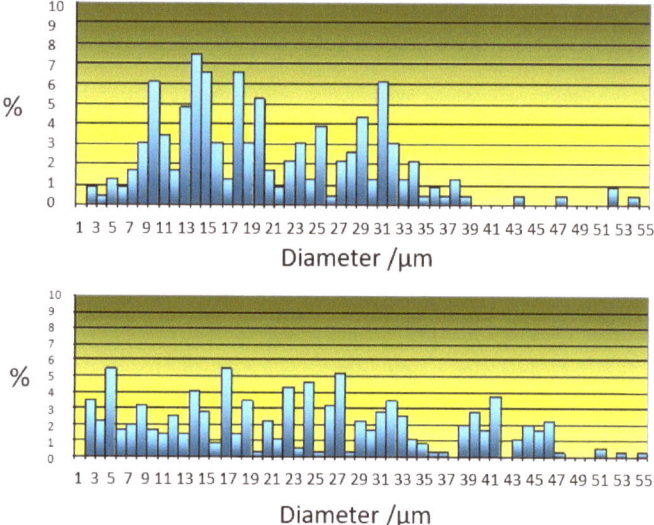

Figure 3. Particle size distribution of alloys A (up) and B (down) milled for 50 h.

Alternative methods for determining particle size distribution are sieve screening or laser diffraction. In this study, measurements made by sieving (three sieves and therefore four values) are consistent with those determined from microscopic observation.

The contamination was measured by ICP. The mechanical alloying process favours contamination from the MA tools. Likewise, the high surface/volume ratio of the particles induces surface oxidation. Nevertheless, the results show only slight contamination from the MA tools (Fe from the container and walls) and oxygen in both alloys after 50 h of MA. Similar results were previously reported [24]. ICP results show that the Fe content increased over the nominal composition and was 1.7 ± 0.3, and 1.6 ± 0.4 at.% for alloys A and B, respectively. Similarly, the oxygen content was 2.1 ± 0.6 and 1.9 ± 0.5 at.% for alloys A and B, respectively; it is probable that a decrease in the iron percentage is associated with a reduction in the oxygen level.

Regarding microstructural analysis, XRD patterns (Figure 4) confirm the formation of the cubic bcc structure with nanocrystalline size. The analysis applying the Williamson–Hall method allows for the determination of both the crystalline size and the microstrain. For alloy A, the crystalline size is 26 ± 2 nm and the microstrain $0.58 \pm 0.06\%$; whereas, for alloy B, the values are 15 ± 2 nm (crystalline size) and $0.61 \pm 0.08\%$ (microstrain). Thus, the increase in boron content favours, as expected, the reduction of the crystalline size, whereas the microstrain (associated with the crystallography defects) is very similar for both samples.

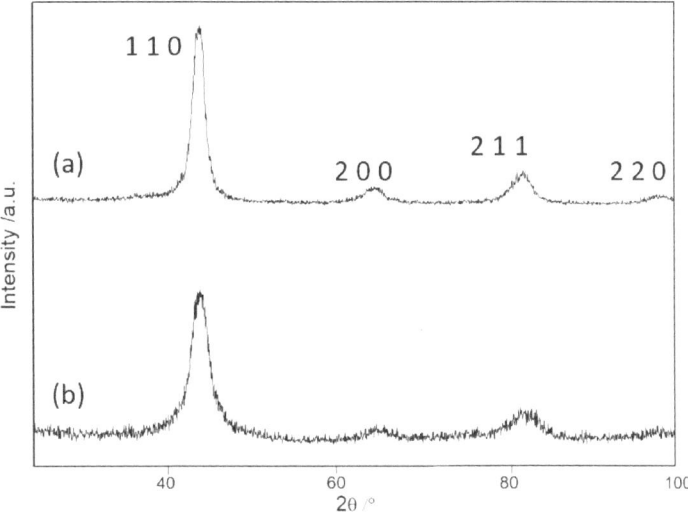

Figure 4. X-Ray diffraction pattern of alloys A (**a**) and B (**b**) milled for 50 h, indicating the Miller indexes of the bcc structure.

3.2. Thermal Analysis

Figure 5 shows the DSC curves of the two alloys milled for 50 h (heating rate: 10 K/min). The general shape of the curves is similar in both alloys. Around 400 K, a large exothermic process begins. This process is typically found in samples produced by mechanical alloying [25]. Its origin is the microstructural relaxation of the high density of crystallographic defects induced during MA grinding. The existence of small exothermic processes of the same character at higher temperatures could be an indicator of a non-completely homogeneous sample. The weak exothermic process around 750 or 780 K in both samples is typical of a partial recrystallisation of the material. The main exothermic process (in the form of a peak, with temperatures around 850 and 880 K in each sample) is due to the crystalline growth of the Fe-rich bcc phase. The development of high-performance soft magnetic alloys is associated with amorphous and low-crystalline-size nanocrystalline alloys. It is necessary that the crystalline size remains smaller than 10 nm. In this case, by applying the random anisotropy model, Hc depends on D^6 because the domain wall effect diminishes, and each grain behaves as a single domain. Thus, crystalline growth should be avoided, and the crystallisation temperature is a limiting temperature for the application of these alloys [26]. It should be remarked that as the Fe/B ratio lowers, the main crystallisation peak is shifted to higher temperatures (about 20 K). Thus, partial substitution of Fe by B increases the thermal stability of the original nanocrystalline phase produced in the mechanical alloying process.

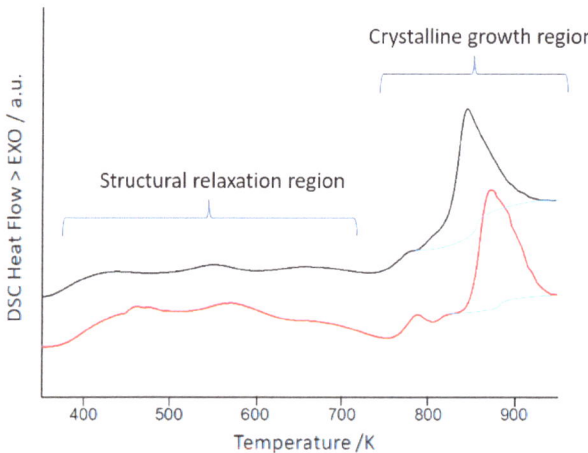

Figure 5. DSC scans at 10 K/min of the as milled alloys A (black) and B (red), indicating the structural relaxation and crystalline growth intervals, as well as the baseline of the main crystallisation process in both samples (blue lines).

One of the most characteristic parameters for the characterisation of crystallisation process is its activation energy. The apparent activation energy of the main crystallisation process was calculated by applying the Kissinger method [27], the typical representation of which (for both analysed alloys) is shown in Figure 6. This method is based on the determination of the peak temperature of the crystallisation process in the experiments carried out at different heating rates. The values are 282 ± 26 and 299 ± 25 kJ/mol for alloys A and B, respectively. These values are consistent with those of the crystallisation of the bcc Fe-rich phase.

The calculated energies of this work are compared with some of those found in the scientific literature (applying Kissinger method) in Table 1.

Table 1. Apparent activation energy of the crystallisation processes (Fe-rich bcc phase).

Composition at.%	Activation Energy kJ mol^{-1}	Initial Structure	Reference
$Fe_{83}P_{16}Cu_1$	238	Amorphous	[28]
$Fe_{68}Nb_6B_{23}Mo_3$	310	Amorphous	[29]
$Fe_{80}Si_{20}$	245	Amorphous	[30]
$Fe_{83}P_{16}Cu_1$	219	Nanocrystalline	[31]
$Fe_{83}P_{14.5}Cu_1Al_{1.5}$	238	Nanocrystalline	[31]
$Fe_{78}Si_{11}B_9$	370	Amorphous	[32]
$Fe_{73.5}Cu_1B_7Si_{15.5}Nb_3$	295	Nanocrystalline	[32]
Fe (99.9% purity)	224	Nanocrystalline	[33]
$Fe_{71}Si_{16}B_9Cu_1Nb_3$	341	Amorphous	[34]
$Fe_{85}Zr_6B_8Cu_1$	282	Nanocrystalline	This Work
$Fe_{80}Zr_5B_{13}Cu_1$	299	Nanocrystalline	This Work

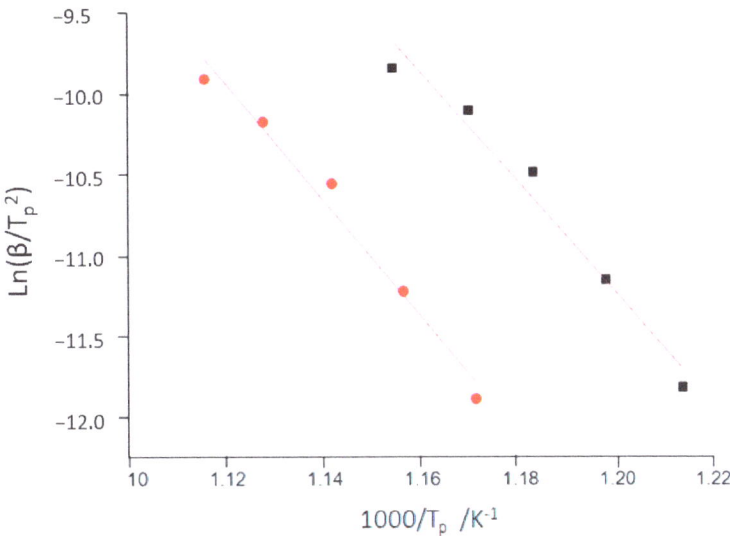

Figure 6. Kissinger plot of the milled alloys A (black) and B (red).

This work's values are in the same relatively wide range [28–34]. The shift in the activation energies can be related to different composition and microstructure (including crystallographic defects). This dispersion is also found in research on alloys of similar composition. Liu et al. [35] found values ranged between 138 and 356 kJ/mol in Fe-Ni-Zr-B alloys produced by mechanical alloying. The addition of Cu facilitates the reduction of (a) the crystallisation temperature and (b) the activation energy [36]. In Fe-based alloys, with a minor addition of Cu, values ranging from 177 to 233 kJ/mol were calculated [37]. Thus, the identification of the process of crystallisation or as crystalline growth is usually performed while taking into account the room temperature nanocrystalline state [26,34]. Likewise, the formation of borides is found at higher temperatures when compared with those of bcc Fe-rich crystallisation [38,39].

Likewise, there are other linear methods (normally based on the determination of the peak temperature). However, the activation energy was similar to those calculated by Kissinger, and the differences are based more on the linear relationships established between the parameters than on real differences in the crystallisation process [29]. For example, representing $\ln(\beta/T_p^2)$ is not the same as representing $\ln(\beta/T_p)$ (in both cases as a function of the inverse of the peak temperature).

The scientific literature on the activation energy of crystallisation processes shows methods in which two energies are determined: that of crystal nucleation and growth [40], the second having a higher value. This approximation is not applicable when there is only crystal growth (for example, in initially nanocrystalline alloys). In recent decades, a set of methods based on the calculation of the apparent activation energy has been extended to different transformation/crystallisation fractions. These methods are defined as isoconversional [41,42]. Figure 7 shows the calculated values (for the main exothermic process) at fractions transformed from 0.1 to 0.9 [42]. A fairly stable value is found, except for high fractions where a slight decrease in activation energy is detected. In the zone of low-medium transformed fraction, the set of calculated values is similar to that determined by applying the Kissinger method, around 288 and 298 kJ/mol for alloys A and B, respectively. For nanocrystalline alloys it is normal for this value to be stable, since the crystal nucleation process is negligible [43]. Thus, the activation energy corresponds to the crystal growth mechanism. In nanocrystalline alloys with similar composition, the same phenomenon was detected [25–34]. At high transformed fractions, the degree of transformation slows down and the local activation energy decreases [44], probably due to the higher influence of the diffusion as main mechanism. There is probably an impingement between the crystal grains. One of the aspects to highlight is that the isoconversional method applied here is not associated with any kinetic model. Therefore, it allows us to parameterise the evolution of the activation energy as a function of the transformed fraction, neglecting the effects of a complex and variable kinetics during the transformation (nucleation, growth, impingement). On the other hand, linear methods are based on concrete hypotheses. For example, the Kissinger equation (used in the analysis of the Figure 6) was obtained assuming: (a) that the fraction transformed at the peak temperature is the same in all experiments, regardless of the heating rate, and (b) the transformation rate (crystal growth in our case) is maximum at the peak temperature.

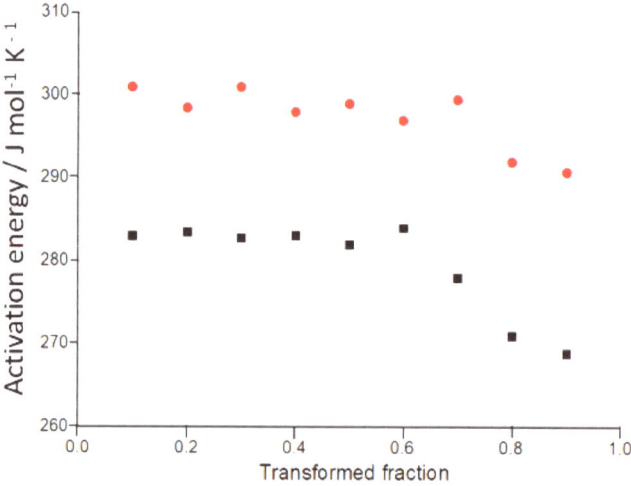

Figure 7. The activation energy as a function of the transformed fraction of the milled alloys A (black) and B (red).

With respect to the thermal stability of both samples after 50 h of grinding, sample B (which has a higher boron content and therefore a lower Fe/B ratio) is the one with the best thermal stability against the main crystallisation process. In general, greater thermal stability is associated with two aspects: (a) a higher transformation temperature and (b) a higher activation energy. In this study, both parameters are better in alloy B. From a technological point of view, what is desirable is a greater thermal stability of the nanocrystalline phase, since crystal growth causes an increase in coercivity and the magnetically soft response.

3.3. Thermomagnetic Analysis

Regarding the magnetic behaviour, one of the most characteristic and fundamental thermomagnetic parameters of magnetically soft alloys is the Curie temperature. This is the temperature that marks the maximum value of applicability (magnetic) of the alloy, since it defines the change from a ferromagnetic to a paramagnetic behaviour. The Curie temperature can be determined from the DSC curves [45]. However, in the case of overlapping processes (as is often the case with alloys produced by mechanical alloying) their determination can be complex [46]. Therefore, on many occasions this temperature is determined directly by magnetic measurements (of variation of magnetisation as a function of temperature). Another alternative, based on thermal analysis techniques, is magnetic thermogravimetry. By adding a small external magnet near the sample zone and performing the thermogravimetry experiment, an apparent variation of the sample mass during the transition from ferromagnetism to paramagnetism is detected. This method, previously applied [47], was used in the present study. Figure 8 shows the thermogravimetric curves recorded. A variation of the apparent mass of the samples around 937 ± 1 K and 971 ± 1 K for alloys A and B, respectively, was found. These temperatures are typical of ferromagnetic alloys with a high iron content.

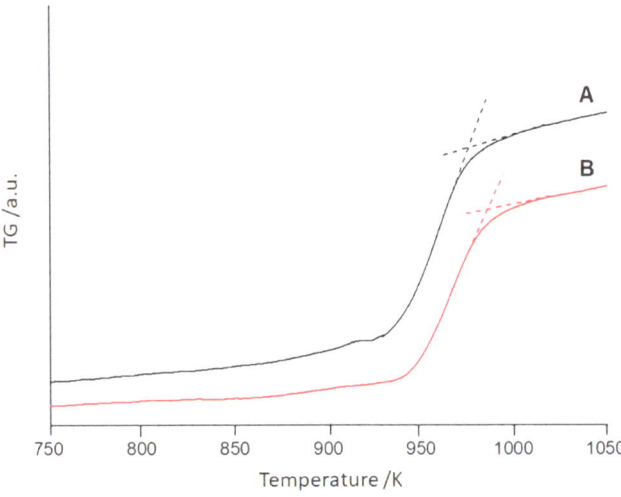

Figure 8. Magnetic thermogravimetry curves of alloys the as milled alloys A (black) and B (red).

The modification made to the thermogravimeter would also allow us to determine the magnetisation of the sample, although for this it would be necessary to always fix the external magnet in the same position and perform a calibration with several magnetic standards (previously analysed with another magnetic measurement technique, such as vibrating sample magnetometry). Consequently, it is better to determine the magnetisation of saturation in magnetic devices (such as a vibrating sample magnetometer). However, in consecutive measurements it is always possible to detect in which samples the change in magnetisation is greater after the transformation associated to the Curie temperature. To make the comparison, it is necessary to normalise taking into account the mass of each sample used in the experiment. In our case, it is detected that the jump is slightly lower in sample A, indicating a somewhat higher initial magnetisation. This result must be confirmed with a magnetic measurement (e.g., by magnetic hysteresis cycles).

As the interest of these alloys is based on their magnetic response, the hysteresis cycles of both alloys (after 50 h of MA) at room temperature were obtained. Figure 9 shows both magnetic cycles (magnetisation M as a function of external magnetic field H). Its analysis allows us to determine the parameters that define its soft magnetic response. The determined values are shown in Table 2. This shows magnetisation of saturation M_S 146 and 139 $A \cdot m^2 \cdot kg^{-1}$, coercivity H_C $12.4 \cdot 10^{-4}$ T and $10.6 \cdot 10^{-4}$ T, and remanence 0.60 and 0.71 $A \cdot m^2 \cdot kg^{-1}$ for alloys A and B, respectively. Thus, the sample with lower nanocrystalline size has lower coercivity.

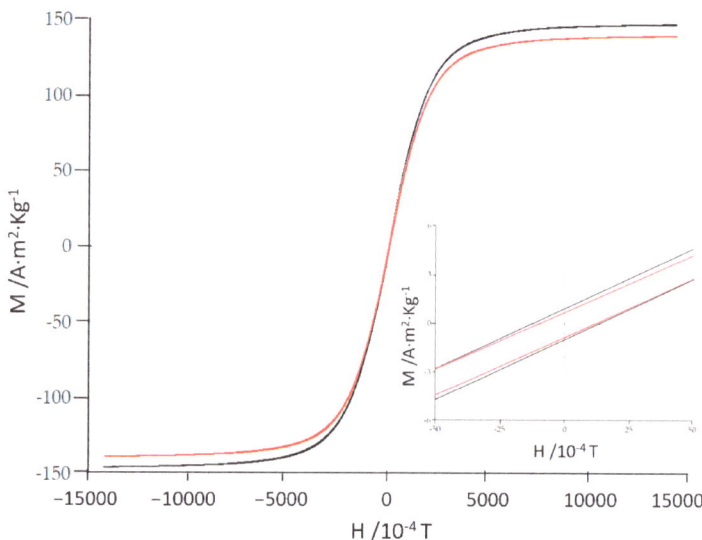

Figure 9. Magnetic hysteresis loops at room temperature of the milled alloys A (black) and B (red). The inset corresponds to the (0,0) region.

Table 2. Magnetic parameters derived from hysteresis cycles at room temperature.

Alloy	H_c 10^{-4} T	M_s $A \cdot m^2 \cdot kg^{-1}$	M_r $A \cdot m^2 \cdot kg^{-1}$	M_r/M_s 10^{-3}
$Fe_{85}Zr_6B_8Cu_1$	12.4	146	0.60	4
$Fe_{80}Zr_6B_{13}Cu_1$	10.6	139	0.71	5

It was found that the magnetisation of saturation has a slightly higher value in sample A. This result is consistent with that detected by magnetic thermogravimetry (Figure 8), where sample A shows a slightly higher variation than that of sample B at the Curie temperature (transition from ferromagnetism to paramagnetism).

The main influence will be related to the Fe atoms' local environment and to Fe-Fe interatomic distance due to the magnetic behaviour of iron. We expected to detect a diminution in the magnetisation of saturation as the Fe/B ratio decreased. Likewise, it is known that the magnetic properties depend strongly on the microstructure evolution, crystalline size, internal stress, particle shape anisotropy, magnetic anisotropy, and magnetostriction of the materials [48]. In our work, higher B content decreases crystalline size and this effect counteracts, partially (favouring soft behaviour), the effect of the reduction of the magnetic element, Fe. However, it is a minor effect on the content of the magnetic element, iron. In both alloys, when changing the percentage of iron from 85 (sample A) to 80 (sample B), a decrease in the magnetic response (saturation magnetisation) of around 6% would be expected. The decrease detected is somewhat smaller, but it is 5%. In relation to this, it should be determined that, from a magnetic point of view, in the two alloys studied the reduction of nanocrystalline size does not counteract the effect of the iron reduction. This phenomenon was to be expected, as the effect on crystal size is considered more important in coercivity. Regarding ulterior compacting to produce bulk samples, the Fe-B bonds dominated the structural compression [23].

The other values of magnetic properties are similar in both samples. The relatively low value of the squareness ratio is typical of alloys obtained in powder form by mechanical alloying [6].

Thus, we can conclude that the alloys produced are magnetically soft at room temperature and that thermal analysis is useful to determine the thermal stability (front crystallisation and front magnetic transition) of this magnetic behaviour.

4. Conclusions

Two ferromagnetic nanocrystalline alloys of the Fe-Zr-B system were produced by mechanical alloying (powder shape). Thermal analysis measurements allowed for the analysis of the thermal stability of the samples. High apparent activation energy, as well as high transition temperatures (crystallisation, Curie), are needed to prevent the loss of the soft behaviour caused by the increase of the crystalline size or the ferro- to paramagnetic transition. The differential calorimetry allowed us to determine (using Kissinger and isoconversional methods) the apparent energy of activation of the main process of crystallisation. The lowest value, 288 kJ/mol, corresponded to the alloy with higher iron content. Regarding the transition temperatures, both the peak crystallisation temperature and the Curie temperature are higher in the alloy with lower Fe/B ratio.

Regarding the magnetic response, the characteristic parameters are similar in both alloys, the magnetisation of saturation being only slightly higher (about 5%) in the alloy with more iron content due to the Fe-Fe magnetic moment atomic interactions.

From a technological point of view, high saturation magnetisation and low coercivity are necessary for the feasibility of the applications of these alloys in devices and systems that must present a soft magnetic behaviour. It is reported that although the saturation magnetisation is slightly lower in the alloy with a lower Fe/B ratio, the decrease in coercivity with increasing boron content (lower Fe/B ratio, alloy labelled as B) is relatively significant due to the smaller size of the nanocrystals of the low Fe/B ratio alloy. This magnetic behaviour, combined with a high thermal stability front crystalline growth (and with the associated loss of the soft magnetic behaviour), of alloy B indicates that the reduction in Fe content does not always provoke a significant decrease in the applicability of this family of alloys. With respect to future outlook and the industrial application of these types of alloys, it is necessary to thoroughly study new compositions and to analyse the consolidation process of these powders and its influence on the final microstructure and magnetic response.

Author Contributions: Conceptualisation, J.-J.S.; methodology, J.S.; formal analysis, W.B.M.; investigation, J.D.; writing—original draft preparation, J.D. and J.-J.S.; supervision, L.E. All authors have read and agreed to the published version of the manuscript.

Funding: Financial support from PID2020-115215RB-C22 project.

Institutional Review Board Statement: Not applicable.

Informed Consent Statement: Not applicable.

Data Availability Statement: Data will be made available upon reasonable requests to the authors.

Conflicts of Interest: The authors declare no conflict of interest.

References

1. Gouasmia, T.; Loudjani, N.; Boulkra, M.; Benchiheub, M.; Belakroum, K.; Bououdina, M. Morphology, structural and microstructural characterizations of mechanically alloyed $Fe_{50}Co_{40}Ni_{10}$ powder mixture. *Appl. Phys. A Sci. Process.* **2022**, *128*, 935. [CrossRef]
2. Panigrahi, M.; Avar, B. Influence of mechanical alloying on structural, thermal and magnetic properties of $Fe_{50}Ni_{10}Co_{10}Ti_{10}B_{20}$ high entropy soft magnetic alloy. *J. Mater. Sci. Mater. Electron.* **2021**, *32*, 21124–21134. [CrossRef]
3. Chaubey, A.K.; Konda Gokuldoss, P.; Wang, Z.; Scudino, S.; Mukhopadhyay, N.K.; Eckert, J. Effect of Particle Size on Microstructure and Mechanical Properties of Al-Based Composite Reinforced with 10 Vol.% Mechanically Alloyed Mg-7.4%Al Particles. *Technologies* **2016**, *4*, 37. [CrossRef]
4. Karthiselva, N.S.; Bakshi, S.R. Reactive Spark Plasma Sintering and Mechanical Properties of Zirconium Diboride–Titanium Diboride Ultrahigh Temperature Ceramic Solid Solutions. *Technologies* **2016**, *4*, 30. [CrossRef]
5. Chebli, A.; Cesnek, M.; Djekoun, A.; Suñol, J.J.; Niznansky, D. Synthesis, characterization and amorphization of mechanically alloyed $Fe_{75}Si_{12}Ti_6B_7$ and $Fe_{73}Si_{15}Ti_5B_7$ powders. *J. Mater. Sci.* **2022**, *57*, 12600–12615. [CrossRef]

6. Carrillo, A.; Daza, J.; Saurina, J.; Escoda, L.; Suñol, J.J. Structural, Thermal and Magnetic Analysis of $Fe_{75}Co_{10}Nb_6B_9$ and $Fe_{65}Co_{20}Nb_6B_9$ Nanostructured Alloys. *Materials* **2021**, *14*, 4542. [CrossRef]
7. Yao, K.F.; Shi, L.X.; Chen, S.Q.; Shao, Y.; Chen, N.; Jia, J.L. Research progress of Fe based soft magnetic amorphous/nanocrystalline alloys. *Acta Phys. Sin.* **2018**, *58*, 17–27. [CrossRef]
8. Zhu, S.; Duan, F.; Ni, J.L.; Feng, S.J.; Liu, X.S.; Lv, Q.R.; Kan, X.C. Soft magnetic composites $FeSiAl/MoS_2$ with high magnetic permeability and low magnetic loss. *J. Alloy. Compd.* **2022**, *926*, 166893. [CrossRef]
9. Yamazaki, T.; Tomita, T.; Uji, K.; Kuwata, H.; Sano, K.; Oka, C.; Sakurai, J.; Hata, S. Combinatorial synthesis of nanocrystalline FeSiBPCuC-Ni-(Nb.;Mo) soft magnetic alloys with high corrosion resistance. *J. Non-Cryst. Solids* **2021**, *563*, 120808. [CrossRef]
10. Miglierini, M.B.; Dekan, J.; Cesnek, M.; Janotova, I.; Svev, P.; Budjos, M.; Kohout, J. Hyperfine interactions in Fe/Co-B-Sn amorphous alloys by Mossbauer spectrometry. *J. Magn. Magn. Mater.* **2020**, *500*, 6417. [CrossRef]
11. Yakin, A.; Simsek, T.; Avar, B.; Simsek, T.; Chattopadhyay, A.K. A review of soft magnetic properties of mechanically alloyed amorphous and nanocrystalline powders. *Emergent Mater.* **2023**, *6*, 453–481. [CrossRef]
12. Yakin, A.; Simsek, T.; Avar, B.; Chattopadhyay, A.K.; Ozcan, S.; Simsek, T. The effect of Cr and Nb addition on the structural, morphological, and magnetic properties of the mechanically alloyed high entropy FeCoNi alloys. *Appl. Phys. A* **2022**, *128*, 686. [CrossRef]
13. Wang, P.; Wei, M.; Dong, Y.; Zhu, Z.; Liu, J.; Pang, J.; Li, X.; Zhang, J. Crystallization evolution behavior of amorphous $Fe_{85.7}Si_{7.9}B_{3.6}Cr_2C_{0.8}$ powder produced by a novel atomization process. *J. Non-Cryst. Solids* **2022**, *594*, 1218254. [CrossRef]
14. Liu, Y.; Yi, Y.; Shao, W.; Shao, Y. Microstructure and magnetic properties of soft magnetic cores of amorphous and nanocrystalline alloys. *J. Magn. Magn. Mater.* **2013**, *330*, 119–133. [CrossRef]
15. Dai, T.; Wang, N. Study of magnetic properties and degradability of gas atomization Fe-based (Fe-Si-B-P) amorphous powder. *J. Supercond. Nov. Magn.* **2019**, *32*, 3699–3702. [CrossRef]
16. Masumoto, H.; Kajiura, Y.; Hosono, M.; Hasegawa, A.; Kumaoka, H.; Yoshimodo, K.; Muri, S. Development of novel Fe based nanocrystalline FeBNbPSi alloy powder with high B_so f 1.41 T by forming stable single amorphous phase. *AIP Adv.* **2022**, *12*, 035312. [CrossRef]
17. Li, G.; Shi, G.; Miao, H.; Liu, D.; Li, Z.; Wang, M.; Wang, L. Effects of the gas-atomization pressure and annealing temperature on the microstructure and performance of FeSiBCuNb nanocrystalline soft magnetic composites. *Materials* **2021**, *16*, 1284. [CrossRef] [PubMed]
18. Afonso, C.R.M.; Kaufman, M.J.; Bolfarini, C.; Botta Filho, W.J.; Kiminami, C.S. Gas atomization of nanocrystalline $Fe_{63}Nb_{10}Al_4Si_3B_{20}$ alloy. *J. Metastable Nanocryst. Mater.* **2004**, *20–21*, 175–182. [CrossRef]
19. Zhao, M.; Pang, J.; Zhang, Y.R.; Zhang, W.; Xiang, Q.C.; Ren, Y.L.; Li, X.Y.; Qiu, K.Q. Optimization of crystallization, microstructure and soft magnetic properties of $(Fe_{0.83}B_{0.11}Si_{0.02}P_{0.03}C_{0.01})_{99.5})Cu_{0.5}$ alloy by rapid annealing. *J. Non-Cryst. Solids* **2022**, *579*, 121380. [CrossRef]
20. Hasiak, M.; Laszcz, A.; Zak, A.; Kaleta, J. Microstructure and Magnetic Properties of NANOPERM-Type Soft Magnetic Material. *Acta Phys. Pol. A* **2019**, *135*, 284–287. [CrossRef]
21. Ozturk, S.; Icin, K.; Gencturk, M.; Gobuluk, M.; Svec, P. Effect of heat treatment process on the structural and soft magnetic properties of Fe38Co38Mo8B15Cu ribbons. *J. Non-Cryst. Solids* **2020**, *527*, 119745. [CrossRef]
22. Nishiyama, N.; Tanimoto, K.; Makino, A. Outstanding efficiency in energy conversion for electric motors constructed by nanocrystalline soft Magnetic Nanomet cores. *AIP Adv.* **2016**, *6*, 055925. [CrossRef]
23. Avar, B.; Chattopadhyay, A.K.; Simsek, T.; Simsek, T.; Ozcan, S.; Kalkan, B. Synthesis and characterization of amorphous-nanocrystalline $Fe_{70}Cr_{10}Nb_{10}B_{10}$ powders by mechanical alloying. *Appl. Phys. A* **2022**, *128*, 537. [CrossRef]
24. Suñol, J.J.; González, A.; Saurina, J.; Escoda, L.; Bruna, P. Thermal and structural characterization of Fe-Nb-B alloys prepared by mechanical alloying. *Mater. Sci. Eng. A* **2004**, *375–377*, 874–880. [CrossRef]
25. Taghvaei, A.H.; Bednarcik, J.; Eckert, J. Influence of annealing on microstructure and magnetic properties of cobalt-based amorphous/nanocrystalline powders synthesized by mechanical alloying. *J. Alloy. Compd.* **2015**, *632*, 296–302. [CrossRef]
26. Daza, J.; Ben Mbarek, W.; Escoda, L.; Suñol, J.J. Characterization and analysis of nanocrystalline soft Magnetic alloys: Fe based. *Metals* **2021**, *11*, 1896. [CrossRef]
27. Vyazovkin, S. Kissinger Method in Kinetics of Materials: Things to Beware and Be Aware of. *Molecules* **2020**, *25*, 2813. [CrossRef]
28. Chen, F.G.; Wang, Y.G.; Mian, X.F.; Hong, Y.; Bi, K. Nanocrystalline $Fe_{83}P_{16}Cu_1$ soft magnetic alloy produced by crystallization of its amorphous precursor. *J. Alloy. Compd.* **2013**, *549*, 26–29. [CrossRef]
29. Liu, Y.; Zhu, M.; Du, Y.; Yao, L.; Jian, Z. Crystallization Kinetics of the $Fe_{68}Nb_6B_{23}Mo_3$ Glassy Ribbons Studied by Differential Scanning Calorimetry. *Crystals* **2022**, *12*, 852. [CrossRef]
30. Zhang, L.l.; Yu, P.F.; Cheng, H.; Zhang, M.D.; Liu, D.J.; Zhou, Z.; Jin, Q.; Liaw, P.K.; Li, G.; Liu, R.P. Crystallization in Fe- and Co-based amorphous alloys studied by in-situ X-ray diffraction. *Metall. Mater. Trans. A* **2016**, *47*, 5859–5862. [CrossRef]
31. Zhu, J.S.; Wang, Y.G.; Xia, G.T.; Dai, J.; Chen, J.K. $Fe_{83}P_{14.5}Cu_1Al_{1.5}$ partial nanocrystalline alloy obtained by one-step melt spinning method. *J. Alloy. Compd.* **2016**, *666*, 243–247. [CrossRef]
32. Li, G.; Li, D.; Ni, X.; Li, Z.; Lu, Z. Effect of copper and niobium addition on crystallization kinetics in Fe-Cu-Nb-Si-B alloys. *Rare Met. Mater. Eng.* **2013**, *42*, 1352–1355. [CrossRef]
33. Malow, T.R.; Koch, C.C. Grain growth in nanocrystalline iron prepared by mechanical attrition. *Acta Mater.* **1997**, *45*, 2177–2186. [CrossRef]

34. Wu, X.; Li, X.; Li, S. Crystallization kinetics and soft magnetic properties of $Fe_{71}Si_{16}B_9Cu_1Nb_3$ amorphous alloys. *Mater. Res. Express* **2020**, *7*, 016118. [CrossRef]
35. Liu, Y.J.; Chang, I.T.H.; Lees, M.R. Thermodynamic and magnetic properties of multicomponent $(Fe,Ni)_{70}Zr_{10}B_{20}$ amorphous alloy powders made by mechanical alloying. *Mater. Sci. Eng. A* **2001**, *304–306*, 992–996. [CrossRef]
36. Warski, T.; Radon, A.; Zackiewicz, P.; Wlodarczyk, P.; Polak, M.; Wojcik, A.; Maziarz, W.; Kolano-Burrian, A.; Hawelek, L. Influence of Cu content on structure, thermal stability and magnetic properties in $Fe_{72-x}Ni_8Nb_4Cu_xSi_2B_{14}$ alloys. *Materials* **2021**, *14*, 726. [CrossRef]
37. Li, D.Y.; Li, X.S.; Zhou, J.; Guo, T.Y.; Tong, X.; Zhang, B.; Wang, C.Y. Development of Fe-based diluted nanocrystalline alloy by substituting C for P in FeSiBCCu system. *J. Alloy. Comp.* **2023**, *952*, 170012. [CrossRef]
38. Manjura Hoque, S.; Hakim, M.A.; Khan, F.A.; Chau, N. Ultra-soft Magnetic properties of devitrified $Fe_{73.5}Cu_{0.6}Nb_{2.4}Si_{13}B_{8.5}$ alloy. *Mater. Chem. Phys.* **2007**, *101*, 112–117. [CrossRef]
39. Chen, Y.B.; Zheng, Z.G.; Wei, J.; Xu, C.; Wang, L.H.; Qiu, Z.G.; Zeng, D.C. effect of Mo addition on thermal stability and magnetic properties in FeSiBPCu nanocrystalline alloys. *J. Non-Cryst. Solids* **2023**, *609*, 122279. [CrossRef]
40. Manchanda, B.; Vimal, K.K.; Kapur, G.S.; Kant, S.; Choudhary, V. Effect of sepiolite on nonisothermal crystallization kinetics of polypropylene. *J. Mater. Sci.* **2016**, *51*, 9535–9550. [CrossRef]
41. Gao, Q.; Jian, Z. Kinetic study on non-isothermal crystallization of $Cu_{50}Zr_{50}$ metallic glass. *Trans. Indian Inst. Mat.* **2017**, *70*, 1879–1885. [CrossRef]
42. Hasani, S.; Rezaei-Shahreza, P.; Seifoddini, A. The effect of Cu minor addition on the non-isothermal kinetic of nano-crystallites formation in $Fe_{41}Co_7Cr_{15}Mo_{14}Y_2C_{15}B_6$ BMG. *J. Therm. Anal. Calorim.* **2021**, *143*, 3365–3375. [CrossRef]
43. Jaafari, Z.; Seifoddini, A.; Hasani, S.; Rezaei-Shahreza, P. Kinetic analysis of crystallization process in $[(Fe_{0.9}Ni_{0.1})_{(77)}Mo_5P_9C_{7.5}B_{1.5}]_{(100-x)}Cu_x$ (x = 0.1at.%) BMG: Non-isothermal condition. *J. Therm. Anal. Calorim.* **2018**, *134*, 1565–1574. [CrossRef]
44. Janovsky, D.; Sveda, M.; Sycheva, A.; Kristaly, F.; Zamborsky, F.; Koziel, T.; Bala, P.; Czel, G.; Kaptay, G. Amorphous alloys and differential scanning calorimetry (DSC). *J. Therm. Anal. Calorim.* **2022**, *147*, 7141–7157. [CrossRef]
45. Kaloshkin, S.; Churyukanova, M.; Tcherdyntsev, V. Characterization of Magnetic Transformation at Curie Temperature in Finemet-type Microwires by DSC. *MRS Online Proc. Libr.* **2012**, *1408*, 107–112. [CrossRef]
46. Alleg, S.; Brahimi, A.; Azzaza, S.; Souilah, S.; Zergoug, M.; Suñol, J.J.; Greneche, J.M. X-ray diffraction, Mössbauer spectroscopy and thermal studies of the mechanically alloyed $(Fe_{1-x}Mn_x)_2P$ powders. *Adv. Powder Technol.* **2018**, *29*, 257–265. [CrossRef]
47. González, A.; Bonastre, A.; Escoda, L.; Suñol, J.J. Thermal analysis of Fe(Co,Ni) based alloys prepared by mechanical alloying. 2007. *J. Therm. Anal. Calorim.* **2007**, *87*, 255–258. [CrossRef]
48. Neamtu, B.V.; Chicinas, H.F.; Gabor, M.; Marinca, T.F.; Lupu, N.; Chicinas, I. A comparative of the Fe-based amorphous alloy prepared by mechanical alloying and rapid quenching. *J. Alloy. Comp.* **2017**, *703*, 19–25. [CrossRef]

Disclaimer/Publisher's Note: The statements, opinions and data contained in all publications are solely those of the individual author(s) and contributor(s) and not of MDPI and/or the editor(s). MDPI and/or the editor(s) disclaim responsibility for any injury to people or property resulting from any ideas, methods, instructions or products referred to in the content.

Article

Features of Metalorganic Chemical Vapor Deposition Selective Area Epitaxy of Al$_z$Ga$_{1-z}$As ($0 \leq z \leq 0.3$) Layers in Arrays of Ultrawide Windows

Viktor Shamakhov [1,*], Sergey Slipchenko [1,*], Dmitriy Nikolaev [1], Ilya Soshnikov [1,2,3], Alexander Smirnov [1], Ilya Eliseyev [1], Artyom Grishin [1], Matvei Kondratov [1], Artem Rizaev [1], Nikita Pikhtin [1] and Peter Kop'ev [1]

1. Ioffe Institute, 26 Politekhnicheskaya, 194021 St. Petersburg, Russia; dim@mail.ioffe.ru (D.N.); ipsosh@mail.ru (I.S.); alex.smirnov@mail.ioffe.ru (A.S.); ilya.eliseyev@mail.ioffe.ru (I.E.); ar.evg.grishin@yandex.ru (A.G.); mikondratov99@gmail.com (M.K.); rizartem@mail.ioffe.ru (A.R.); nike@mail.ioffe.ru (N.P.); ps@kopjev.ioffe.ru (P.K.)
2. Nanotechnology Center, Epitaxial Nanotechnology Laboratory at Alferov University, 194021 St. Petersburg, Russia
3. Institute for Analytical Instrumentation, 31-33 Ivana Chernykh, 190103 St. Petersburg, Russia
* Correspondence: shamakhov@mail.ioffe.ru (V.S.); serghpl@mail.ioffe.ru (S.S.)

Citation: Shamakhov, V.; Slipchenko, S.; Nikolaev, D.; Soshnikov, I.; Smirnov, A.; Eliseyev, I.; Grishin, A.; Kondratov, M.; Rizaev, A.; Pikhtin, N.; et al. Features of Metalorganic Chemical Vapor Deposition Selective Area Epitaxy of Al$_z$Ga$_{1-z}$As ($0 \leq z \leq 0.3$) Layers in Arrays of Ultrawide Windows. *Technologies* **2023**, *11*, 89. https://doi.org/10.3390/technologies11040089

Academic Editors: Sergey N. Grigoriev, Marina A. Volosova and Anna A. Okunkova

Received: 23 May 2023
Revised: 27 June 2023
Accepted: 29 June 2023
Published: 7 July 2023

Copyright: © 2023 by the authors. Licensee MDPI, Basel, Switzerland. This article is an open access article distributed under the terms and conditions of the Creative Commons Attribution (CC BY) license (https://creativecommons.org/licenses/by/4.0/).

Abstract: Al$_z$Ga$_{1-z}$As layers of various compositions were grown using metalorganic chemical vapor deposition on a GaAs substrate with a pattern of alternating SiO$_2$ mask/window stripes, each 100 μm wide. Microphotoluminescence maps and thickness profiles of Al$_z$Ga$_{1-z}$As layers that demonstrated the distribution of the growth rate and z in the window were experimentally studied. It was shown that the layer growth rate and the AlAs mole fraction increased continuously from the center to the edge of the window. It was experimentally shown that for a fixed growth time of 10 min, as z increased from 0 to 0.3, the layer thickness difference between the center of the window and the edge increased from 700 Å to 1100 Å, and the maximum change in z between the center of the window and the edge reached Δz 0.016, respectively. Within the framework of the vapor-phase diffusion model, simulations of the spatial distribution of the layer thickness and z across the window were carried out. It was shown that the simulation results were in good agreement with the experimental results for the effective diffusion length D/k: Ga—85 μm, Al—50 μm.

Keywords: selective area epitaxy; mocvd; AlGaAs; photoluminescence; profilometry

1. Introduction

To date, one of the topical tasks in the field of photonic integrated circuits technology is the monolithic integration of electro-optical elements [1–6] that perform various functions, including the control and generation of both optical radiation (multiwave laser sources, modulators, low-loss waveguides, splitters, combiners, etc.) and electrical signals. One of the ways to effectively address this challenge is the selective area epitaxy (SAE) technique [7–10]. The main feature of this growth method is that epitaxial growth occurs on a pre-prepared substrate with a passivating mask deposited on the surface. This mask forms regions which suppress growth, while epitaxial growth occurs in unprotected regions (windows). At present, single-mode lasers with monolithically integrated modulators [11] and couplers [12], multiwavelength single-mode laser systems [13], monolithic semiconductor sources of femtosecond laser pulses [14], and tunable semiconductor lasers with ultra-wide tuning ranges [15] are fabricated using SAE. The state-of-the-art formation of nano-objects such as quantum dots [16] and nanowires [17] is also implemented using SAE. To date, the main SAE techniques are metalorganic chemical vapor deposition (MOCVD) [10,12], molecular beam epitaxy (MBE) [18], and chemical beam epitaxy (CBE) [19].

One of the characteristic features of SAE is associated with the influence of the geometric dimensions of the mask and windows on the composition and properties of the grown epitaxial layers [20,21]. The fundamental reason for this behavior is related to mass conservation during the growth process. As a result, the reduction in the growth area associated with growth inhibition in the mask region leads to an increase in the growth rate in the window. In the SAE process, precursors can reach either the mask or the window [9] through migration inside the boundary layer of the gas phase. Precursors that reach the window region undergo a pyrolysis reaction and participate in the growth of the epitaxial layer. Precursors reaching the mask can either be adsorbed onto the mask surface and migrate to the window area through surface diffusion, or they can be desorbed from the mask surface within a short time. The desorbed precursors then return to the gas phase and diffuse towards the window due to the resulting concentration gradient between the mask and the window. The growth process is determined by the total contribution of these two diffusion processes. In addition, the diffusion in the gas phase is the dominant process in MOCVD according to references [22,23]. This is due to the fact that the diffusion length in the gas phase is much larger than the surface diffusion length. According to references [24,25], the diffusion length in the gas phase can reach up to 100 µm, while the surface diffusion length remains below 1 µm. As a result, the growth rate of the epitaxial layer increases, and its distribution becomes inhomogeneous within the window region.

It follows from the above discussion that the ability to predict the properties of deposited materials is important for SAE. To date, the most widely used model for MOCVD is the vapor-phase diffusion model [9,10,20,25,26]. In this model, the mask surface diffusion is neglected. Surface migration plays an important role only when the mask geometry is close to the surface diffusion length. This model has demonstrated its effectiveness, as it gives good agreement between the simulated and experimental results for binary layers [27].

However, there are no data in the literature on how the simulation results agree if several different values are taken for z. There are also almost no data for periodic structures in which neighboring elements are located (close enough) such that they can influence each other. An element refers to two stripes of the mask and the window between them. Isolated elements are typically studied, i.e., the neighboring element is most often located at a distance of more than 300 µm from the studied element.

The novelty of the results is in the fact that, for the first time, experimental and theoretical studies have been carried out on the growth of layers of AlGaAs/GaAs solid solutions obtained using the selective area epitaxy technique in ultrawide windows. The results are important to the development of selective area epitaxy techniques for the multilayer heterostructure growth used in many optoelectronic devices. Section 1 of this paper describes the features of the technology used for obtaining bulk layers of solid solutions of various compositions (composition of $Al_zGa_{1-z}As$ ($0 \leq z \leq 0.3$)) grown using selective area epitaxy on a GaAs substrate with a periodic structure consisting of alternating 100 µm wide stripes with and without SiO_2. In Section 2, the vapor-phase diffusion model is considered, which describes the selective area epitaxy of $Al_zGa_{1-z}As$ layers in ultrawide windows. Section 3 presents the results of experimental studies on the spatial distribution of the thickness and composition of the selective area epitaxy-grown $Al_zGa_{1-z}As$ layers, as well as the simulation results. Within the framework of the vapor-phase diffusion model, the effective diffusion lengths for Ga and Al were chosen to ensure good agreement between the simulation and experiment on thickness and composition variation across the window.

2. Materials and Methods

2.1. Fabrication of Experimental Samples

An EMCORE GS3100 (EMCORE Corp., Somerset, NJ, USA) setup with a vertical reactor and resistive heating of the substrate holder was used for epitaxial growth. The work pressure in the reactor was maintained about 77 Torr. The substrate temperature and rotation speed were 750 °C and 1000 rpm, respectively. Trimethylgallium (TMGa) and trimethylaluminum (TMAl) (Elma-Chem, Zelenograd, Russia) were used as sources of group 3 atoms, and arsine (AsH_3) (Salyut, Nizhny Novgorod, Russia) was used as the source of group 5 atoms. The carrier gas was hydrogen (H_2). In this study, the growth time was fixed at 10 min for all samples. Two types of samples were fabricated. Both types of samples were grown on a precisely oriented n-GaAs (100) substrate 2 inches in diameter (Wafer Technology Ltd., Milton Keynes, UK). For the first type of samples (SE—standard epitaxy), standard epitaxial growth was carried out on a substrate without a mask. Four $Al_{z0}Ga_{1-z0}As$ samples were grown with the following compositions (z_0) and growth rates (V_{planar}): sample SE1—$z_0 = 0$, $V_{planar} = 200$ Å/min; sample SE2—$z_0 = 0.11$, $V_{planar} = 225$ Å/min; sample SE3—$z_0 = 0.19$, $V_{planar} = 247$ Å/min; sample SE4—$z_0 = 0.3$, $V_{planar} = 286$ Å/min. Samples of the second type were SAE-grown. To do this, at the initial stage, a pattern of alternating stripes (100 μm wide dielectric mask/100 μm wide window without a dielectric) oriented in the direction [011] was made on the substrate. The 1000 Å thick SiO_2 mask was deposited by ion-plasma sputtering. The pattern was formed using lithography and an etching process in a buffered oxide solution (BOE 5:1). Before the AlGaAs layer growth, the substrate with a pattern was annealed at a temperature of 750 °C for 20 min in the arsine flow, followed by the SAE process. As a result, four samples of the second type were grown with different z_0 at the same fluxes as for standard epitaxy: $z_0 = 0$ for SAE1; $z_0 = 0.11$ for SAE2; $z_0 = 0.19$ for SAE3; $z_0 = 0.3$ for SAE4. The actual compositions of the layers and their distribution in the window for all samples of the second type will be discussed below. It should be noted that the deposition of polycrystals on the mask was observed for the SAE4 sample under the given growth conditions (Figure 1a), but their density was not high, so the sample was not excluded from the studies. Figure 1a shows a segment located in the center of the mask. As can be seen from Figure 1a, for this area (12 × 9 μm), the number of polycrystals is about 400 pieces, which is an average density of about 3.7 pcs/μm². In this case, the linear dimensions of polycrystals do not exceed 100 nm. It should be noted that the density of polycrystals slightly decreases towards the mask/window interface. Also, one can observe a negligible number of polycrystals in the mask region within 1 μm from the mask/window interface. This behavior can be explained based on the study in [28]. For each type of reactant species, there is a certain threshold concentration on the mask surface, above which a heterogeneous nucleation occurs. This threshold is higher for ideal than for non-ideal areas of the mask surface (for example, roughness or defects on the mask surface). Polycrystals precipitate when one of the threshold concentrations is exceeded. Moreover, there is a region at the mask/window interface where nucleation does not occur. For comparison, Figure 1b shows an image of the mask surface for the SAE3 sample on which there are no polycrystals.

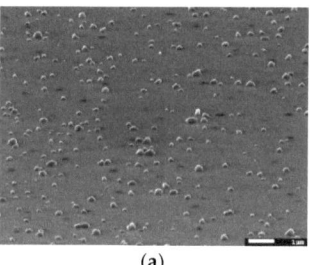
(a)
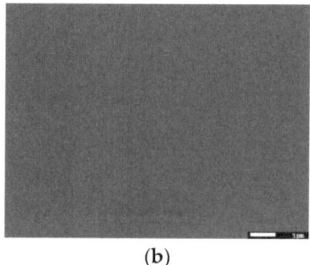
(b)

Figure 1. Scanning electron microscope (SEM) JOEL JSM-7001F (JOEL Ltd., Tokyo, Japan) image of the surface in the center of the mask for SAE4 (**a**) and SAE3 (**b**) samples.

Further, the microphotoluminescence (μPL) spectra for samples of the second type (SAE2–SAE4) were studied with a spatial resolution in the window region. The μPL measurements were performed at room temperature (300 K) using a Horiba LabRAM HR Evolution UV-VIS-NIR (Horiba Jobin Yvon, Longjumeau, France) spectrometer equipped with a confocal microscope. The spectra were measured using the continuous-wave (cw) excitation at 532 nm (2.33 eV) of a Nd:YAG laser (Torus, Laser Quantum, Stockport, UK) with a power on the samples as low as ~40 µW. The spectra were recorded using a 600 lines/mm grating and Peltier-cooled electron-multiplying charge-coupled device (EMCCD) detector, while a Leica PL FLUOTAR 50 × NIR (NA = 0.55) long working-distance objective lens was used to focus the incident beam onto a spot of ~2 µm diameter. The measurements were carried out with point-to-point scanning with a step of 1 µm. The z-distribution was determined for the selectively grown epitaxial layer from the obtained data. As an alternative to the proposed technique for determining the composition, energy-dispersive X-ray spectroscopy (EDS) can be used [29,30]. However, we consider it more labor-intensive and less accurate in resolution for spatial scanning.

Then, the mask was removed from the samples (SAE1–SAE4) using a buffer etchant. After that, the thickness distribution of the selectively grown epitaxial layer was measured across the window stripe for each sample. An AmBios XP-1 (Ambios Technology Inc., Santa Cruz, CA, USA) surface profilometer was used for the study.

2.2. SAE Simulation Model

The selective epitaxy process was quantitatively described using the vapor-phase diffusion model [9,10,20,25,26]. The simulation model is based on the calculation of the concentration profile of precursors in the gas phase above the substrate surface. The profile was determined by solving the Laplace equation in the boundary layer window of width F and height M. Figure 2 shows a schematic explaining the boundary conditions for the selective epitaxy process simulation within the vapor-phase diffusion model. In our case, we used a 2D model, because the length of the mask stripes was much greater than its width, which allowed us to neglect diffusion along the mask stripes. Then, the Laplace equation took the form:

$$\frac{\partial^2 N}{\partial x^2} + \frac{\partial^2 N}{\partial y^2} = 0, \quad (1)$$

where N is the precursor concentration, x is the coordinate in the direction across the window, and y is the coordinate in the direction perpendicular to the growth plane.

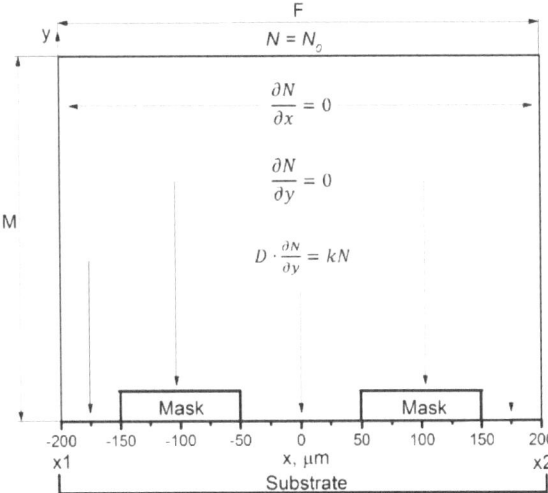

Figure 2. Two-dimensional simulation domain of the vapor-phase diffusion model. F and M are the width and height of the boundary layer. $x1$ and $x2$ are the boundaries of the simulation that takes place along the x axis.

The boundary conditions at the edges of the window and the boundary layer can be written as follows:

1—the upper part of the boundary layer is at a sufficiently large distance from the substrate to avoid perturbations introduced by the mask.

$$N|_{y=M} = N_0, \qquad (2)$$

where N_0 is the precursor concentration at the upper boundary of the boundary layer, which is constant.

2—precursor concentration does not change in lateral directions within the boundary layer.

$$\frac{\partial N}{\partial x}|_{x=x1,x2} = 0 \qquad (3)$$

3—precursors are not deposited on the surface of the mask.

$$\frac{\partial N}{\partial y}|_{y=0} = 0, \qquad (4)$$

4—precursors are deposited on the surface of the semiconductor.

$$D \cdot \frac{\partial N}{\partial y}|_{y=0} = kN, \qquad (5)$$

where k is a surface reaction rate constant; D is a mass diffusion constant.

The precursor concentration profile was determined by the D/k ratio, which can be considered as the effective diffusion length. D/k can be estimated either by theoretical calculation [9,26] or by fitting to experimental data.

To estimate the layer thickness distribution across the window, the concept of growth-rate enhancement (GRE) was used. GRE characterizes the layer growth-rate change during selective epitaxy relative to the growth rate during standard epitaxy, i.e., deposition on a substrate without a mask. GRE is calculated by:

$$GRE = \frac{H}{H_{planar}} = \frac{V}{V_{planar}} = \frac{N}{N_0} \cdot \left(1 + \frac{M}{\frac{D}{k}}\right), \qquad (6)$$

where H and H_{planar} are the thicknesses of the selectively grown layer and the standard-grown layer, respectively; V and V_{planar} are the growth rates of the selectively grown layer and the standard-grown layer, respectively. It is worth noting here that the values of H_{planar} and V_{planar} are constant in the growth plane for standard epitaxy with the selected regimes, while there is a dependence on the x coordinate characterizing the lateral position in the window in the case of selective epitaxy.

These calculations are applicable for binary compounds. For ternary solid solutions of the $A_zB_{1-z}C$ type (where A and B are elements of the 3rd group), GRE is determined using a linear relationship between the binary compounds AC and BC forming a ternary solid solution [10]:

$$GRE_{ABC} = z_0 \cdot GRE_{AC} + (1 - z_0) \cdot GRE_{BC}, \qquad (7)$$

where GRE_{ABC}, GRE_{AC}, and GRE_{BC} are GREs for the ABC solid solution and the binary compounds that form it, respectively; z_0 is the AC mole fraction under the same growth conditions at standard epitaxy.

According to [10], the z distribution of the $A_zB_{1-z}C$ solid solution across the window at SAE is determined by the GRE_{ABC} distribution and the composition z_0 of the $A_{z0}B_{1-z0}C$ solid solution at standard epitaxy according to the relation:

$$z = \frac{z_0 \cdot GRE_{AC}}{GRE_{ABC}}, \qquad (8)$$

3. Results and Discussion

3.1. Studies of the SAE-Grown AlGaAs Layers' Thickness Profiles across the Window

We examined the SE samples to control the layer composition and the deposition rate prior to studying the SAE samples. To test the growth rate, SE1–SE4 samples were diced at their centers to obtain chips for scanning electron microscope (SEM) studies. Then, the thicknesses of the deposited layers were determined using SEM images. From the data obtained, the growth rate was determined, which coincided with the preset one with good accuracy. The PL spectra were also measured for these samples, from which the z_0 was determined, which corresponded to the specified composition with a high accuracy.

In the first part of the analysis of experimental results, let us consider the thickness profiles and determine the change in the thickness of the AlGaAs epitaxial layers across the window for samples SAE1–SAE4. For each sample, four neighboring windows were scanned. Figure 3 shows the measurement results (solid lines), and it is clear that with an increase in the AlAs mole fraction in the deposited AlGaAs layer (from SAE1 to SAE4), its thickness increases. It can be seen there is a fairly good reproducibility of the layer thicknesses in the windows for each sample. For the SAE1 sample, the thickness in the window center has values in the range of 3730–3860 Å, and at the window edge, 4430–4600 Å; for the SAE2 sample, it is of 4030–4080 Å in the center and 4820–4900 Å at the edge; for the SAE3 sample: 4270–4390 Å in the center and 5150–5400 Å at the edge; for the SAE4 sample: 4980–5150 Å in the center and 6090–6250 Å at the edge. It can be seen that the difference in the thickness of the selectively grown layer in the window center with respect to the edge increases with an increase in the AlAs mole fraction.

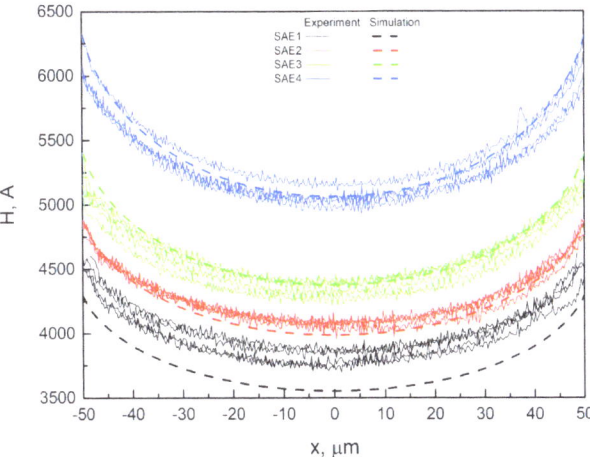

Figure 3. Profiles of the layer thickness across the window for SAE1–SAE4 samples. Solid lines are experimental data; dotted lines are simulation results.

Next, in order to compare the simulation model with the experimental results, the change in the layer thickness across the window during SAE for different z_0 corresponding to the SAE1–SAE4 samples was calculated. First, we chose the D/k for Ga, which provides the best agreement between the calculated thickness and the experimentally measured thickness for the SAE1 sample (GaAs layer) in the window center. GRE values were calculated for D/k ranging from 70 to 150 μm. In the simulation, the boundary layer height M was 1500 μm. Figure 4a shows the calculated GREs; it is clear that the GRE increases as D/k increases to 85 μm at the window center. A further increase in D/k leads to a decrease in GRE in the window center. For different values of D/k (μm), the GRE values (center/window edge) are as follows: 70 μm—1.775/2.219; 85 μm—1.776/2.146; 95 μm—1.775/2.107; 100 μm—1.774/2.09; 110 μm—1.771/2.058; 150 μm—1.752/1.963.

Also for the SAE1 sample, which has an experimental difference in layer thickness between the center and edge of the window of 700–740 Å, the theoretical differences for various D/k values are as follows: 70 µm—888 Å; 85 µm—740 Å; 95 µm—664 Å; 100 µm—632 Å; 110 µm—574 Å; 150 µm—422 Å. The cumulative evidence is that the results closest to the experiment were obtained at D/k for Ga equal to 85 µm. For Al, we could not choose the D/k based on experimental data because of the active growth of the AlAs layer on the surface of the mask at SAE. In [21], the value of D/k for Ga coincides with the value chosen in the present study (85 µm); therefore, the D/k for Al of 50 µm also indicated in [21] was used in the simulation of SAE2–SAE4 samples. Figure 4b shows the GRE change over the window width for GaAs and AlAs binary compounds, as well as the AlGaAs GRE calculated using Equation (7) for the compositions z_0 corresponding to the SAE2–SAE4 samples. The resulting GRE values for the center/edge of the window are 1.776/2.146, 1.775/2.17, 1.774/2.188, and 1.772/2.212 for SAE1, SAE2, SAE3, and SAE4, respectively. It can be seen that the GRE difference between the center and the edge increases with increasing AlAs mole fraction.

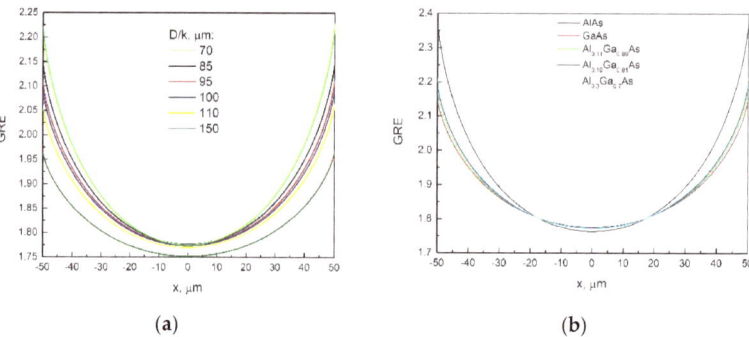

Figure 4. Simulated GRE distribution across the window: (**a**) for D/k ranging from 70 to 150 µm; (**b**) for AlAs and GaAs binary compounds, as well as for $Al_{z0}Ga_{1-z0}As$ solid solutions with z_0 (at standard epitaxy) of 0.11, 0.19, and 0.3, obtained at D/k ratios of 85 µm and 50 µm for Ga and Al, respectively.

Figure 4b shows that a decrease in the D/k ratio leads to an increase in the GRE difference between the edge and the center of the window. A lower D/k ratio indicates that the precursors are adsorbed onto the substrate at a higher rate. Consequently, most of the precursors were deposited in the window near the window/mask interface, and the GRE decreased more rapidly with distance from the window/mask interface. As a result, the AlGaAs solid solutions turned out to be enriched in Al near the window/mask interface.

The layer thickness distribution across the window was calculated for samples SAE1–SAE4 using the obtained GRE profiles (Figure 4b). The layer thickness was calculated based on Equation (6):

$$H = GRE \cdot H_{planar} = GRE \cdot V_{planar} \cdot t, \qquad (9)$$

where t is the layer growth time.

Figure 3 (dashed curves) shows the calculated layer thickness distribution across the window for SAE1–SAE4 samples and it is seen that the calculated curves are in good agreement with the experimental data. The change in the layer thickness in the center and at the edge of the window (experiment/simulation) was also compared, which was: 700–740/740 Å, 790–820/890 Å, 880–1010/1020 Å, and 1100–1110/1250 Å for SAE1, SAE2, SAE3, and SAE4 samples, respectively. It can be seen that the experimental difference in the layer thickness between the center and the edge of the window is in good agreement with that of the simulation. From the obtained results, we can conclude that the simulation is in good agreement with the experimental results for the composition range $0 \leq z_0 \leq 0.3$ of $Al_{z0}Ga_{1-z0}As$ solid solutions at D/k ratios of 85 µm and 50 µm for Ga and Al, respectively.

3.2. Studies of the SAE-Grown AlGaAs Layers' Composition Profiles across the Window

In the second part of the experimental results analysis, let us compare the simulated and experimental profiles across the window of SAE-grown AlGaAs layers with different values of z. To this end, the µPL spectra were measured for SAE2–SAE4 samples at 300 K with a scanning step of 1 µm across the window. Figure 5 shows µPL spectra maps for SAE2–SAE4 samples. For all samples, a blue shift is observed when moving from the center to the edge of the window. For SAE2 (Figure 5a), a local peak of low intensity is observed in the 820–880 nm wavelength range, which was caused by the emission of the n-GaAs substrate. It can also be noted that the intensity of the PL spectrum increases towards the edge of the window relative to the center in the SAE2 and SAE3 samples, which coincides with the behavior of the PL spectrum measured by our group for AlGaAs at a temperature of 80 K in [26]. The SAE4 sample shows the opposite behavior compared with the SAE2 and SAE3 samples, i.e., the intensity of the PL spectrum at the edge of the window is lower than that at the center of the window. However, the reason for this behavior is not yet clear.

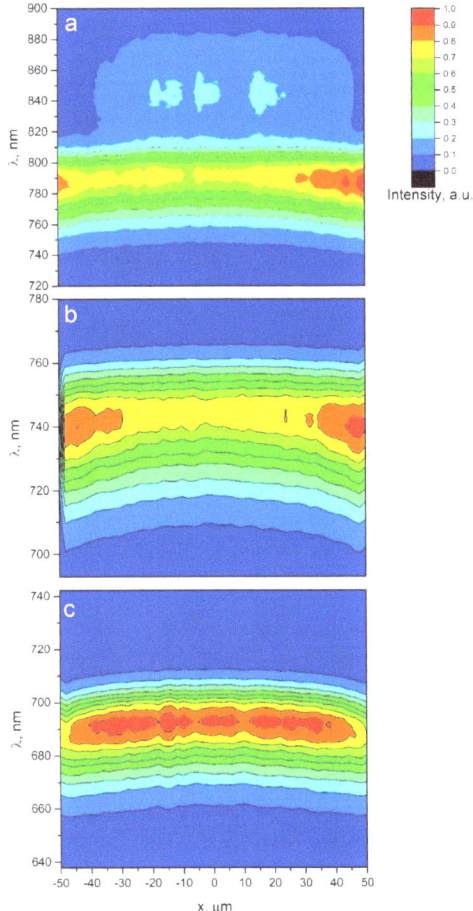

Figure 5. Maps of the µPL spectra measured at 300 K showing the spatial distribution of the spectral position and intensity of the PL spectra for the samples: (**a**)—SAE2, (**b**)—SAE3, (**c**)—SAE4.

From the µPL spectra obtained, z profiles across the window were plotted for the SAE2–SAE4 samples. To calculate z of the $Al_zGa_{1-z}As$ solid solution, the following relation between the band gap E and z was used [31]:

$$E = 1.424 + 1.247 \cdot z \ (z \leq 0.45) \tag{10}$$

Figure 6 (solid curves) shows z profiles across the window calculated using µPL spectra and Equation (10) for SAE2–SAE4 samples. Figure 6 (dashed lines) shows z profiles calculated using Equation (8) based on the vapor-phase diffusion model. Figure 6 shows that the simulation and experiment are in good agreement in the middle of the window for the SAE2 and SAE3 samples, and as the distance from the center to the edge of the window increases, the experimental values of z become slightly higher than the calculated ones. The experimental curve for the SAE4 sample is below the calculated one, which may be due to the formation of a small amount of polycrystals on the mask. Polycrystals can contain Al mainly, which leads to a decrease in the amount of Al diffusing towards the window in the gas phase. The change in z between the center and edge of the window for SAE-grown $Al_zGa_{1-z}As$ layers (experiment/simulation) was: 0.111/0.109 and 0.124/0.12, 0.191/0.189 and 0.205/0.206, and 0.293/0.298 and 0.309/0.321 in the center and at the edge of the window for the SAE2, SAE3, and SAE4 samples, respectively. We can conclude that the simulation model allows us to estimate the change in z across the window for $Al_zGa_{1-z}As$ solid solutions up to $z = 0.3$ with good accuracy. It should be noted that the experimental z variation between the center and edge of the window ranges from 0.013 to 0.016 as z_0 increases from 0.11 to 0.3, and the simulation gives an increase in this variation from 0.011 to 0.023.

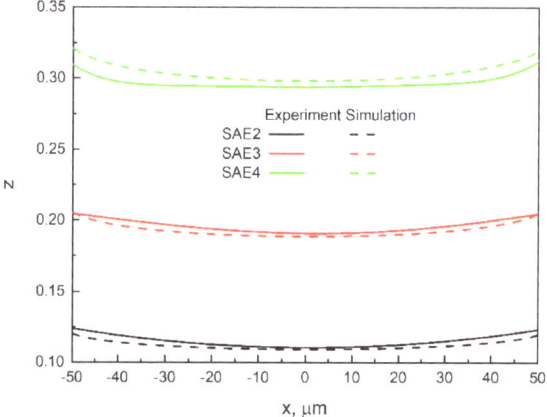

Figure 6. Profiles of the composition of $Al_zGa_{1-z}As$ layers across the window for samples SAE2–SAE4. Solid curves were obtained from the experimental data on µPL; dotted curves were simulated based on the vapor-phase diffusion model.

4. Conclusions

The behavior of the spatial variation of the main characteristics of $Al_zGa_{1-z}As$ epitaxial layers ($0 \leq z \leq 0.3$) grown via selective area epitaxy in arrays of ultra-wide windows (the growth rate, the layer thickness, and the layer composition distribution) have been determined on the basis of experimental studies of their properties. Comparison of experimental results with simulation in the framework of the vapor-phase diffusion model demonstrates a satisfactory agreement at D/k ratios of 85 µm and 50 µm for Ga and Al, respectively, which implies that the proposed simulation model is suitable for predicting the properties of layers in the development of multilayer structures and devices based on them. In the future, the results obtained will be used in the study of strained quantum wells, which is crucial in the field of light-emitting structures.

Author Contributions: Conceptualization, V.S. and S.S.; methodology, V.S. and S.S.; formal analysis, V.S.; investigation, D.N., I.S., A.S. and I.E.; writing—original draft preparation, V.S. and S.S.; writing—review and editing, V.S. and S.S.; visualization, A.G., M.K. and A.R.; supervision, N.P. and P.K.; project administration, N.P.; funding acquisition, N.P. All authors have read and agreed to the published version of the manuscript.

Funding: This research received no external funding.

Institutional Review Board Statement: Not applicable.

Informed Consent Statement: Not applicable.

Data Availability Statement: All the data taken in this work have been published in this paper.

Acknowledgments: The authors would like to thank Ilya Shashkin for his assistance in the preparation of the manuscript.

Conflicts of Interest: The authors declare no conflict of interest.

References

1. Williams, K.A.; Bente, E.A.J.M.; Heiss, D.; Jiao, Y.; Ławniczuk, K.; Leijtens, X.J.M.; van der Tol, J.J.G.M.; Smit, M.K. InP photonic circuits using generic integration [Invited]. *Photon. Res.* **2015**, *3*, B60–B68. [CrossRef]
2. Zhang, C.; Li, X. III–V Nanowire Transistors for Low-Power Logic Applications: A Review and Outlook. *IEEE Trans. Electron. Devices* **2016**, *63*, 223–234. [CrossRef]
3. Dietrich, C.P.; Fiore, A.; Thompson, M.G.; Kamp, M.; Höfling, S. GaAs integrated quantum photonics: Towards compact and multi: Functional quantum photonic integrated circuits. *Laser Photonics Rev.* **2016**, *10*, 870–894. [CrossRef]
4. Zhang, C.; Miao, X.; Chabak, K.D.; Li, X. A review of III–V planar nanowire arrays: Selective lateral VLS epitaxy and 3D transistors. *J. Phys. D Appl. Phys.* **2017**, *50*, 393001. [CrossRef]
5. Bogdanov, S.; Shalaginov, M.Y.; Boltasseva, A.; Shalaev, V.M. Material platforms for integrated quantum photonics. *Opt. Mater. Express.* **2017**, *7*, 111–132. [CrossRef]
6. Smit, M.; Williams, K.; van der Tol, J. Past, present, and future of InP-based photonic integration. *APL Photonics* **2019**, *4*, 050901. [CrossRef]
7. Heinecke, H.; Milde, A.; Baur, B.; Matz, R. Selective-area growth of III/V semiconductors in chemical beam epitaxy. *Semicond. Sci. Technol.* **1993**, *8*, 1023–1031. [CrossRef]
8. Coleman, J.J. Metalorganic chemical vapor deposition for optoelectronic devices. *Proc. IEEE* **1997**, *85*, 1715–1729. [CrossRef]
9. Kim, J.D.; Chen, X.; Coleman, J.J. 10 Selective Area Masked Growth (Nano to Micro). In *Handbook of Crystal Growth*, 2nd ed.; Kuech, T.F., Ed.; Elsevier: Amsterdam, The Netherlands, 2015; Volume 3A, pp. 441–481. [CrossRef]
10. Wang, B.; Zeng, Y.; Song, Y.; Wang, Y.; Liang, L.; Qin, L.; Zhang, J.; Jia, P.; Lei, Y.; Qiu, C.; et al. Principles of Selective Area Epitaxy and Applications in III–V Semiconductor Lasers Using MOCVD: A Review. *Crystals* **2022**, *12*, 1011. [CrossRef]
11. Delprat, D.; Ramdane, A.; Silvestre, L.; Ougazzaden, A.; Delorme, F.; Slempkes, S. 20-Gb/s integrated DBR laser-EA modulator by selective area growth for 1.55-μm WDM applications. *IEEE Photon. Technol. Lett.* **1997**, *9*, 898–900. [CrossRef]
12. Osowski, M.L.; Lammert, R.M.; Coleman, J.J. A dual-wavelength source with monolithically integrated electroabsorption modulators and Y-junction coupler by selective-area MOCVD. *IEEE Photon. Technol. Lett.* **1997**, *9*, 158–160. [CrossRef]
13. Fujii, T.; Takeda, T.; Nishi, H.; Diamantopoulos, N.-P.; Sato, T.; Kakitsuka, T.; Tsuchizawa, T.; Matsuoand, S. Multiwavelength membrane laser array using selective area growth on directly bonded InP on SiO2/Si. *Optica* **2020**, *7*, 838–846. [CrossRef]
14. Xu, J.; Liang, S.; Liu, S.; Qiao, L.; Sun, S.; Deng, Q.; Zhu, H. Passively mode-locked quantum-well laser with a saturable absorber having gradually varied bandgap. *IEEE Photon. Technol. Lett.* **2017**, *29*, 889–892. [CrossRef]
15. Lemaitre, F.; Latkowski, S.; Fortin, C.; Lagay, N.; Pajkovic, R.; Smalbrugge, E.; Decobert, J.; Ambrosius, H.; Williams, K. Selective area growth in generic integration for extended range tunable laser source. In Proceedings of the 2018 IEEE Photonics Conference (IPC), Reston, VA, USA, 30 September–4 October 2018; pp. 1–2. [CrossRef]
16. Kim, H.; Wei, W.; Kuech, T.F.; Gopalan, P.; Mawst, L.J. Room temperature operation of InAs quantum dot lasers formed by diblock-copolymer lithography and selective area MOCVD growth. In Proceedings of the 2017 IEEE Photonics Conference (IPC), Orlando, FL, USA, 1–5 October 2017; pp. 405–406. [CrossRef]
17. Barrigón, E.; Heurlin, M.; Bi, Z.; Monemar, B.; Samuelson, L. Synthesis and applications of III–V nanowires. *Chem. Rev.* **2019**, *119*, 9170–9220. [CrossRef]
18. Nishinaga, T.; Bacchin, G. Selective area MBE of GaAs, AlAs and their alloys by periodic supply epitaxy. *Thin Solid Films* **2000**, *367*, 6–12. [CrossRef]
19. Zannier, V.; Li, A.; Rossi, F.; Yadav, S.; Petersson, K.; Sorba, L. Selective-Area Epitaxy of InGaAsP Buffer Multilayer for In-Plane InAs Nanowire Integration. *Materials* **2022**, *15*, 2543. [CrossRef]

20. Gibbon, M.; Stagg, J.P.; Cureton, C.G.; Thrush, E.J.; Jones, C.J.; Mallard, R.E.; Pritchard, R.E.; Collis, N.; Chew, A. Selective-area low-pressure MOCVD of GaInAsP and related materials on planar InP substrates. *Semicond. Sci. Technol.* **1993**, *8*, 998–1010. [CrossRef]
21. Decobert, J.; Dupuis, N.; Lagreeb, P.Y.; Lagay, N.; Ramdane, A.; Ougazzaden, A.; Poingt, F.; Cuisin, C.; Kazmierski, C. Modeling and characterization of AlGaInAs and related materials using selective area growth by metal-organic vapor-phase epitaxy. *J. Cryst. Growth* **2007**, *298*, 28–31. [CrossRef]
22. Colas, E.; Shahar, A.; Soole, B.D.; Tomlinson, W.J.; Hayes, J.R.; Caneau, C.; Bhat, R. Lateral and longitudinal patterning of semiconductor structures by crystal growth on nonplanar and dielectric-masked GaAs substrates: Application to thickness-modulated waveguide structures. *J. Cryst. Growth* **1991**, *107*, 226–230. [CrossRef]
23. Kayser, O. Selective growth of InP/GaInAs in LP-MOVPE and MOMBE/CBE. *J. Cryst. Growth* **1991**, *107*, 989–998. [CrossRef]
24. Yamaguchi, K.-I.; Okamoto, K. Lateral Supply Mechanisms in Selective Metalorganic Chemical Vapor Deposition. *Jpn. J. Appl. Phys.* **1993**, *32*, 1523–1527. [CrossRef]
25. Sugiyama, M. Selective area growth of III-V semiconductors: From fundamental aspects to device structures. In Proceedings of the 22nd International Conference on Indium Phosphide and Related Materials (IPRM), Kagawa, Japan, 31 May–4 June 2010; pp. 1–6. [CrossRef]
26. Shamakhov, V.; Nikolaev, D.; Slipchenko, S.; Fomin, E.; Smirnov, A.; Eliseyev, I.; Pikhtin, N.; Kop'ev, P. Surface Nanostructuring during Selective Area Epitaxy of Heterostructures with InGaAs QWs in the Ultra-Wide Window. *Nanomaterials* **2021**, *11*, 11. [CrossRef] [PubMed]
27. Slipchenko, S.; Shamakhov, V.; Nikolaev, D.; Fomin, E.; Soshnikov, I.; Bondarev, A.; Mitrofanov, M.; Pikhtin, N.; Kop'ev, P. Basics of surface reconstruction during selective area metalorganic chemical vapour-phase epitaxy of GaAs films in the stripe-type ultra-wide window. *Appl. Surf. Sci.* **2022**, *588*, 152991. [CrossRef]
28. Yamaguchi, K.-I.; Okamoto, K. Analysis of Deposition Selectivity in Selective Epitaxy of GaAs by Metalorganic Chemical Vapor Deposition. *Jpn. J. Appl. Phys.* **1990**, *29*, 2351–2357. [CrossRef]
29. Eraky, M.S.; Sanad, M.M.S.; El-Sayed, E.M.; Shenouda, A.Y.; El-Sherefy, E.-S. Phase transformation and photoelectrochemical characterization of Cu/Bi and Cu/Sb based selenide alloys as promising photoactive electrodes. *AIP Adv.* **2019**, *9*, 115115. [CrossRef]
30. Eraky, M.S.; Sanad, M.M.S.; El-Sayed, E.M.; Shenouda, A.Y.; El-Sherefy, E.-S. Influence of the electrochemical processing parameters on the photocurrent–voltage conversion characteristics of copper bismuth selenide photoactive films. *Eur. Phys. J. Plus* **2022**, *137*, 907. [CrossRef]
31. Casey, H.C.; Panish, M.B. *Heterostructure Lasers Part B: Materials and Operating Characteristics*; Academic Press: San Francisco, CA, USA; London, UK, 1978; 344p.

Disclaimer/Publisher's Note: The statements, opinions and data contained in all publications are solely those of the individual author(s) and contributor(s) and not of MDPI and/or the editor(s). MDPI and/or the editor(s) disclaim responsibility for any injury to people or property resulting from any ideas, methods, instructions or products referred to in the content.

Article

Optical Properties of AgInS₂ Quantum Dots Synthesized in a 3D-Printed Microfluidic Chip

Konstantin Baranov [1,*], Ivan Reznik [1,2,*], Sofia Karamysheva [1], Jacobus W. Swart [2], Stanislav Moshkalev [3] and Anna Orlova [1,*]

1. International Laboratory Hybrid Nanostructures for Biomedicine, ITMO University, Saint Petersburg 199034, Russia; spkaramysheva@itmo.ru
2. Faculty of Electrical Engineering and Computing, University of Campinas, Campinas 13083-970, Brazil; jacobus@unicamp.br
3. Centre for Semiconductor Components and Nanotechnology, University of Campinas, Campinas 13083-870, Brazil; stanisla@unicamp.br
* Correspondence: baranov.const@mail.ru (K.B.); ivanreznik1993@mail.ru (I.R.); a.o.orlova@gmail.com (A.O.)

Abstract: Colloidal nanoparticles, and quantum dots in particular, are a new class of materials that can significantly improve the functionality of photonics, electronics, sensor devices, etc. The main challenge addressed in the article is modification of the syntheses of colloidal NP to launch them into mass production. It is proposed to use an additive printing method of chips for microfluidic synthesis, and it is shown that our approach allows to offer a cheap, easily scalable and automated synthesis method which allows to increase the product yield up to 60% with improved optical properties of AgInS₂ quantum dots.

Keywords: microfluidic synthesis; 3D printing; colloidal luminescent quantum dots; optical properties

Citation: Baranov, K.; Reznik, I.; Karamysheva, S.; Swart, J.W.; Moshkalev, S.; Orlova, A. Optical Properties of AgInS₂ Quantum Dots Synthesized in a 3D-Printed Microfluidic Chip. *Technologies* **2023**, *11*, 93. https://doi.org/10.3390/technologies11040093

Academic Editors: Sergey N. Grigoriev and Nam-Trung Nguyen

Received: 19 May 2023
Revised: 25 June 2023
Accepted: 10 July 2023
Published: 12 July 2023

Copyright: © 2023 by the authors. Licensee MDPI, Basel, Switzerland. This article is an open access article distributed under the terms and conditions of the Creative Commons Attribution (CC BY) license (https://creativecommons.org/licenses/by/4.0/).

1. Introduction

Quantum dots (QD) are a striking representative of a new class of luminophores and colloidal nanoparticles simultaneously. The challenge today is to develop new synthesis methods that are easily scalable and automated. The temperature and mixing rate of precursors during the formation of structures have a huge impact on their optical properties such as the position, shape and width of the photoluminescent (PL) band and the PL quantum yield (PLQY) of quantum dots [1,2]. It is well known that reproducing the synthesis of nanoparticles using the same technique with classical hydrothermal synthesis might cause unstable results due to human factors, limiting the nanostructures production on an industrial scale and their application in medicine [3]. Microfluidic synthesis is the introduction of reagents through programmable syringe pumps, which allows a continuous supply of precursors at a certain rate and automation of the system to produce the required particles [4]. In addition, a large number of microfluidic chip channels can ensure efficient mixing of reagents. These advantages of microfluidic methods over traditional ones have motivated the development, fabrication, and use of inexpensive devices that allow precise temperature control and efficient reagent mixing [5]. All the characteristics mentioned above make microfluidics very advantageous for various applications, from chemical, biological, and material industries to pharmacy and clinical diagnostics [6,7].

Most research groups are focused on using a soft lithography method in order to manufacture microfluidic devices [8]. However, this technology is time-consuming and expensive. For example, in [9], the chip fabrication for QD synthesis took about 30 h. Three-dimensional printing for the formation of microfluidic chips is an emerging trend, and the growing number of publications on microfluidic synthesis confirms its relevance [10]. To the best of our knowledge, this is the first time when quantum dots were synthesized in exclusively 3D-printed microfluidic chips. Here, we suggest a new approach to manufacturing microfluidic chips for QD syntheses that can simplify the formation, reduce the cost of materials used, and significantly save time on commissioning the final product. The yield of the production of $AgInS_2$ (AIS) QD was 1.5 times higher than that of the classical hydrothermal method of synthesis in a flask.

Among different types of quantum dots, the most studied group is CdX (X = S, Se, Te), but the high cytotoxicity of cadmium hinders its wider application in biology and biomedicine and limits its mass use in photonic and electronic devices due to utilization problems [11]. Ternary quantum dots, including AIS QD, are the representatives of nanostructured luminophores with well-tuned optical properties, containing no heavy metal atoms [12]. Therefore, the development of approaches to their mass production will make it possible in the future to produce environmentally friendly nanostructured materials for use in photovoltaics, imaging, and sensorics [13]. In addition, it is well known that the AIS QD are capable of generating superoxide anion when exposed to UV- and visible range radiation due to their energy structure [14]. Interaction of such reactive oxygen species (ROS) with bacteria can lead to the destruction of the latter [15]. Therefore, the AIS QDs are potential candidates for the photodynamic therapy of bacteria [16]. Furthermore, such QD are characterized by a relatively simple hydrothermal synthesis procedure that could be easily transferred to continuous microfluidic synthesis [17].

In this work, it is shown that the use of additive printing to produce chips for microfluidic synthesis makes it possible to synthesize biocompatible colloidal AIS QDs whose luminescence properties are better than those of similar QDs synthesized at the same macroparameters (temperature, pressure and velocity of precursor pumping) by classical colloidal synthesis. Optical properties of AIS QDs synthesized by the microfluidic method and classical hydrothermal method in a flask were compared. It was demonstrated that the PLQY of the samples synthesized by the microfluidic method for 18 and 180 s was found to be 2.5 times higher than that of the samples synthesized in a flask. In addition, the increased yield of the product compared to the classical hydrothermal method, combined with full automation of the synthesis process, can significantly reduce the cost of the synthesized $AgInS_2$ quantum dots.

2. Materials and Methods

2.1. Chemicals

Indium (III) chloride ($InCl_3$), silver nitrate ($AgNO_3$), sodium sulfide ($Na_2S \times 9H_2O$), ammonia hydrate (NH_4OH), mercaptoacetic acid (MAA), 2-propanol, and thioglycolic acid (TGA) were purchased from Sigma-Aldrich (St. Louis, MO, USA). All chemicals were used without further purification.

2.2. Quantum Dot $AgInS_2$ Synthesis

The reagents for the synthesis of the aqueous AIS QDs were mixed according to the procedure described in [17]. A mixture of precursors was formed and alternately added to 5 mL H_2O in the following order with respective concentrations: 0.052 mL $AgNO_3$ (0.1 M), 0.104 mL TGA (1M), 0.0334 mL NH_4OH (5M), 0.0364 mL $InCl_3$ (1M), 0.08 mL NH_4OH (5M), 0.052 mL ($Na_2S \times 9H_2O$) (1M). Then, AIS QD syntheses were performed in a microfluidic chip and by the classical hydrothermal method in a flask, which is described in detail in Section 3.

2.3. Characterization Methods

The Anycubic Photon Mono 3D printer (Anycubic, Shenzhen, China) was used to fabricate the chip. The absorption spectra of the samples were recorded using a UV Probe 3600 spectrophotometer (Shimadzu, Kyoto, Japan). PL spectra were examined with a Cary Eclipse spectrofluorometer (Varian, Belrose, NSW, Australia). PL kinetics was analyzed using the Micro-Time100 single-photon counting microscope (PicoQuant, Berlin, Germany). Sample concentrations of AIS QDs were estimated using the methodology presented in [18].

The calculation of concentration for the studied samples is presented in Appendix A and Section 3.

The PLQY of the synthesized QDs was estimated relative to the PL quantum yield of rhodamine 6G according to Formula (1):

$$\varphi = \varphi_R \frac{I \cdot D_R \cdot n^2}{I_R \cdot D \cdot n^2}, \qquad (1)$$

where n is the refractive index of the solvent, I is the integral luminescence intensity, and D is the optical density at the excitation wavelength of the sample. Index R refers to the parameters of the reference sample (rhodamine in ethanol). It should be noted that when calculating the integral luminescence intensity, the spectra cut at 800 nm were interpolated into the long-wavelength region, symmetrical to the short-wavelength region, in the Origin software package using the inversion operation.

The QDs luminescence decay times and amplitudes were estimated by approximating the experimental curves with Formula (2):

$$y = y_o + \sum_{i=1}^{n} A_i \exp\left(\frac{-(x - x_0)}{\tau_i}\right), \qquad (2)$$

where A_i is the amplitude of the ith component and τ_i is the characteristic decay time of the ith component.

3. Results and Discussion

3.1. Microfluidic Chip Formation

Photopolymer resin-based chips were prepared for nano- and microparticle microfluidic fusion. The microfluidic chip is shown in Figure 1 and is a continuous serpentine channel with a cross-section of 1 mm².

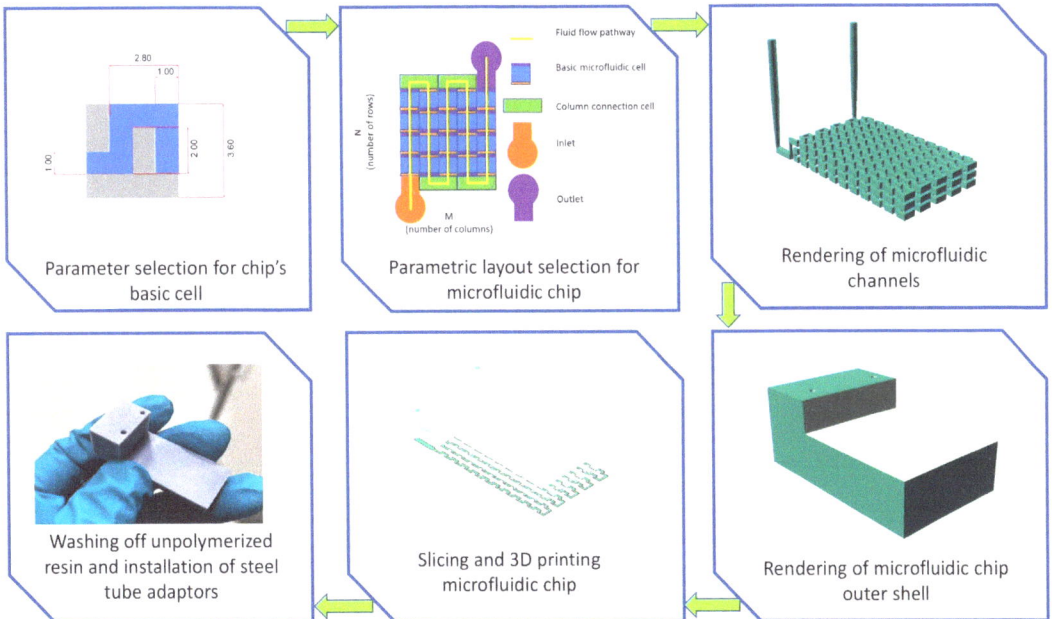

Figure 1. Schematic of formation of three-dimensional microfluidic chips.

One of the requirements for the microfluidic chip to be created is the ability to reproduce a large number of syntheses. Attempts to reduce the cross-sectional size caused destruction of the channels. Therefore, a cross-section of 1 mm^2 was chosen in order to avoid channel plugging and increase the chip's lifetime. The length of internal channels and volume of the chip are 1080 mm and 1.8 mL, respectively. In this work, it is planned to change the thermal treatment time by changing the rate of precursor injection into the microfluidic chip. The time for classical QD AIS synthesis in the flask is approximately 30 min (1800 s). Since the microfluidic chip should be compact, internal channels with a total volume of 1.8 mL were created. It allows 1 mL of precursors to flow through an internal volume in 30 min. To ensure the compactness of the chip, the base cells composing the channel were arranged in a three-dimensional matrix of ($6 \times 10 \times 5$) elements. An inlet channel and an outlet channel were connected to the main channel for fluid input and to the outlet channels of the microfluidic chip.

Figure 1 shows a 3D model of the microfluidic chip in a section and photos of chips formed using photopolymer 3D printing. To form physical copies of the chips using additive manufacturing techniques, the chip design was subtracted from the solid block, thus forming the cavities of the microfluidic channels. Microfluidic chips were printed on a 3D printer using Phrozen TR250LV photopolymer resin (Phrozen, Taiwan, China). The choice of this photopolymer resin is justified by its outstanding physical and mechanical characteristics compared with the usual photopolymer resins used in 3D printing. In particular, after photopolymerization, this resin is characterized by an increased mechanical tensile strength in the temperature range from 0 to 100 °C, which is ideal for use in hydrothermal types of nanoparticles synthesis, where the temperature inside the reactor reaches 90 °C.

The chip printing parameters for this photopolymer resin were as follows: the thickness of one layer and irradiation time per layer were 50 µm and 2 s, respectively. After printing finalization, the channels of the chip and its surface contained a large amount of unpolymerized resin. To remove it, the chip was placed in a bath filled with isopropanol and exposed to ultrasound for 10 min. After that, the internal channels of the chip were flushed with pure isopropanol using a syringe pump connected to the chip inlet.

The syringe pumps were connected to the microfluidic chip using Teflon tubes with inner and outer diameters of 1 and 1.5 mm, and steel adapters with inner and outer diameters of 0.9 and 1.1 mm, respectively. The microfluidic chip connection itself was performed in two stages. At the first stage, the steel adapter was inserted into the Teflon tube with one end, and the other end was inserted into the inlet or outlet port of the microfluidic chip. At the second stage, to seal the chip, all joints were smeared with several layers of photopolymer resin followed by polymerization of the joints using UV light irradiation with a wavelength of 405 nm for 10–30 s.

3.2. Synthesis of AIS Quantum Dots in a Microfluidic Chip

The amount of precursors and their ratio was the same for microfluidic and classical syntheses (Section 2). In the case of classical hydrothermal synthesis, the precursor mixture was placed in a three-head flask, which was further subjected to stirring and thermal treatment. In the second case, to heat the reagents, the microfluidic chip was placed in a water bath mounted on a heating element. The temperature of water in the flask and in the chip was set to 40, 60, and 90 °C and was maintained by feedback from the heating element and the temperature sensor located in the water bath. These temperature regimes were chosen to compare the optical properties of AIS QD. According to the synthetic protocol [17], the duration of thermal treatment synthesis is about 30 min. Therefore, a 1800 s treatment time was used as a starting point for both classical and microfluidic synthesis for an explicit comparison in efficiencies between these two types of synthesis. Additionally, 180 and 18 s treatment times were used to establish the dependence of the optical properties of quantum dots on the precursor treatment time. In order to sustain such treatment time in microfluidic devices, various precursor feeding rates were used, thereby precisely adjusting the temperature exposure time of the precursor solution. The precursor mixture was fed using a syringe pump at 0.001, 0.01 and 0.1 mL/s. It allows the same volume of precursor mixture, i.e., 1 mL, to pass through our chip in 18, 180, and 1800 s, respectively. After synthesis, QDs solutions were washed with the addition of isopropyl alcohol in a 1:2 ratio and centrifuged for 3 min at 12,000 min^{-1}. Then, they were redissolved in water. It is noteworthy that all syntheses were carried out in one chip. After the eight syntheses mentioned in this work, no external macromechanical damage was found. All connections of Teflon hoses that supply and divert precursor solution to the microfluidic reactor remained sealed. This confirms the possibility of extended utilization of such reactors in syntheses carried out at the temperature of up to 90 °C.

3.3. Dependence of $AgInS_2$ QD Optical Properties on Synthesis Temperature

Figure 2 shows the absorption and photoluminescence (PL) spectra of the AIS QDs synthesized in the microfluidic chip at 40 °C, 60 °C, and 90 °C for 1800 s.

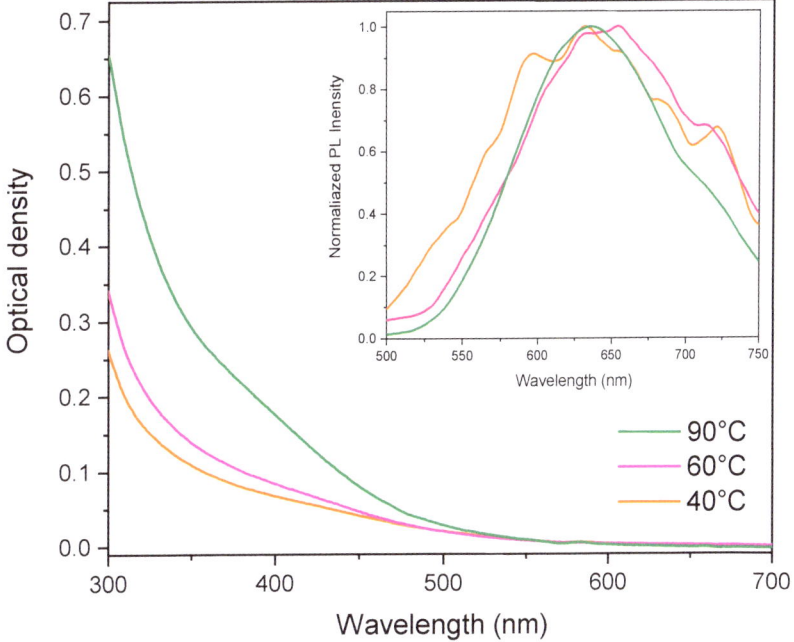

Figure 2. Absorption and normalized PL spectra (the inset) of $AgInS_2$ quantum dots synthesized in a microfluidic chip; PL excitation wavelength is 405 nm.

Figure 2 shows that the microfluidic synthesized ensembles of AIS quantum dots demonstrate typical absorption and PL spectra for ternary QDs [19,20]. It is well postulated that the complex shape of the PL band of AIS QDs is attributed to several radiative transitions [17]. Luminescence in $AgInS_2$ and $CuInS_2$ QDs is primarily attributed to defects in the crystal lattice of sulfur, indium, silver, and copper atoms, rather than excitonic processes [21]. The presence of a long-wave luminescence shoulder is a result of the recombination of the electron–hole pair at defective indium sites [22].

It is not surprising that increasing the temperature of QD synthesis leads to increased optical density in the QD absorption spectrum because of increase in precursor reaction rates [23] and improvement in QD crystallinity [24]. PLQY of QD ensembles estimated in a comparative method with Rhodamine 6G as the reference demonstrates the linear dependence on synthesis time, as shown in Figure 3.

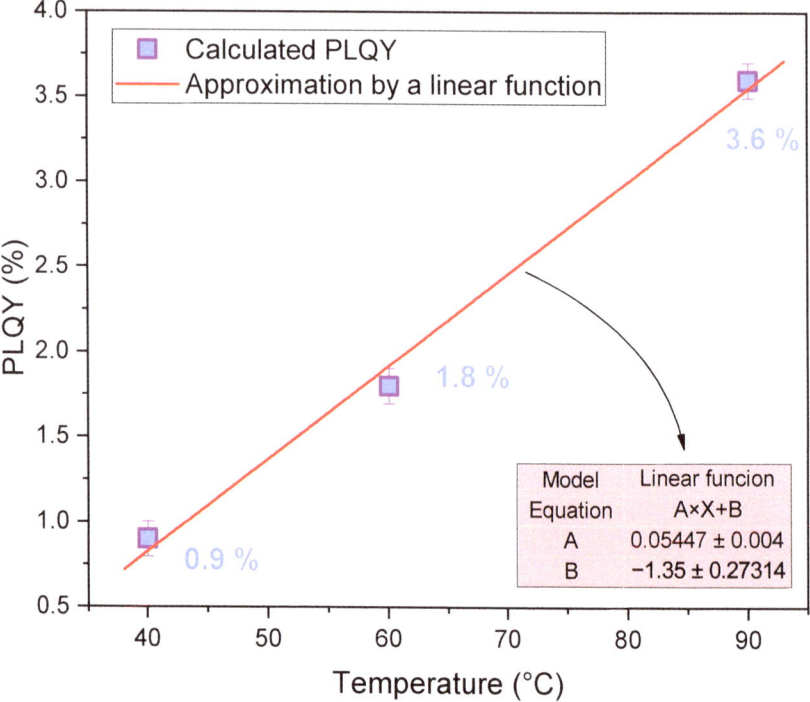

Figure 3. The dependence of the PLQY of the AIS QDs synthesized for 1800 s on the synthesis temperature and its linear fitting.

Figure 3 shows that increase in the synthesis temperature of AgInS$_2$ QDs leads to a linear increase in PLQY. The PL quantum yield of samples synthesized in the microfluidic chip at 40, 60, and 90 °C was 0.9, 1.8, and 3.6%, respectively. Since the microfluidic synthesis of the AIS QDs at 90 °C allows achieving the best PLQY in the range of 40–90 °C, this temperature was used for 18 and 180 s QDs synthesis and for the study on how synthesis time impacts QD optical properties and their yield.

3.4. Impact of Synthesis Time on the Optical Properties of AgInS$_2$ QDs

Figure 4 presents the absorption and photoluminescence excitation (PLE) spectra of AgInS$_2$ QD synthesized in a microfluidic chip and in a flask at 90 °C for precursor thermal treatment times of 18, 180, and 1800 s in the reactor.

AIS QDs synthesized in our microfluidic chip demonstrate a more pronounced optical absorption edge band in comparison with QDs synthesized in a flask (Figure 4a,b). Typically, electronic absorption bands of QDs are more sharp in their PLE spectra [25]. It allows using PLE spectra (Figure 4c,d) of QDs to estimate the absorption edge band position of QDs more correctly in comparison with their absorption spectra. Our data clearly show that increasing the synthesis time of up to 1800 s in the microfluidic chip leads to the shift in the QD absorption band to higher energy because of decreasing QD size [26].

Figure 5 presents normalized PL spectra of AgInS$_2$ quantum dots synthesized in a microfluidic chip and in a flask at 90 °C for precursor thermal treatment times of 18, 180, and 1800 s in the reactor.

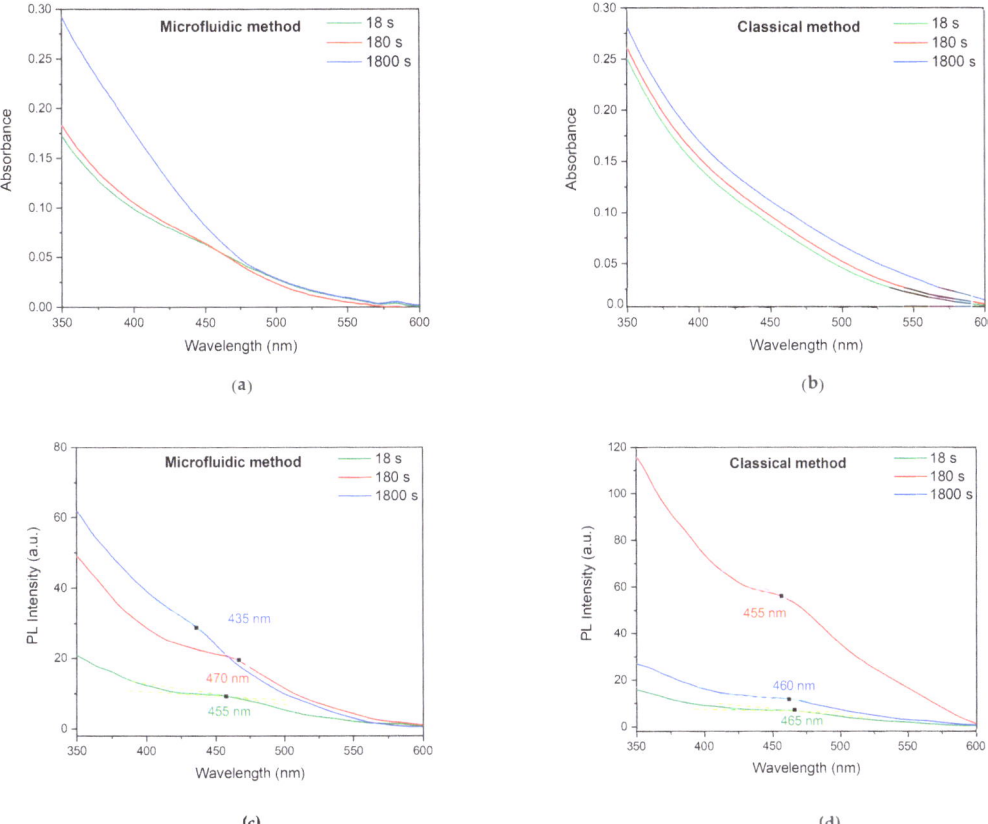

Figure 4. Absorption (**a**,**b**) and PLE (**c**,**d**), spectra of the AIS QDs synthesized in a microfluidic chip and by classical hydrothermal method in a flask at 90 °C and different thermal treatment times. PL acquisition wavelength is 610 nm.

Figure 5 shows that QDs synthesized in a flask exhibit classical dynamics of nanocrystal growth accompanied by a gradual increase in luminescence intensity and a long-wave shift of the PL band [27]. The PL band shift to the higher energy region because of increasing microfluidic synthesis time correlates very well with the shift of PL band absorption spectrum (Figure 4) and suggests the decrease in QD size.

Thermal treatment times of 18, 180, and 1800 s correspond to precursor injection rates of 0.1, 0.01, and 0.001 mL, respectively. Such rates ensure the laminar fluid flow [28]; therefore, the reagents in the channel mix mainly by diffusion and convection of molecules under the influence of ambient temperature. According to [19,26,29], the shift of the QD PL band to the higher-energy region indicates a decrease in the proportion of silver atoms in the AIS QDs and disappearance of the shoulder in the PL band of AIS QDs located at 700 nm.

A comprehensive study of the optical and photophysical properties of the AIS quantum dots in [26] allows us to suggest that the average size of our QDs is varied from 1.4 to 2 nm because of the QD absorption (Figure 4) and PL spectra (Figure 5).

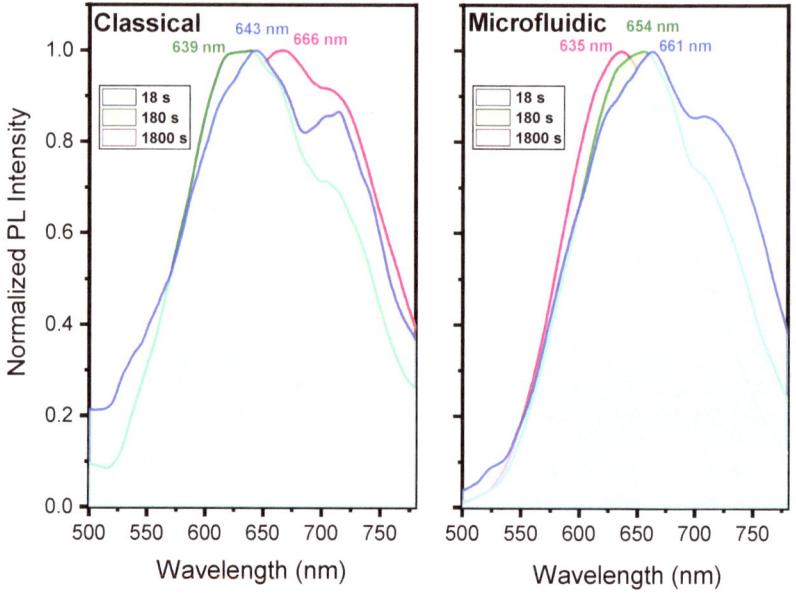

Figure 5. Normalized PL spectra of the AIS QDs synthesized in a microfluidic chip and by classical hydrothermal method in a flask at 90 °C and different thermal treatment times, PL excitation wavelength is 405 nm.

Figure 6 shows the dependence of the PL QY of the classical and microfluidic syntheses of AIS QDs on the synthesis time.

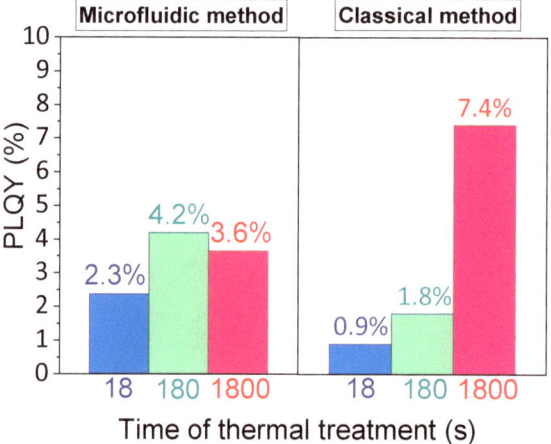

Figure 6. Dependence of PLQY of AIS QDs synthesized at 90 °C on the synthesis time in the microfluidic chip and in a flask.

It was found that PLQY of AIS QDs synthesized with the classical approach reaches 7.4%. It correlates very well with the PLQY of typical hydrophilic AIS QDs [30]. The best QY PL for our AIS QDs synthesized in the microfluidic chip is close to 4%. The increase in microfluidic synthesis time up to 1800 s not only reduces the QD size, shifts absorption and PL bands to higher energy, and decreases PL bandwidth, but also slightly decreases their

PLQY. Our data demonstrate that PLQY of QDs synthesized in the microfluidic chip for 18 and 180 s is 2.5 times higher than for QDs synthesized in a flask for the same time (Figure 6). It is assumed that the difference in dependencies of PLQY of QDs on synthesis time for microfluidic and classical syntheses is related to the high coefficient of mass temperature transfer in the reagent mixture in the microfluidic chip [31].

Figure 7 shows the dependences of the AIS QD PL kinetics of QD synthesized by hydrothermal and microfluidic methods on the synthesis time. The experimental data were approximated with three-exponential functions according to Formula (2).

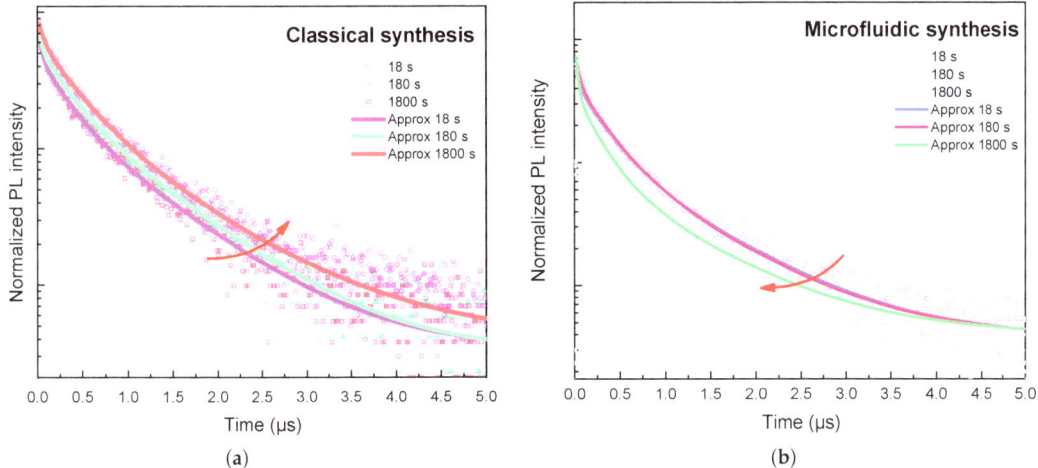

Figure 7. PL decay curves of AIS QDs synthesized at 90 °C during 18 s, 180 s, and 1800 s: (**a**) classical and (**b**) microfluidic syntheses. Red arrows show the dynamics of decay curves with changes in heat treatment time.

PL decay curves for all AIS QD samples shown in Figure 7 were fitted with three-exponential functions which is typical for ternary QDs [32]. The fitting parameters of PL decay curves for all AIS QD samples are presented in Table 1.

Table 1. The fitting parameters of PL decay curves for AIS QD samples.

Type of Synthesis	Time of Synthesis, s	τ_1, ns	τ_2, ns	τ_3, ns	A_1, %	A_2, %	A_3, %
Classical	18	35 ± 3	273 ± 13	818 ± 21	39.3	41.5	39.2
Classical	180	42 ± 3	219 ± 9	810 ± 13	31.5	46.1	22.4
Classical	1800	55 ± 5	370 ± 11	1026 ± 35	24.3	52.6	23.1
Microfluidic	18	32 ± 2	276 ± 10	911 ± 25	36.4	43.9	19.8
Microfluidic	180	40 ± 3	287 ± 6	915 ± 12	35.3	45.6	19.1
Microfluidic	1800	28 ± 1	249 ± 5	993 ± 15	46.2	38.6	15.3

The analysis of PL decay curves of classical synthesized AIS QDs shows that an increase in the synthesis time from 18 to 1800 s leads to an increase in all characteristic PL decay times. This trend correlates very well with the increase in PLQY because of increasing synthesis time (see Figure 6). The correlation of characteristic PL decay times with PLQY also remains for microfluidic-synthesized AIS QDs for the shortest (30–40 ns) and the middle (250–290 ns) components, but not for the longest component (900–1000 ns). The increase in characteristic PL decay times can indicate an improved surface quality or an increase in size of the synthesized QDs [33]. The shortest PL decay time belongs to defects of the crystal lattice [34]. Therefore, it can be an effective marker of the quality of the QD surface. The data presented in Table 1 clearly indicate a gradual decrease in the number of defects on the surface of QDs synthesized by the microfluidic method due to a decrease in the contribution of the short-lived component of QDs luminescence. It is worth noting that, according to the already published works, the long-lived components of the three-component QDs luminescence (about 100–900 ns) belong to the luminescence at the donor–acceptor pair transitions [22].

The ratio of yield of QDs synthesized at different regimes in a microfluidic chip and a flask can be estimated because the same volume of precursor mixtures with the same precursor concentrations was used in all our syntheses (for details, see Appendix A). QDs are nanocrystals whose absorption spectra depend on QD size because of the quantum confinement effect [35]. This fact is widely used to estimate the Cd-based QD size and extinction coefficients [36]. Although absorption spectra of ternary QDs based on silver or copper contain no sharp bands, this approach has also been applied to estimate their sizes and extinction coefficients. It was demonstrated that the position of the fundamental edge in CIS QD absorption spectra allows estimating QD size and extinction coefficient [18]. It is well known that stoichiometry, electronic structure and optical properties of CIS and AIS QDs are very similar [37]. Therefore, the concentration of our QD samples was estimated using the absorption, PL and PLE spectra of our spectra. For details of QD concentration estimation of our samples, refer to Appendix A. Table 2 accumulates the absorption band position (ABP), molar extinction coefficient (MEC), and QD concentration.

Table 2. Calculations of the concentration of AIS QD solutions prepared by microfluidic synthesis and by classical hydrothermal synthesis at 90 °C with treatment times of 18, 180 and 1800 s.

Type of Synthesis	Time of Synthesis, s	ABP, nm	MEC, 10^5 M^{-1}cm^{-1}	Optical Density	Concentration, 10^{-6} M
Classical	18	465	0.12692	0.063	4.9
Classical	180	455	0.10944	0.088	8.6
Classical	1800	460	0.13753	0.097	7.1
Microfluidic	18	455	0.10944	0.057	5.2
Microfluidic	180	470	0.13217	0.049	3.7
Microfluidic	1800	435	0.09269	0.105	11.3

According to the data presented in Table 2 and Appendix A, the relative QD yield ratio of classical and microfluidic synthesis strongly depends on their synthesis time. The relative yield of the AIS quantum dot product is about 1.6, 0.4, 1 for heat treatment times of 1800, 180, and 18 s, respectively. It is intriguing that the QD yields in classical and microfluidic syntheses are the same for the shortest synthesis time, i.e., 18 s. It means that the efficiency of precursor mixing is about the same for both classical and microfluidic syntheses and there are no advantages to using the latter one. In contrast, our data demonstrate that increasing the synthesis time up to 180 s leads to a strong decrease of the QD yield ratio. Therefore, despite the relatively good PLQY of AIS QDs synthesized with microfluidic approach, it is suggested that 180 s is not optimal for microfluidic synthesis of AIS QDs in our chip. At the same time, our data strongly confirm the increase in QD synthesis yield up to 60% in comparison with classical one if typical classical AIS QD synthesis time [17] is used.

4. Conclusions

This study introduces a novel technique for the flow hydrothermal synthesis of three-component AgInS$_2$ quantum dots using additive manufacturing to form microfluidic chips. It demonstrates the capability of operating photopolymer resin-based chips for prolonged durations without developing any structural defects that could compromise the mechanical properties of the microfluidic chip channels. Based on our analysis of the photophysical properties of the AIS QDs, the microfluidic chip exhibited a significantly higher coefficient of mass and heat transfer compared to the conventional flask reactor. The photoluminescence quantum yield of samples synthesized by the microfluidic method for 18 and 180 s is found to be about 2.5 times higher than that for QDs synthesized in a flask for the same time. It is demonstrated that the product yield of AIS quantum dots synthesized in the microfluidic chip is up to 60% higher than in a flask for the treatment time, which is typical for AIS QD classical synthesis, i.e., 1800 s.

Author Contributions: Conceptualization, I.R., K.B. and A.O.; methodology, I.R.; writing—original draft preparation and investigation, K.B., I.R. and S.K.; formal analysis, J.W.S. and S.M.; validation and supervision, A.O. All authors have read and agreed to the published version of the manuscript.

Funding: This work was financially supported by Ministry of Education and Science of the Russian Federation: State assignment, Passport 2019-1080 (Goszadanie 2019-1080), and by RPMA grant of School of Physics and Engineering of ITMO University (N° 621317).

Institutional Review Board Statement: Not applicable.

Informed Consent Statement: Not applicable.

Data Availability Statement: Data is available upon request.

Conflicts of Interest: The authors declare no conflict of interest.

Appendix A

Since the optical properties and structure of the AIS and CIS QDs are similar, and the number of works aimed at studying the optical properties of the CIS QDs prevails, the observed optical properties can be explained in the first approximation by generalizing the existing results about the CIS QDs to the AIS QDs. To determine the concentration of the solutions, this approximation was used. In work [18], the dependence of the CIS QD size on the wavelength of the fundamental absorbance edge (WFAE) and the dependence of the CIS QD size on the molar extinction coefficient (MEC) were experimentally obtained. Thus, the known location of the absorbtion band position of AIS QD allows us to determine the concentration of nanoparticles, given their size, via the Bouguer–Lambert–Beer Formula (A1):

$$D = C \cdot \varepsilon \cdot l, \qquad (A1)$$

where D is optical density; ε is a molar extinction coefficient; l is distance; C is concentration.

In order to determine the WFAE, two tangents to the graph of the luminescence excitation intensity spectra were plotted (Figure 4c,d). The obtained wavelengths were compared with the particle size [18] and are written in Table 2. The resulting AIS QD sizes were compared with the extinction coefficients in work [18] used to determine the concentration. Then, using the absorption spectra (Figure 4a,b), the optical densities were determined. Using Formula (A1), concentrations were obtained, provided that l is 1 cm.

To determine the relative yield of the AIS quantum dot product, the ratio of concentrations of samples obtained by different methods for the same heat treatment time was calculated. Thus, the microfluidic synthesis/classical synthesis ratio is about 1.6, 0.4, 1 for heat treatment times of 1800, 180, and 18 s, respectively.

References

1. Mi, W.; Tian, J.; Tian, W.; Dai, J.; Wang, X.; Liu, X. Temperature dependent synthesis and optical properties of CdSe quantum dots. *Ceram. Int.* **2012**, *38*, 5575–5583. [CrossRef]
2. Bian, F.; Sun, L.; Cai, L.; Wang, Y.; Zhao, Y. Quantum dots from microfluidics for nanomedical application. *Wiley Interdiscip. Rev. Nanomed. Nanobiotechnol.* **2012**, *11*, 1567. [CrossRef]
3. Abedini-Nassab, R.; Pouryosef Miandoab, M.; Şaşmaz, M. Microfluidic Synthesis, Control, and Sensing of Magnetic Nanoparticles: A Review. *Micromachines* **2021**, *12*, 768. [CrossRef]
4. Shepherd, S.J.; Issadore, D.; Mitchell, M.J. Microfluidic formulation of nanoparticles for biomedical applications. *Biomaterials* **2021**, *274*, 120826. [CrossRef]
5. Niculescu, A.-G.; Chircov, C.; Bîrcă, A.C.; Grumezescu, A.M. Nanomaterials Synthesis through Microfluidic Methods: An Updated Overview. *Nanomaterials* **2021**, *11*, 864. [CrossRef] [PubMed]

6. Ma, J.; Lee, S.M.-Y.; Yi, C.; Li, C.-W. Controllable synthesis of functional nanoparticles by microfluidic platforms for biomedical applications—A review. *Lab Chip* **2017**, *17*, 209–226. [CrossRef]
7. Lai, X.; Lu, B.; Zhang, P.; Zhang, X.; Pu, Z.; Yu, H.; Li, D. Sticker Microfluidics: A Method for Fabrication of Customized Monolithic Microfluidics. *Acs Biomater. Sci. Eng.* **2019**, *5*, 6801–6810. [CrossRef] [PubMed]
8. Kajtez, J.; Buchmann, S.; Vasudevan, S.; Birtele, M.; Rocchetti, S.; Pless, C.J.; Heiskanen, A.; Barker R.A.; Martínez-Serrano, A.; Parmar, M.; et al. 3D-Printed Soft Lithography for Complex Compartmentalized Microfluidic Neural Devices. *Adv. Sci.* **2020**, *7*, 2198–3844. [CrossRef] [PubMed]
9. Ma, H.; Pan, L.; Wang, J.; Zhang, L.; Zhang, Z. Synthesis of AgInS2 QDs in droplet microreactors: Online fluorescence regulating through temperature control. *Chin. Chem. Lett.* **2019**, *30*, 79–82. [CrossRef]
10. Weisgrab, G.; Ovsianikov, A.; Costa, P.F. Functional 3D Printing for Microfluidic Chips. *Adv. Mater. Technol.* **2019**, *4*, 1900275. [CrossRef]
11. Cheng, Y.; Ling, S.D.; Geng, Y.; Wang, Y.; Xu, J. Microfluidic synthesis of quantum dots and their applications in bio-sensing and bio-imaging. *Nanoscale Adv.* **2021**, *3*, 2180–2195. [CrossRef] [PubMed]
12. Girma, W.M.; Fahmi, M.Z.; Permadi, A.; Abate, M.A.; Chang, J.-Y. Synthetic strategies and biomedical applications of I–III–VI ternary quantum dots. *J. Mater. Chem.* **2021**, *5*, 6193–6216. [CrossRef]
13. May, B.M.; Bambo, M.F.; Hosseini, S.S.; Sidwaba, U.; Nxumalo, E.N.; Mishra, A.K. A review on I–III–VI ternary quantum dots for fluorescence detection of heavy metals ions in water: Optical properties, synthesis and application. *RSC Adv.* **2022**, *12*, 11216–11232. [CrossRef] [PubMed]
14. Baranov, K.; Kolesova, E.; Baranov, M.; Orlova, A. Generation of Reactive Oxygen Species by AgInS$_2$/TiO$_2$ Nanocomposites upon Exposure to UV and Visible Radiation. *Opt. Spectrosc.* **2022**, *130*, 1562–6911. [CrossRef]
15. Wang, G.; Jin, W.; Qasim, A.M.; Gao, A.; Peng, X.; Li, W.; Feng, H.; Chu, P.K. Antibacterial effects of titanium embedded with silver nanoparticles based on electron-transfer-induced reactive oxygen species. *Biomaterials* **2017**, *124*, 25–34. [CrossRef]
16. Mir, I.A.; Radhakrishanan, V.S.; Rawat, K.; Prasad, T.; Bohidar, H.B. Bandgap Tunable AgInS based Quantum Dots for High Contrast Cell Imaging with Enhanced Photodynamic and Antifungal Applications. *Sci. Rep.* **2018**, *8*, 9322. [CrossRef]
17. Raevskaya, A.; Lesnyak, V.; Haubold, D.; Dzhagan, V.; Stroyuk, O.; Gaponik, N.; Zahn, D.R.T.; Eychmüller, A. A Fine Size Selection of Brightly Luminescent Water-Soluble Ag-In-S and Ag-In-S/ZnS Quantum Dots. *J. Phys. Chem.* **2017**, *121*, 9032–9042. [CrossRef]
18. Qin, L.; Li, D.; Zhang, Z.; Wang, K.; Ding, H.; Xie, R.; Yang, W. The determination of extinction coefficient of CuInS$_2$, and ZnCuInS$_3$ multinary nanocrystals. *Nanoscale* **2012**, *4*, 6360. [CrossRef]
19. Hu, X.; Chen, T.; Xu, Y.; Wang, M.; Jiang, W.; Jiang, W. Hydrothermal synthesis of bright and stable AgInS$_2$ quantum dots with tunable visible emission. *J. Lumin.* **2018**, *200*, 189–195. [CrossRef]
20. May, B.M.M.; Parani, S.; Oluwafemi, O.S. Detection of Ascorbic Acid using Green Synthesized AgInS$_2$ Quantum Dots. *Mater. Lett.* **2018**, *236*, 432–435. [CrossRef]
21. Zang, H.; Li, H.; Makarov, N.S.; Velizhanin, K.A.; Wu, K.; Park, Y.-S.; Klimov, V.I. Thick-Shell CuInS$_2$/ZnS Quantum Dots with Suppressed "Blinking" and Narrow Single-Particle Emission Line Widths. *Nano Lett.* **2017**, *17*, 1787–1795. [CrossRef]
22. Fu, M.; Luan, W.; Tu, S.-T.; Mleczko, L. Optimization of the recipe for the synthesis of CuInS$_2$/ZnS nanocrystals supported by mechanistic considerations. *Green Process. Synth.* **2017**, *6*, 133–146. [CrossRef]
23. Zikalala, N.; Parani, S.; Tsolekile, N.; Oluwafemi, O.S. Facile green synthesis of ZnInS quantum dots: Temporal evolution of their optical properties and cell viability against normal and cancerous cells. *J. Mater. Chem.* **2020**, *8*, 9329–9336. [CrossRef]
24. Mahmoud, W.E.; Yaghmour, S.J. The temporal evolution of the structure and luminescence properties of CdSe semiconductor quantum dots grown at low temperatures. *J. Lumin.* **2012**, *132*, 2447–2451. [CrossRef]
25. Tonti, D.; van Mourik, F.; Chergui, M. On the Excitation Wavelength Dependence of the Luminescence Yield of Colloidal CdSe Quantum Dots. *Nano Lett.* **2014**, *4*, 2483–2487. [CrossRef]
26. Stroyuk, O.; Raievska, O.; Kupfer, C.; Solonenko, D.; Osvet, A.; Batentschuk, M.; Brabec, C.J.; Zahn, D.R.T. High-Throughput Time-Resolved Photoluminescence Study of Composition- and Size-Selected Aqueous Ag-In-S Quantum Dots. *J. Phys. Chem.* **2021**, *125*, 12185–12197. [CrossRef]
27. Cheng, O.H.-C.; Qiao, T.; Sheldon, M.T.; Son, D.H. Size- and Temperature-dependent Photoluminescence Spectra of Strongly Confined CsPbBr3 Quantum Dots. *Nanoscale* **2020**, *12*, 13113–13118. [CrossRef]
28. Song, Y.; Hormes, J.; Kumar, C.S.S.R. Size- and Temperature-dependent Microfluidic Synthesis of Nanomaterials. *Small* **2008**, *4*, 698–711. [CrossRef]
29. Moodelly, D.; Kowalik, P.; Bujak, P.; Pron, A.; Reiss, P. Synthesis, photophysical properties and surface chemistry of chalcopyrite-type semiconductor nanocrystals. *J. Mater. Chem.* **2019**, *7*, 11665–11709. [CrossRef]
30. Kang, X.; Huang, L.; Yang, Y.; Pan, D. Scaling up the Aqueous Synthesis of Visible Light Emitting Multinary AgInS$_2$/ZnS Core/Shell Quantum Dots. *J. Phys. Chem.* **2015**, *119*, 7933–7940. [CrossRef]
31. Wang, J.; Shao, C.; Wang, Y.; Sun, L.; Zhao, Y. Microfluidics for Medical Additive Manufacturing. *Engineering* **2020**, *6*, 1244–1257. [CrossRef]
32. Lazareva, A.A.; Reznik, I.A.; Dubavik, A.Y.; Veniaminov A.V.; Orlova A.O. Investigation of photoluminescence kinetics CuInS$_2$/ZnS quantum dots. *J. Phys. Conf. Ser.* **2021**, *2058*, 012007. [CrossRef]

33. Sun, J.; Ikezawa, M.; Wang, X.; Jing, P.; Li, H.; Zhao, J.; Masumoto, Y. Photocarrier recombination dynamics in ternary chalcogenide CuInS$_2$ quantum dots. *Phys. Chem. Chem. Phys.* **2015**, *17*, 11981–11989. [CrossRef]
34. Motevich, I.G.; Zenkevich, E.I.; Stroyuk, A.L.; Raevskaya, A.E.; Kulikova, O.M.; Sheinin, V.B.; Koifman, O.I.; Zahn, D.R.T.; Strekal, N.D. Effect of pH and Polyelectrolytes on the Spectral-Kinetic Properties of AIS/ZnS Semiconductor Quantum Dots in Aqueous Solutions. *J. Appl. Spectrosc.* **2021**, *87*, 1057–1066. [CrossRef]
35. Empedocles, S.A.; Norris, D.J.; Bawendi, M.G. Photoluminescence Spectroscopy of Single CdSe Nanocrystallite Quantum Dots. *Phys. Rev. Lett.* **1996**, *77*, 3873–3876. [CrossRef]
36. Yu, W.W.; Qu, L.; Guo, W.; Peng, X. Experimental Determination of the Extinction Coefficient of CdTe, CdSe, and CdS Nanocrystals. *Chem. Mater.* **2003**, *15*, 2854–2860. [CrossRef]
37. Liu, L.; Hu, R.; Law, W.-C.; Roy, I.; Zhu, J.; Ye, L.; Hu, S.; Zhang, X.; Yong, K.-T. Optimizing the synthesis of red- and near-infrared CuInS$_2$ and AgInS$_2$ semiconductor nanocrystals for bioimaging. *Analyst* **2013**, *138*, 6140. [CrossRef]

Disclaimer/Publisher's Note: The statements, opinions and data contained in all publications are solely those of the individual author(s) and contributor(s) and not of MDPI and/or the editor(s). MDPI and/or the editor(s) disclaim responsibility for any injury to people or property resulting from any ideas, methods, instructions or products referred to in the content.

Communication

Moisture Condensation on Epitaxial Graphene upon Cooling

Muhammad Farooq Saleem [1,2], Niaz Ali Khan [3], Muhammad Javid [4], Ghulam Abbas Ashraf [5], Yasir A. Haleem [6], Muhammad Faisal Iqbal [7], Muhammad Bilal [8], Peijie Wang [9,*] and Lei Ma [1,*]

[1] Tianjin International Center for Nanoparticles and Nanosystems, Tianjin University, Tianjin 300072, China
[2] GBA Branch of Aerospace Information Research Institute, Chinese Academy of Sciences, Guangzhou 510700, China
[3] Hubei Key Laboratory of Advanced Textile Materials & Application, Wuhan Textile University, Yangguang Road 1, Wuhan 430200, China
[4] Institute of Advanced Magnetic Materials, College of Materials and Environmental Engineering, Hangzhou Dianzi University, Hangzhou 310012, China
[5] Key Laboratory of Integrated Regulation and Resources Development on Shallow Lake, Ministry of Education, College of Environment, Hohai University, Nanjing 210098, China
[6] Department of Physics, Khwaja Fareed University of Engineering and Information Technology, Rahim Yar Khan 64200, Pakistan
[7] Department of Physics, Riphah International University, Faisalabad 38000, Pakistan
[8] College of Materials Science and Technology, Nanjing University of Aeronautics and Astronautics, Nanjing 210016, China
[9] The Beijing Key Laboratory for Nano-Photonics and Nano-Structure, Department of Physics, Capital Normal University, Beijing 100048, China
* Correspondence: pjwan@cnu.edu.cn (P.W.); maleixinjiang@tju.edu.cn (L.M.)

Citation: Saleem, M.F.; Khan, N.A.; Javid, M.; Ashraf, G.A.; Haleem, Y.A.; Iqbal, M.F.; Bilal, M.; Wang, P.; Ma, L. Moisture Condensation on Epitaxial Graphene upon Cooling. *Technologies* 2023, 11, 30. https://doi.org/10.3390/technologies11010030

Academic Editors: Sergey N. Grigoriev, Marina A. Volosova and Anna A. Okunkova

Received: 25 November 2022
Revised: 8 February 2023
Accepted: 11 February 2023
Published: 13 February 2023

Copyright: © 2023 by the authors. Licensee MDPI, Basel, Switzerland. This article is an open access article distributed under the terms and conditions of the Creative Commons Attribution (CC BY) license (https://creativecommons.org/licenses/by/4.0/).

Abstract: Condensation of moisture on the epitaxial graphene on 6H-SiC was observed below room temperature despite continuous nitrogen flow on the graphene surface. Raman peaks associated with ice were observed. A combination of peaks in the frequency range of 500–750 cm^{-1}, along with a broad peak centered at ~1327 cm^{-1}, were also observed and were assigned to airborne contaminants. The latter is more important since its position is in the frequency range where the defect-associated D band of graphene appears. This band can be easily misunderstood to be the D band of graphene, particularly when the Raman spectrum is taken below room temperature. This peak was even observed after the sample was brought back to room temperature due to water stains. This work highlights the importance of careful Raman investigation of graphene below room temperature and its proper insulation against moisture.

Keywords: epitaxial graphene; Raman scattering; condensation; adsorption

1. Introduction

The 2D materials have drawn considerable interest in the last few decades for their unique properties and important device applications [1]. Particularly, graphene has been studied widely compared with any other 2D material since its discovery [2]. Its exceptional carrier mobility and favorable electronic properties have attracted much interest of device fabrication technology [3,4]. Graphene can be produced by physical and chemical methods [5]. Graphene of a precise number of layers can be achieved by optimizing the growth recipes on metal, semiconducting and insulating substrates by physical growth methods. CVD and epitaxial growth methods on solid surfaces have proved to be the most promising methods to obtain large area high-quality graphene. Graphene growth on technologically important SiC substrate has been achieved on both the Si and C-faces. The properties of graphene on two faces are affected by the type of plane on which it is grown. Graphene on the Si-face is particularly different due to an unavoidable underlying buffer layer of carbon that grows between the graphene and silicon face during the growth process. Charge transfer and substrate interaction of graphene on the Si-face affect its properties more compared with the C-face. However, poor control over the number of graphene layers is the key issue on the C-face that usually yields thick graphene layers. Although the focus of this research is on graphene growth on the Si-face for its controllability on the number of layers, graphene on the C-face exhibits higher carrier mobility that makes it superior [6]. The carrier mobility of graphene is highly influenced by the number of layers; therefore, growth control is technologically very important [7,8].

The change in ambient temperature and interaction of environmental species can strongly affect the properties of atomically thin and sensitive graphene layers, which make it the material of choice for sensing applications. Graphene's wetting properties have been studied extensively and a clear change in its wetting properties has been observed simply by surface functionalizing. Many studies have given controversial and inconsistent results about graphene properties and the performance of graphene based devices so far [9]. The surface charge, defects and adsorbates are considered to be the sources of inconsistent graphene properties [9,10]. The intentional adsorption of known molecular species on graphene surfaces results in controlled doping and modification of its various electrical and optical properties. However, the unintentional adsorption of unknown ambient species can lead to degradation, inconsistent device behaviors, and poor reliability of graphene-based devices. The single-atom thick graphene surface exposed to air can have bulk aerosol and moisture deposits. Such materials trapped in moisture are more likely to adsorb on open surfaces, particularly below room temperature. It is indeed very important to keep the surface clear to be able to benefit from the intrinsic properties of highly sensitive graphene layers. In this study, we observed a change in the Raman spectrum of epitaxial graphene upon cooling due to moisture condensation on its surface that is significant and can result in the misleading evaluation of its quality if dealt with carelessly. For quality assurance and stable graphene-based device behaviors, proper surface insulation is necessary below room temperature. Raman spectroscopy is one of the most sensitive techniques that has successfully detected the contamination deposition of the graphene surface. Further investigation with similar characterization techniques can allow the detection of contaminants and their effects on the properties of graphene-based devices.

2. Experimental

2.1. Materials

Vanadium-doped semi-insulating 6H-SiC substrate was purchased from Wolfspeed (4600 Silicon Drive Durham, North Carolina 27703, USA). The thickness of 6H-SiC substrate was 369 µm. The substrates were polished on both Si and C faces.

2.2. Epitaxial Graphene Growth

The 6H-SiC substrates were ultrasonically cleaned in acetone and ethanol prior to graphene growth. Graphene was grown in a home-made RF-furnace at 1550 °C on the atomically flat 6H-SiC substrate for 25 min in a high vacuum (<10^{-10} Torr). At sufficiently high temperature, Si evaporates from the substrate leaving C atoms behind that form graphene. The main components of the furnace include a large copper coil, sample stage, quartz tube, a protective shield and an AC power supply. An alternating current is produced using the power supply in the middle RF range. It heats the sample placed in the tube. A turbo pump is used to evacuate the tube.

2.3. Measurements

Optical images of the graphene surface grown on the carbon face of the 6H-SiC substrate were taken using an optical microscope attached with RH13325 (R-2000) Raman spectrometer in the complete temperature cycle between room temperature and −180 °C using a 50× objective lens. The 532 nm excitation laser with 2 µm spot size of the same Raman spectrometer was used to obtain the in situ temperature-dependent Raman spectra of the graphene in the same temperature range. A 50× objective lens was used. The sample was placed in a hot/cold cell and the liquid nitrogen was passed on it while taking temperature-dependent Raman spectra and optical images of the graphene surface at atmospheric pressure. The Raman spectra and optical images were taken after every 30 °C change in temperature continuously in the complete temperature cycle. The temperature was allowed to stabilize before each spectrum was obtained. Low room temperature and the humid lab environment allowed increased moisture condensation on the sample surface as shown in Figure 1. High vacuum annealing of the graphene sample was performed at 500 °C before Raman spectra and optical images were taken to clean any contaminants from the graphene sample surface.

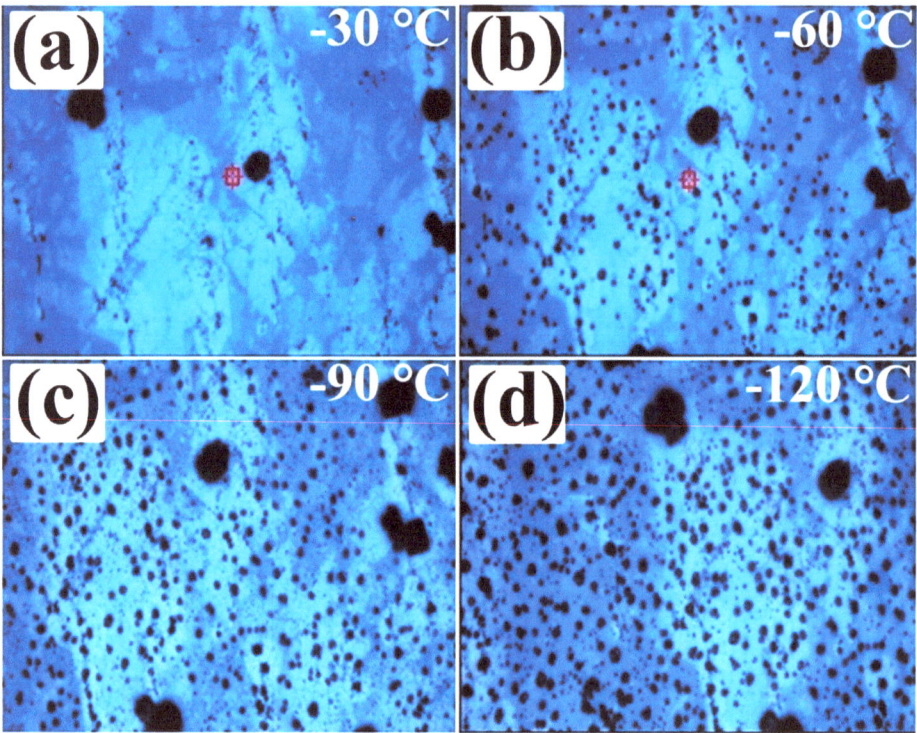

Figure 1. Optical microscopy images of the epitaxial graphene at (**a**) −30 °C, (**b**) −60 °C, (**c**) −90 °C, and (**d**) −120 °C temperatures.

3. Results and Discussion

The epitaxial graphene on 6H-SiC was annealed in a vacuum for 5 h to get rid of any moisture or contaminants before the temperature-dependent Raman spectra were obtained. However, graphene has been reported to be more chemically active for adsorption after vacuum annealing [11]. It is assumed that the adsorbates can be removed by annealing leaving active sites for more adsorbates to come which makes graphene more lipophilic. Graphene on the C-face of SiC contains wrinkles, ridges and ripples that can further facilitate adsorption [12].

Temperature-dependent Raman spectra were obtained in a complete cycle between room temperature and −180 °C. The surface of the sample was continuously observed through the optical microscope attached with Raman spectroscopy. The moisture starts to condense on the surface of the sample upon cooling despite the continuous flow of nitrogen. When the temperature reaches near −30 °C, the water droplets become noticeable on the surface as shown in optical microscopy images in Figure 1. Only big and distant droplets appear at first. The smaller droplets start to appear on the surface between the big droplets with further cooling.

An apparently clear area between the droplets was chosen to take the temperature-dependent Raman spectra but the continuous moisture condensation attenuated the Raman signal of the sample. The Raman signal was lost continuously, and the focus adjustments were made after every few degrees centigrade change in temperature. A strong background in Raman spectra was observed upon cooling that was subtracted from all the spectra for clarity. Moisture condensation and contaminants can lead to fluorescence background [13]. The contaminations resulted in some peaks other than normal Raman peaks of graphene and SiC substrate. The peaks at 227, 293, 414, 550–750, 1327, and 3200 cm^{-1} are observed upon cooling below 0 °C and remain observable during the complete temperature cycle as shown in Figure 2a,b. The peaks are assigned to the moisture and contaminants it carries [14–17]. The broad feature in the range 550–750 cm^{-1} is a combination of peaks that is due to the complex form of contaminants in the moisture. The contaminants can be airborne aerosols such as dust, soot, smoke, pollen, etc., or a combination of them. The droplets were not observable through the microscope once the sample was brought back to room temperature. Figure 2c shows the comparison of the Raman spectra for the sample before and after the complete cooling cycle. The peaks related to moisture and contaminants remain evident even when the sample is heated back to room temperature. This confirms that complete desorption of moisture and contaminants does not occur upon heating the sample back to room temperature. The hydrated contaminants can only be cleaned by high vacuum annealing of the sample. The graphene stored at ambient pressure should be annealed in a vacuum once it faces temperatures below room temperature to restore its actual properties.

The peak near 1327 cm^{-1} is significant due to its position near the D band of graphene. Low-temperature Raman spectroscopy of graphene has been reported many times [18–20]. Intentional graphene surface functionalization and molecule adsorption has also been studied extensively [21–23]. The activation of D band only by cooling or contaminating/functionalizing the surface of graphene has never been reported. Thus, the possibility of this peak being a D band can be ruled out. The peak is assigned to the contaminants that moisture carries.

In many parts of the world, -30 °C is room temperature at some point. A low temperature is often maintained in laboratories to avoid excessive heating of equipment. Graphene exposed to such environments can easily attract pollutants present in the air. A peak near 1327 cm^{-1} can be easily misinterpreted as a D band of graphene in the graphene sample exposed to open air below room temperature. The D band is used as the most reliable measure of graphene quality [24]. Graphene surface should be carefully cleaned by high vacuum annealing before Raman analysis once the surface has faced a temperature below 0 °C in open air.

To further confirm the origin of the observed peaks, the Raman laser was focused on a big droplet on the graphene surface, and temperature-dependent Raman spectra were obtained as shown in Figure 3a. Intense and clear ice signals were observed in the range of 2800–3800 cm^{-1} with peaks at 227 and 1327 cm^{-1} [25]. The Raman peaks of graphene and SiC were also observed. The ice-related broad feature in the range of 2800–3800 cm^{-1} was fitted and gave the well-known four peaks associated with ice at 3205, 3113, 3221 and 3364 cm^{-1} as shown in Figure 3b. Interestingly, the previously observed features in the range of 550–750 cm^{-1} on a surface between the droplets were not observed this time. Only a small peak near 550 cm^{-1} was observed. The water droplets and contaminants may have their preferential adsorption sites. The type of contaminants that prefer to adsorb between the water droplets and give rise to Raman features in the range of 550–750 cm^{-1} do not adsorb on water droplet formation sites. Careful graphene surface characterization and proper insulation are, hence, important in low-temperature environments.

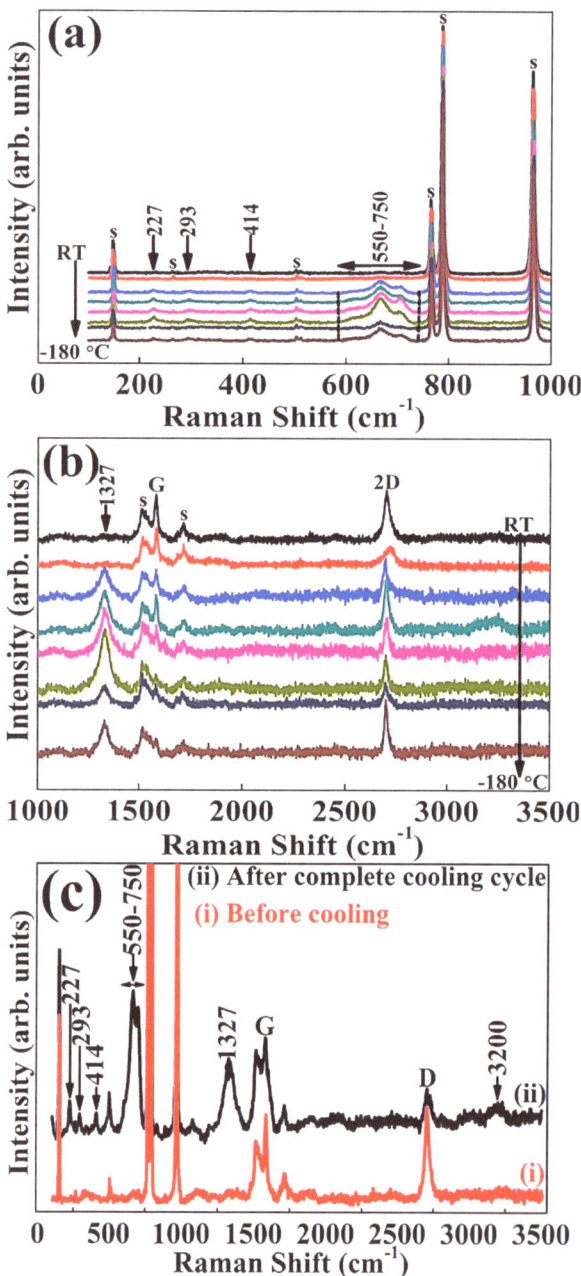

Figure 2. (**a**) Temperature-dependent Raman spectra recorded from room temperature to −180 °C in Raman frequency range of 1000–3500 cm^{-1}. (**b**) Temperature-dependent Raman spectra were recorded from room temperature to −180 °C in the Raman frequency range of 100–1000 cm^{-1}. (**c**) Raman spectra of epitaxial graphene were taken at room temperature before and after the cooling cycle.

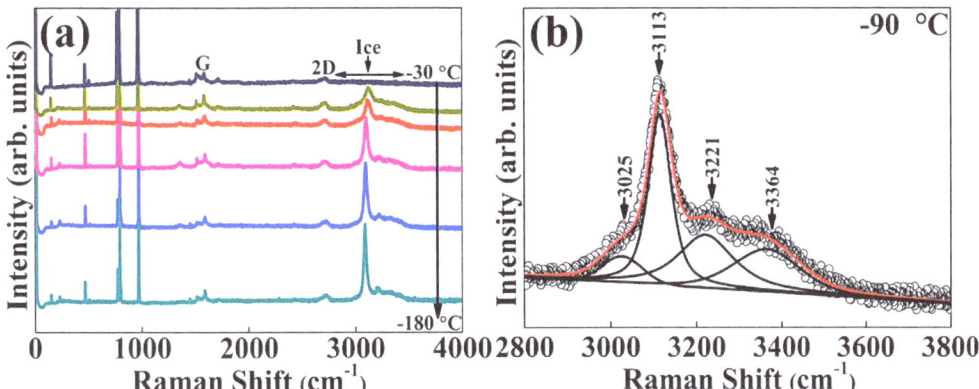

Figure 3. (**a**) Temperature-dependent Raman spectra obtained by focusing the laser to a frozen water droplet (**b**) fitting of the water-related peak obtained at −90 °C.

4. Conclusions

Condensation of moisture on the epitaxial graphene on 6H-SiC was investigated by Raman spectroscopy below room temperature. The Raman spectra showed the Raman peaks associated with ice and contaminants below 0 °C. A peak near 1327 cm^{-1} associated with contaminants is particularly important for its position since it is in the frequency range where the D band of graphene occurs. The D band is usually used as the most reliable measure of graphene quality. Careful Raman investigation of graphene and its insulation against moisture is, therefore, important below 0 °C. Proper investigation of the abundant species in the air can provide a clear understanding of the contaminants and their possible effects on graphene's intrinsic properties that should be explored further by using sensitive characterization techniques similar to Raman spectroscopy.

Author Contributions: Conceptualization, M.F.S. and L.M.; methodology, M.F.S.; software, G.A.A., M.J. and Y.A.H.; validation, L.M., M.F.I. and M.B.; formal analysis, P.W., M.F.S. and L.M.; investigation, Y.A.H., M.F.S., M.J. and G.A.A.; resources, L.M., N.A.K. and P.W.; data curation, M.F.S., Y.A.H. and M.B.; writing—original draft preparation, M.F.S., L.M., M.F.I. and M.J.; writing—review and editing, M.B. and M.F.I.; visualization, M.B., M.F.I. and M.J.; supervision, L.M., P.W. and M.F.S.; project administration, L.M. and P.W.; funding acquisition, L.M. and P.W. All authors have read and agreed to the published version of the manuscript.

Funding: This research received no external funding.

Institutional Review Board Statement: Not applicable.

Informed Consent Statement: Not applicable.

Data Availability Statement: All the data taken in this work has been published in this paper.

Conflicts of Interest: The authors declare that they have no conflict of interest.

References

1. Mondal, B.; Gogoi, P.K. Nanoscale Heterostructured Materials Based on Metal Oxides for a Chemiresistive Gas Sensor. *ACS Appl. Electron. Mater.* **2022**, *4*, 59–86. [CrossRef]
2. Novoselov, K.S.; Geim, A.K.; Morozov, S.V.; Jiang, D.; Zhang, Y.; Dubonos, S.V.; Grigorieva, I.V.; Firsov, A.A. Electric field effect in atomically thin carbon films. *Science* **2004**, *306*, 666–669. [CrossRef] [PubMed]
3. Ferrari, A.C.; Meyer, J.C.; Scardaci, V.; Casiraghi, C.; Lazzeri, M.; Mauri, F.; Piscanec, S.; Jiang, D.; Novoselov, K.S.; Roth, S.; et al. Raman Spectrum of Graphene and Graphene Layers. *Phys. Rev. Lett.* **2006**, *97*, 187401. [CrossRef] [PubMed]
4. Amanatiadis, S.; Zygiridis, T.; Kantartzis, N. Radiation Efficiency Enhancement of Graphene Plasmonic Devices Using Matching Circuits. *Technologies* **2021**, *9*, 4. [CrossRef]

5. Li, J.; Zeng, H.; Zeng, Z.; Zeng, Y.; Xie, T. Promising Graphene-Based Nanomaterials and Their Biomedical Applications and Potential Risks: A Comprehensive Review. *ACS Biomater. Sci. Eng.* **2021**, *7*, 5363–5396. [CrossRef] [PubMed]
6. Bouhafs, C.; Stanishev, V.; Zakharov, A.A.; Hofmann, T.; Kühne, P.; Iakimov, T.; Yakimova, R.; Schubert, M.; Darakchieva, V. Decoupling and ordering of multilayer graphene on C-face 3C-SiC(111). *Appl. Phys. Lett.* **2016**, *109*, 203102. [CrossRef]
7. Brzhezinskaya, M.; Kononenko, O.; Matveev, V.; Zotov, A.; Khodos, I.I.; Levashov, V.; Volkov, V.; Bozhko, S.I.; Chekmazov, S.V.; Roshchupkin, D. Engineering of Numerous Moiré Superlattices in Twisted Multilayer Graphene for Twistronics and Straintronics Applications. *ACS Nano* **2021**, *15*, 12358–12366. [CrossRef]
8. Dien, N.T.; Hirai, Y.; Koshiba, J.; Sakai, S. Factors affecting multiple persistent organic pollutant concentrations in the air above Japan: A panel data analysis. *Chemosphere* **2021**, *277*, 130356. [CrossRef]
9. Zhang, J.; Jia, K.; Huang, Y.; Liu, X.; Xu, Q.; Wang, W.; Zhang, R.; Liu, B.; Zheng, L.; Chen, H.; et al. Intrinsic Wettability in Pristine Graphene. *Adv. Mater.* **2022**, *34*, 2103620. [CrossRef]
10. Belyaeva, L.A.; van Deursen, P.M.G.; Barbetsea, K.I.; Schneider, G.F. Hydrophilicity of Graphene in Water through Transparency to Polar and Dispersive Interactions. *Adv. Mater.* **2018**, *30*, 1703274. [CrossRef]
11. Ni, Z.H.; Wang, H.M.; Luo, Z.Q.; Wang, Y.Y.; Yu, T.; Wu, Y.H.; Shen, Z.X. The effect of vacuum annealing on graphene. *J. Raman Spectrosc.* **2010**, *41*, 479–483. [CrossRef]
12. Yazdi, G.R.; Akhtar, F.; Ivanov, I.G.; Schmidt, S.; Shtepliuk, I.; Zakharov, A.; Iakimov, T.; Yakimova, R. Effect of epitaxial graphene morphology on adsorption of ambient species. *Appl. Surf. Sci.* **2019**, *486*, 239–248. [CrossRef]
13. Kögler, M.; Heilala, B. Time-gated Raman spectroscopy—A review. *Meas. Sci. Technol.* **2020**, *32*, 012002. [CrossRef]
14. Craig, R.L.; Bondy, A.L.; Ault, A.P. Surface Enhanced Raman Spectroscopy Enables Observations of Previously Undetectable Secondary Organic Aerosol Components at the Individual Particle Level. *Anal. Chem.* **2015**, *87*, 7510–7514. [CrossRef] [PubMed]
15. Doughty, D.C.; Hill, S.C. Raman spectra of atmospheric particles measured in Maryland, USA over 22.5 h using an automated aerosol Raman spectrometer. *J. Quant. Spectrosc. Radiat. Transfer* **2020**, *244*, 106839. [CrossRef]
16. Doughty, D.C.; Hill, S.C. Raman spectra of atmospheric aerosol particles: Clusters and time-series for a 22.5 hr sampling period. *J. Quant. Spectrosc. Radiat. Transfer* **2020**, *248*, 106907. [CrossRef]
17. Moorchilot, V.S.; Aravind, U.K.; Menacherry, S.P.M.; Aravindakumar, C.T. Single-Particle Analysis of Atmospheric Aerosols: Applications of Raman Spectroscopy. *Atmosphere* **2022**, *13*, 1779. [CrossRef]
18. Liu, Y.; Shi, Y.; Zhou, W.; Shi, W.; Dang, W.; Li, X.; Liang, B. The split-up of G band and 2D band in temperature-dependent Raman spectra of suspended graphene. *Opt. Laser Technol.* **2021**, *139*, 106960. [CrossRef]
19. Yang, M.; Wang, L.; Qiao, X.; Liu, Y.; Shi, Y.; Wu, H.; Liang, B.; Li, X.; Zhao, X. Temperature Dependence of G and D' Phonons in Monolayer to Few-Layer Graphene with Vacancies. *Nanoscale Res. Lett.* **2020**, *15*, 189. [CrossRef]
20. Linas, S.; Magnin, Y.; Poinsot, B.; Boisron, O.; Förster, G.D.; Martinez, V.; Fulcrand, R.; Tournus, F.; Dupuis, V.; Rabilloud, F.; et al. Interplay between Raman shift and thermal expansion in graphene: Temperature-dependent measurements and analysis of substrate corrections. *Phys. Rev. B* **2015**, *91*, 075426. [CrossRef]
21. Cunha, R.; Perea-López, N.; Elías, A.L.; Fujisawa, K.; Carozo, V.; Feng, S.; Lv, R.; dos Santos, M.C.; Terrones, M.; Araujo, P.T. Probing the interaction of noble gases with pristine and nitrogen-doped graphene through Raman spectroscopy. *Phys. Rev. B* **2018**, *97*, 195419. [CrossRef]
22. Valeš, V.; Kovaříček, P.; Fridrichová, M.; Ji, X.; Ling, X.; Kong, J.; Dresselhaus, M.; Kalbáč, M. Enhanced Raman scattering on functionalized graphene substrates. *2D Mater.* **2017**, *4*, 025087. [CrossRef]
23. Hung, Y.-J.; Hofmann, M.; Cheng, Y.-C.; Huang, C.-W.; Chang, K.-W.; Lee, J.-Y. Characterization of graphene edge functionalization by grating enhanced Raman spectroscopy. *RSC Adv.* **2016**, *6*, 12398–12401. [CrossRef]
24. Yang, G.; Li, L.; Lee, W.B.; Ng, M.C. Structure of graphene and its disorders: A review. *Sci. Technol. Adv. Mater.* **2018**, *19*, 613–648. [CrossRef] [PubMed]
25. Mazumder, M.; Das, R.; Sajib, M.S.J.; Gomes, A.J.; Islam, M.; Selvaratnam, T.; Rahman, A. Comparison of Different Hydrotalcite Solid Adsorbents on Adsorptive Desulfurization of Liquid Fuel Oil. *Technologies* **2020**, *8*, 22. [CrossRef]

Disclaimer/Publisher's Note: The statements, opinions and data contained in all publications are solely those of the individual author(s) and contributor(s) and not of MDPI and/or the editor(s). MDPI and/or the editor(s) disclaim responsibility for any injury to people or property resulting from any ideas, methods, instructions or products referred to in the content.

Article

Improvement of β-SiC Synthesis Technology on Silicon Substrate

Yana Suchikova [1,*], Sergii Kovachov [1], Ihor Bohdanov [1], Artem L. Kozlovskiy [2,3], Maxim V. Zdorovets [2,3] and Anatoli I. Popov [2,4,*]

[1] The Department of Physics and Methods of Teaching Physics, Berdyansk State Pedagogical University, 71100 Berdyansk, Ukraine; essfero@gmail.com (S.K.); bogdanovbdpu@gmail.com (I.B.)
[2] Engineering Profile Laboratory, L.N. Gumilyov Eurasian National University, Satpaev Str. 5, Astana 010008, Kazakhstan; kozlovskiy.a@inp.kz (A.L.K.); mzdorovets@gmail.com (M.V.Z.)
[3] Laboratory of Solid State Physics, Institute of Nuclear Physics, Almaty 050032, Kazakhstan
[4] Institute of Solid State Physics, University of Latvia, 8 Kengaraga Str., LV-1063 Riga, Latvia
* Correspondence: yanasuchikova@gmail.com (Y.S.); popov@latnet.lv (A.I.P.)

Abstract: This article presents an enhanced method for synthesizing β-SiC on a silicon substrate, utilizing porous silicon as a buffer layer, followed by thermal carbide formation. This approach ensured strong adhesion of the SiC film to the substrate, facilitating the creation of a hybrid heterostructure of SiC/por-Si/mono-Si. The surface morphology of the SiC film revealed islands measuring 2–6 μm in diameter, with detected micropores that were 70–80 nm in size. An XRD analysis confirmed the presence of spectra from crystalline silicon and crystalline silicon carbide in cubic symmetry. The observed shift in spectra to the low-frequency zone indicated the formation of nanostructures, correlating with our SEM analysis results. These research outcomes present prospects for the further utilization and optimization of β-SiC synthesis technology for electronic device development.

Keywords: silicon carbide; silicon; electrochemical etching; solar cells; carbonization; annealing; heterostructures

1. Introduction

Silicon has long been the primary material for solar energy applications, largely owing to its natural abundance and low toxicity [1,2]. However, recent studies have highlighted certain limitations in the energy conversion efficiency of silicon-based photovoltaic cells [3,4], catalyzing the exploration of alternative semiconducting materials. In this vein, semiconductors such as cadmium telluride (CdTe) [5,6], gallium arsenide (GaAs) [7,8], indium phosphide (InP) [9,10], and complex ternary compounds like $Cd_xTe_yO_z$ [11,12], $Al_xGa_{1−x}As$ [13,14], and $CuGa_xIn_{1−x}Se_2$ [15,16] have garnered significant attention. Despite the promising attributes of these materials, they are often compromised by high production costs [17]. To ameliorate this, nanostructuring the surfaces of these semiconductors has been suggested as a method to enhance their light absorption coefficients [18–20]. Nonetheless, the scarcity of elements like indium and gallium restricts the large-scale application of such materials to space-based solar cells [21,22], emphasizing the need for efficient and cost-effective materials for terrestrial use.

Emerging as a viable option is silicon carbide (SiC), a material comprising silicon and carbon, two of Earth's most abundant elements [23–25]. The well-characterized semiconductor properties of SiC make it conducive for industrial-scale utilization [26–35], akin to silicon. In addition to its robust thermal conductivity, electric breakdown voltage, and current density, SiC can operate over a broad temperature range without undergoing degradation in its monocrystalline structure or phase transitions. Through doping, the semiconductor properties of SiC can be tailored to yield both n-type and p-type materials [36,37]. These defects have been reported to impede the electrical and mechanical properties of the material, as corroborated by extensive studies [38,39]. Such defects are not just distinctive to silicon carbide but are a recurring theme across the broader category of semiconductors, each requiring its unique set of mitigative strategies.

Another formidable obstacle lies in the lattice mismatch when SiC films are grown on non-lattice-matched substrates. The repercussions of this discrepancy are often the formation of strain-induced defects that can severely compromise the crystalline quality of SiC films, thereby limiting their applicability in electronic devices [40,41].

To ameliorate these complexities, prior research has established the efficacy of incorporating intermediate buffer layers as a palliative measure [42,43]. Notably, porous domains or interstitial spaces generated on the semiconductor surface via electrochemical procedures have been shown to serve as effective buffer layers, considerably reducing defect densities and lattice mismatch-induced stresses [44–46]. Additionally, it has been demonstrated repeatedly that the nanostructuring of silicon leads to a change in the properties of the monocrystalline counterpart, allowing for a significant improvement in the characteristics of solar cells fabricated based on it [47–49].

The present study introduces a cost-effective approach for synthesizing high-quality β-SiC films on silicon substrates, utilizing a porous silicon (por-Si) intermediary buffer layer.

Using the electrochemical etching technique, the methodology begins by establishing a porous layer on the monocrystalline silicon surface. This method possesses distinct economic advantages such as reduced material wastage, lower energy consumption, and a minimal requirement of specialized equipment. Furthermore, the electrochemical etching process can be precisely controlled to produce uniform pores, thus reducing the need for additional post-processing or treatments. Considering these benefits, electrochemical etching can be up to 40–60% more cost-effective compared to other pore-formation methods.

The subsequent growth of SiC is facilitated by rapid thermal vacuum carbonization. Including the buffer porous layer is pivotal for attaining a pristine silicon carbide film.

Regarding material safety, both silicon and silicon carbide stand out for their benign nature. Silicon, a primary component of the Earth's crust, is non-toxic and abundantly available. Silicon carbide, similarly, is considered environmentally friendly and non-toxic compared to other semiconductors that are frequently employed in solar cells, such as cadmium telluride or copper indium gallium selenide, which have raised environmental and health concerns. This inherent non-toxicity of the β-SiC/por-Si/mono-Si heterostructure materials emphasizes their suitability for sustainable and eco-friendly solar cell applications.

Given the economic and environmental merits, the β-SiC/por-Si/mono-Si heterostructure demonstrates significant potential as an advanced and viable material for use in solar cell applications. The novelty of our work lies in using a porous buffer layer, which not only facilitates the growth of high-quality silicon carbide films but also enhances the adhesion and structural integrity of the resulting heterostructure. Consequently, this new approach reduces production costs and yields a high-quality heterostructure with excellent electronic properties, making the β-SiC/por-Si/mono-Si heterostructure a promising material for application in solar cells.

2. Materials and Methods

2.1. Samples for the Experiment

A monocrystalline Si of n-type conductivity was utilized for the experiment, doped with phosphorus to a concentration of 1.8×10^{17} cm^{-1}. Plates oriented as (111) with dimensions of $1 \times 2 \times 0.2$ cm were employed.

2.2. Electrolytes, Precursors, and Equipment

For the electrochemical modification of the silicon substrate, an initial step involved the removal of the native oxide passivating layer. This procedure was accomplished through electrochemical etching using a hydrochloric acid (HCl) solution. The experimental setup included a standard three-electrode electrochemical cell. Within this configuration, a platinum electrode served as the counter electrode, while a silver–silver chloride electrode acted as the reference.

To augment the efficiency of the etching process, the electrochemical cell was equipped with two additional subsystems. Firstly, an air-blowing module was incorporated to eliminate the formation of oxygen bubbles on the sample surface that could otherwise hinder the etching process. Second, an electrolyte stirring mechanism was added to ensure uniform electrolyte distribution and consistent etching across the substrate.

After removing the oxide layer, the next phase involved the formation of a porous silicon layer. This was achieved through electrochemical etching in a hydrofluoric acid (HF) solution within the same electrochemical cell. The electrolyte was maintained at room temperature during this stage of the procedure. The etching was conducted under ambient lighting conditions and did not require specialized environmental controls.

Post-etching, the samples underwent a drying process within a SNOL 58/350 muffle furnace to remove any residual moisture and to prepare their surfaces for subsequent procedures.

Finally, the silicon carbide layer was synthesized using a Jipelec JetFirst-100 oven. A methane (CH_4) gas precursor was deployed for this purpose, facilitating the formation of Si-C bonds and thereby culminating in the growth of a high-quality SiC layer.

All chemicals utilized were procured from Spetsprompostach LLC. The hydrochloric HCl and HF acids were of analytical grade (ARG, 99.8%), indicating a high purity level suitable for analytical applications and ensuring minimal interference from impurities during the electrochemical etching processes. The silicon (Si) substrate was of semiconductor grade, a designation for materials with controlled and specified impurity levels to ensure optimal electrical conductivity and functionality in semiconductor applications. Methane (CH_4), which was utilized to synthesize the silicon carbide layer, was of ultra-high purity (UHP) grade with a purity level of 99.995%, ensuring the high-quality synthesis of the SiC layer.

2.3. Experiment Steps

The experiment was conducted in several stages, specifically:

- Stage 1—Formation of the porous Si layer;
- Stage 2—Removal of reaction products from the sample surface;
- Stage 3—Formation of the SiC layer;
- Stage 4—Post-processing of the samples.

Detailed experimental stages are provided below.

2.3.1. Stage 1. Electrochemical Etching in Acid Solutions

The preliminary phase of the experimental procedure focused on the meticulous preparation of the monocrystalline silicon (mono-Si) surface to render it amenable for silicon carbide (SiC) layer formation. A point of concern in semiconductor processing is the natural tendency of the silicon surface to oxidize following mechanical operations such as grinding and polishing. This oxidation forms a thin, amorphous oxide layer documented to expedite the device degradation rate and pose challenges in further fabrication steps [50,51]. Therefore, it is imperative to effectively remove this passivating oxide layer to preserve the monocrystalline silicon substrate's intrinsic properties and facilitate subsequent process steps [52,53].

To further facilitate the quality of SiC films, a strategic approach involves formulating a textured or porous silicon layer on the monocrystalline silicon surface. These specially engineered buffer layers act as a soft substrate, addressing and alleviating the perennial lattice mismatch between the SiC film and the mono-Si substrate. The resulting architecture diminishes the manifestation of elastic stresses at the interface, thereby potentially enhancing the SiC films' overall quality and reprehensive understanding and repeatability. The specific conditions under which this initial phase of the experiment was conducted are systematically documented in Table 1, while the accompanying Figure 1 provides a visual representation to aid interpretation.

Table 1. Conditions and sub-stages of Si sample preparation.

№	Purpose	Electrolyte	Potential, V	Time, min
1.1	Removal of oxide	5% HCl	2	3
1.2	Formation of a porous layer	50% HF	5	7
1.3	Removal of reaction products from the surface	H_2O	0	2

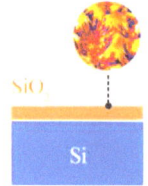 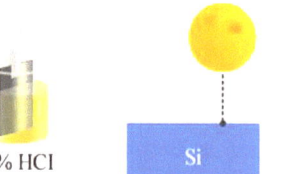   

Figure 1. Schematic diagram of the formation of the Si porous layer.

A standard three-electrode cell ensemble was employed in the electrochemical etching process utilized for both the removal of oxide and the formation of a porous layer on the monocrystalline silicon substrate. In the initial sub-stage (1.1), a potential of 2V was applied for 3 min in a 5% HCl solution to facilitate the removal of the oxide layer. Subsequently, in the second sub-stage (1.2), the porous silicon layer was formed by applying a potential of 5V for 7 min in a 50% HF solution. This higher potential drove the electrochemical etching of the silicon substrate, inducing porosity. The specific current during these processes was not fixed but could fluctuate to maintain the set potential, ensuring that the desired electrochemical reactions occurred. The final sub-stage (1.3) involved rinsing the silicon substrate in water to remove any reaction products from the surface, setting the stage for subsequent procedures. Through this systematic electrochemical etching process, the silicon substrate was adequately prepared to synthesize the silicon carbide layer, addressing the inherent challenges associated with the lattice mismatch and the passivating oxide layer.

2.3.2. Stage 2. Removal of Moisture Residues and Reaction Products from the Sample Surface

The second stage of the experimental process was engineered to meticulously remove moisture residues and byproducts from the treated surface of the porous silicon (por-Si) sample. Such procedural steps are integral to ensure an uncontaminated substrate for subsequent silicon carbide (SiC) film formation. To achieve this, samples were subjected to a tightly controlled drying sequence in a SNOL 58/350 oven, characterized by three discrete sub-stages delineated by specific durations and temperatures as outlined in Table 2.

Table 2. Conditions and sub-stages for removing moisture residues from the por-Si surface.

№	Time, min	Temperature, °C
2.1	180	150
2.2	60	300
2.3	60	400

Table 2 comprehensively summarizes the conditions and sub-stages: starting with a 180-min treatment at 150 °C, followed by 60 min at 300 °C and then 400 °C. The judiciously selected time and temperature parameters were optimized to remove any vestigial moisture residues without adversely affecting the integrity of the por-Si substrate.

Initiating the SiC film formation immediately after this stage was paramount to mitigate the potential of the sample oxidation processes post-drying. Thus, the transition to the SiC film formation was executed without delay to maintain a contiguous processing flow.

2.3.3. Stage 3. Thermal Deposition and SiC Film Formation

Employing rapid thermal vacuum carbonization as our deposition technique, we used methane (CH_4) as the precursor material (Figure 2). The procedure was bifurcated into two temporal series, each encapsulating a duration of 60 s and punctuated by a 2-min hiatus, a design choice strategically deployed to allow for the chamber's re-equilibration. As can be seen in Table 3, the process underwent a rapid temperature ramp from an initial 50 °C to a final state of 900 °C, conducted under a controlled vacuum pressure of 1×10^{-2} Pa.

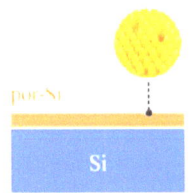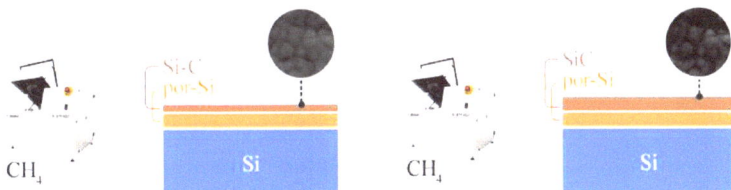

Figure 2. Schematic diagram of the formation of the heterostructure β-SiC/por-Si/mono-Si.

Table 3. Parameters for por-Si surface carbonization.

№	Time, min	Temperature, °C	Pressure in the Reaction Chamber, P, Pa
3.1	1	50→900 [1]	1×10^{-2}
3.2	1	50→900	1×10^{-2}

[1] The notation (50→900) indicates that the temperature was gradually increased within the specified ranges. The interval between processing batches was 1 min. These short carbonization batches ensure uniform carbon distribution on the surface, which is essential for forming a high-quality thin silicon carbide layer.

The rapid thermal vacuum carbonization technique was chosen for its proven ability to produce high-fidelity SiC films with minimized defect density, offering a potential solution to inherent fabrication challenges. This process involves an instantaneous elevation in temperature, designed to expedite deposition while suppressing the possible formation of undesirable phases or defects. The constant pressure conditions in the reaction chamber also serve as a critical parameter to inhibit unwanted chemical reactions, thereby promoting stoichiometric fidelity in the resulting SiC films.

2.3.4. Stage 4. Post-Processing of the β-SiC/Por-Si/Mono-Si Heterostructure

For Stage 4, post-processing procedures were initiated to stabilize the formed β-SiC/por-Si/mono-Si heterostructures. The post-processing primarily aimed to eliminate weakly adhered atoms and extraneous byproducts of the deposition process. Table 4 outlines the conditions: firstly, chemical etching was applied using a 2% HCl solution for 20 min at ambient temperature to remove oxides and other residual compounds. Secondly, thermal annealing was performed in a nitrogen atmosphere for 90 min at 150 °C to relieve residual stresses and improve crystallinity.

Table 4. Parameters for post-processing the surface of the formed β-SiC/por-Si/mono-Si heterostructure.

№	Method	Reagent	Time, min	Temperature, °C
4.1	Chemical etching	2% HCl	20	20
4.2	Thermal annealing	N_2	90	150

After completing the aforementioned stages, the samples were allowed to rest in an open-air environment for 90 days. This time frame was sufficient for fully assessing the samples' long-term structural stability and integrity.

2.4. Characterization

The structural features of the SiC films were comprehensively analyzed using scanning electron microscopy (SEM), carried out with an SEO-SEM Inspect S50-B microscope (Fei Company, Hillsboro, OR, USA). X-ray diffraction methods were also employed, utilizing a Dron–3M device with CuKα radiation (Sumy State University, Sumy, Ukraine). Energy dispersive spectroscopy (EDS) was executed to elucidate the elemental composition, facilitated by an AZtecOne spectrometer (Oxford Instruments, Abingdon, UK).

3. Results

3.1. Subsection

As illustrated in Figure 3, the surface morphology of the β-SiC/por-Si/mono-Si heterostructure is comprehensively explored. Remarkably, the heterostructure exhibits the formation of distinct island-like agglomerates that collectively merge to form a densely packed film. Within the context of the inset associated with Figure 3, a closer microscopic observation reveals that each of these island agglomerates is interspersed with minute pores, the dimensions of which have been carefully measured. These pores exhibit a size distribution that falls within approximately 70 to 80 nm.

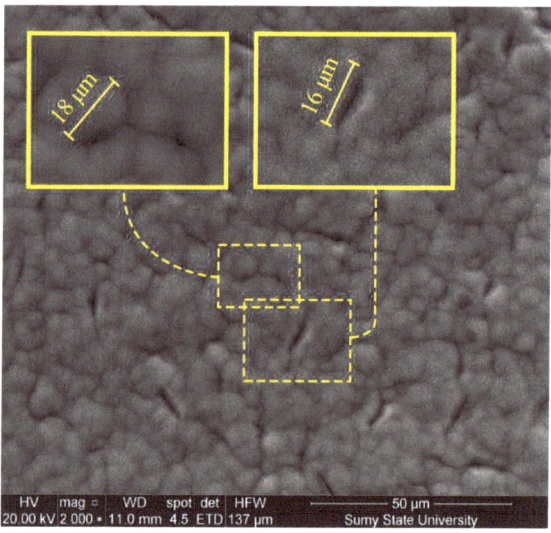

Figure 3. Surface morphology of the β-SiC/por-Si/mono-Si heterostructure.

These pores are likely resultant artifacts from the chemical etching process conducted on the silicon carbide film, specifically during the hydrochloric acid chemical treatment stage (identified as step 4 in the post-processing sequence of the heterostructure fabrication). The interspersing channels separating these island agglomerates have dimensions extending up to approximately 10 μm.

Figure 4 is devoted to a statistical investigation into the distribution of these islands based on their respective diameters. The analytical quantification was executed utilizing the robust analytical capabilities of the ImageJ 1.53q software product. According to the compiled data, the islands exhibit diameters that are dispersed across a range extending from 2 to 6 μm. Most notably, the statistical preponderance is observed for islands with a diameter of approximately $d \approx 3$ μm.

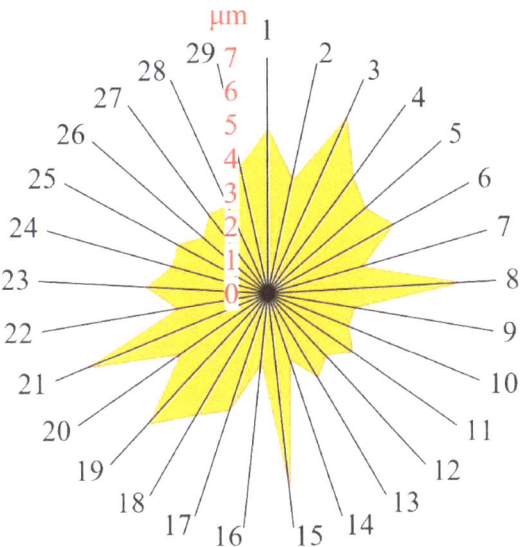

Figure 4. Diameter distribution diagram of the SiC film islands.

Moving to Figure 5, a cross-sectional analysis of the synthesized β-SiC/por-Si/mono-Si film is presented. The measured thickness of the film is precisely 2 μm. Upon microscopic examination, the film manifests a conspicuous and highly uniform overlay upon the porous silicon (por-Si) layer, devoid of any discernible imperfections such as pores, cracks, or other types of morphological defects. Moreover, the top SiC layer appears to maintain a high degree of homogeneity and compositional uniformity throughout the span of the film. Importantly, the adhesion between the SiC film and the underlying por-Si substrate is noted to be exceptionally robust, as evidenced by the absence of any zones that are indicative of film separation or detachment from the substrate.

Figure 5. Cross-section of the β-SiC/por-Si/mono-Si film.

3.2. EDX Analysis

As elucidated in Figure 6, a detailed compositional mapping of the surface of the SiC film, as adhered to the por-Si/mono-Si substrate, has been conducted utilizing energy-dispersive X-ray (EDX) analysis mapping technology. The results from this analytical technique unambiguously reveal that the surface composition of the film is constituted exclusively of silicon (Si) and carbon (C) atoms. Intriguingly, carbon is observed to be uniformly and homogeneously distributed across the surface—a notable feature that substantiates the high quality of the film in question.

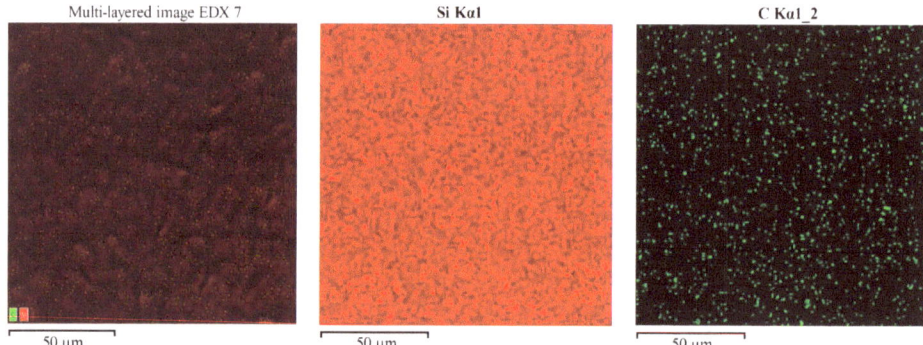

Figure 6. EDX analysis of the surface of the heterostructure β-SiC/por-Si/mono-Si obtained using mapping technology.

Further, the silicon concentration appears to be relatively higher than that of carbon within the mapped region. This phenomenon is postulated to be the result of the scattering or reflection of X-rays of the underlying por-Si/mono-Si substrate. Such a reflection phenomenon is a strong indicator that the silicon carbide (SiC) film may be of a relatively thin character, thus enabling the substrate's features to influence the EDX compositional map.

Proceeding to Figure 7, an EDX spectroscopic analysis is performed on the surface of the aforementioned β-SiC/por-Si/mono-Si heterostructure. The resulting spectrum provides quantitative insights into the percentage compositions of the constituent elements. Again, it becomes evident that silicon (Si) is present in a significantly higher concentration than carbon. This observation is consistent with the aforementioned hypothesis regarding X-ray reflection from the substrate layer, thereby reinforcing the idea that the SiC film may be relatively thin.

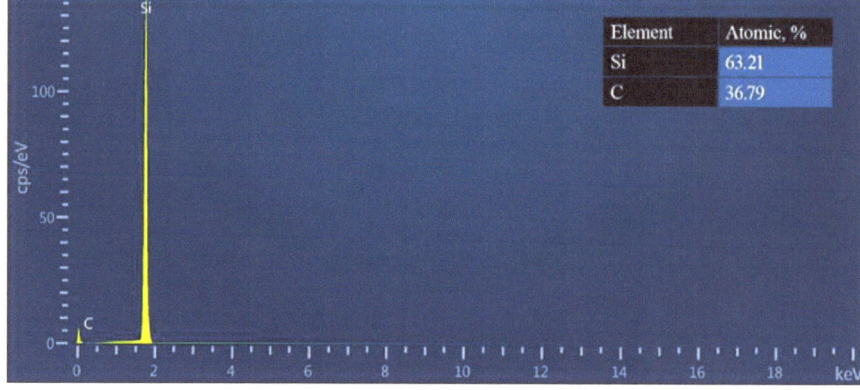

Figure 7. EDX spectrum of the surface of the β-SiC/por-Si/mono-Si heterostructure; inset shows the percentage composition of elements.

Importantly, the EDX analysis did not detect the presence of any extraneous elements on the surface, thereby confirming the purity of the material. Additionally, a key positive outcome that should be highlighted is the absence of any surface oxidation. This is an important metric for the quality of the film and underscores its potential for a wide range of applications where surface purity is a critical parameter.

3.3. XRD Analysis

The X-ray diffraction (XRD) pattern of the fabricated β-SiC/por-Si/mono-Si heterostructure is presented in Figure 8. A close inspection reveals the presence of three prominent peaks, whose angular positions (2θ) and corresponding crystallographic orientations (hkl) are systematically tabulated in Table 5. The peaks correspond to crystalline structures of silicon (Si) and silicon carbide (SiC) in its β-modification state. These spectral features can be mapped to specific lattice planes, thus affirming the underlying crystallography.

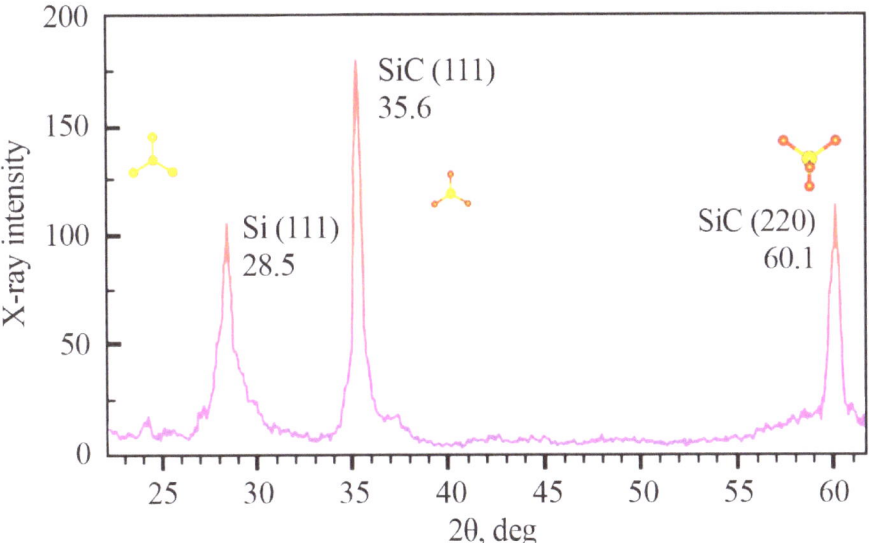

Figure 8. XRD pattern of β-SiC/por-Si/mono-Si heterostructure.

Table 5. Crystal structures determined by XRD spectrophotometry.

№	2θ, °	hkl	Crystal	Crystal System
1	28.5	(111)	Si	Cubic
2	35.6	(111)	Si	Cubic
3	60.1	(220)	SiC	Cubic

Intriguingly, a noticeable peak shift toward the left-hand side of the diffractogram is observed when compared to the standard peak positions for crystalline silicon. This spectral shift can be rationalized by the quantum size effects that manifest in nanometer-sized features within the material. More specifically, these nanometer-sized features are the interstitial spaces between the pores in the silicon layer. These spaces are engineered into the material by the etching process of the monocrystalline silicon substrate, a process that was conducted as an initial step (Step 1) in the fabrication protocol.

Of particular interest is the intense peak located at an angular position of $2\vartheta = 35.6°$, which is characteristically representative of β-SiC crystallizing in a zinc-blende-type lattice [54,55]. The prominence of this peak and the absence of other intense peaks corroborate the excellent crystallinity of the resultant SiC layers, suggesting that the synthesized heterostructure manifests high structural fidelity and integrity.

In summary, the XRD analysis offers compelling evidence of the high-quality crystalline nature of the obtained β-SiC/por-Si/mono-Si heterostructure. The observed spectral features support the notion that the material is not only well-crystallized but that it also possesses unique characteristics enabled by quantum size effects, underscoring its potential utility in advanced applications.

4. Discussion

One of the formidable challenges encountered in the epitaxial growth of silicon carbide (SiC) films on silicon (Si) substrates lies in the substantial presence of packing defects, as corroborated by the existing literature [56,57]. Such defects can be particularly detrimental, as they precipitate a breakdown under the application of high electric fields, thereby undermining the functional integrity of the device. The etiology of these packing defects can be primarily ascribed to the lattice mismatch between the SiC film and the silicon substrate [58].

To offer a nuanced understanding of this phenomenon, Table 6 presents the lattice parameters for both monocrystalline silicon and silicon carbide [59]. While both of these semiconductors crystallize in the cubic crystal system, their lattice parameters and unit-cell volumes exhibit pronounced disparities. These differences, although subtle, often translate into significant mechanical stress during the growth of SiC films on monocrystalline silicon substrates. The inherent tension culminates in defect formations and compromises the film's overall quality (as illustrated in Figure 9a).

Table 6. Structural parameters of Si and SiC.

Parameter	Si	SiC
Crystal system	Cubic	Cubic
Space group name	P1	F-43m
Lattice parameters, Å	a = 4.348	a = 5.6608
Unit-cell volume, 10^{-6} pm^{-3}	82.20	181.39

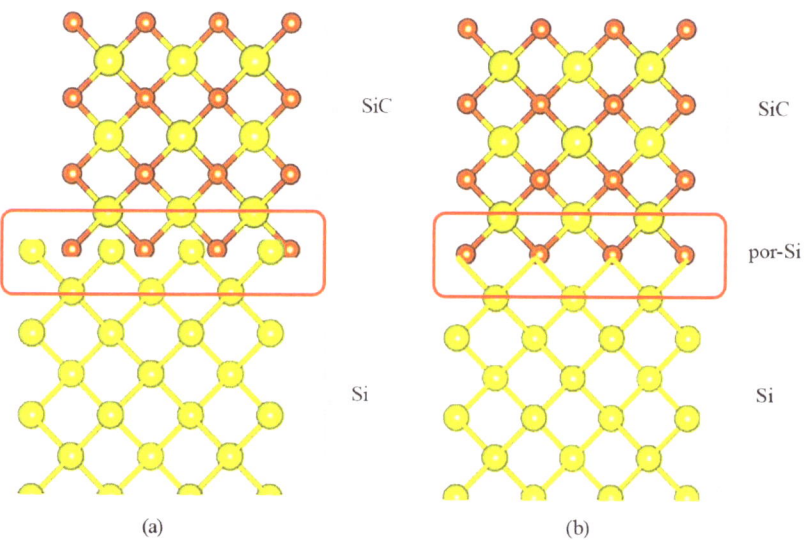

Figure 9. Schematic representation of the formation of SiC/Si heterostructures (**a**) and SiC/por-Si/Si (**b**).

A promising approach to alleviate these issues involves creating a porous layer on the silicon substrate, before SiC deposition. As depicted in Figure 9b, this porous layer can serve as a mechanical buffer, effectively mitigating the excessive elastic pressures that typically arise due to lattice mismatches. The porous architecture essentially functions as a "soft" substrate, facilitating more harmonious integration of SiC with the underlying silicon. Figure 10 represents the SiC film growth on mono-Si (Figure 10a) and por-Si (Figure 10b) substrates. In Figure 10a, the progression from pure mono-Si to a rougher β-SiC film signifies strain due to lattice mismatches. Conversely, in Figure 10b, the por-Si begins with a structured porous surface, and as SiC deposition occurs, it shows a smoother β-SiC film, underscoring the stress-buffering capacity of the porous architecture. This improved consistency in Figure 10b illustrates enhanced film–substrate adhesion and provides empirical support to the viability of the por-Si approach in addressing lattice mismatch challenges.

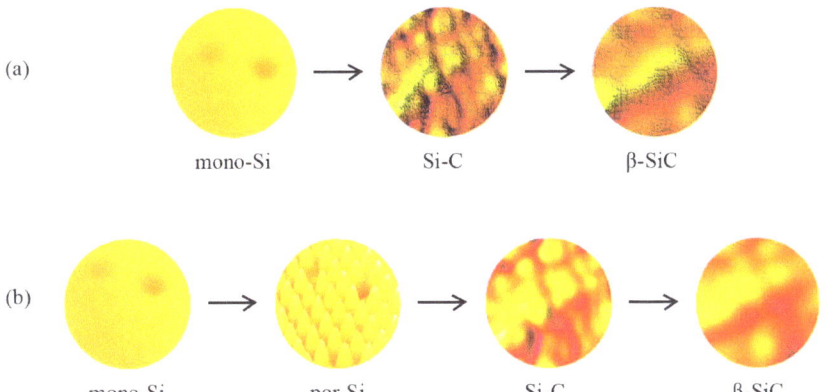

Figure 10. Simulation of the SiC film growth process on the surface of mono-Si (**a**) and por-Si (**b**).

The lattice mismatch between SiC and Si serves as a fundamental bottleneck in achieving high-quality epitaxial films. However, incorporating a porous buffer layer presents a feasible solution to this problem, as it allows for better stress distribution and enhanced adhesion, ultimately forming superior-quality SiC films.

In conventional thermal methods for forming silicon carbide (SiC), carbon atoms are chemically bonded to silicon atoms by forming Si-C covalent bonds [60,61]. However, a challenge arises when considering the surface properties of monocrystalline silicon (mono-Si), which inherently possesses a smooth topology. Due to the natural occurrence of auto-passivation facilitated by silicon dioxide, monocrystalline silicon surfaces tend to have minimal atoms with dangling or broken bonds [62]. Consequently, this surface morphology impinges upon the efficacy of most traditional techniques for the epitaxial growth of SiC films on mono-Si substrates.

The SiC films grown on such smooth mono-Si substrates are often fraught with many imperfections, both on the surface and within the bulk material. These defects manifest in various forms, including dislocations, stacking faults, and film cracking, as illustrated in Figure 10a. The cracks compromise the mechanical and electronic properties of the material, significantly limiting its application potential.

Contrastingly, when pre-formed on the mono-Si surface, a porous silicon (por-Si) layer offers a more amenable substrate for SiC film growth. The surface of porous silicon is characterized by a rough, textured morphology, depicted in Figure 10b. This textural variation in por-Si is crucial: it engenders micro-protrusions or elevated zones interspersed between the pores. These micro-protrusions act as localized sources of silicon atoms, which can readily bond with incoming carbon atoms during SiC film formation.

The intricate porous architecture thus plays a dual role; first, it provides a multitude of nucleation sites that better facilitate the formation of Si-C bonds, and second, it introduces a level of surface roughness that enhances the mechanical interlocking between the SiC film and the underlying silicon substrate. The result is the formation of a high-quality crystalline β-SiC film, which benefits from reduced defects and superior bond strength.

It is imperative to highlight that the properties and applications of porous silicon as a substrate have been the subject of extensive academic investigations [63–65]. These studies underscore the porous structure's instrumental role in mitigating challenges associated with SiC film growth on monocrystalline silicon, thereby contributing to improved SiC/mono-Si heterostructures.

In summary, while the surface characteristics of monocrystalline silicon have traditionally posed limitations on the growth of defect-free SiC films, introducing a porous silicon layer offers an effective countermeasure. This adaptability can pave the way for developing high-quality SiC films, an area of ongoing research and technological importance.

Many mechanisms have been elucidated in film formation on semiconductor substrates to govern the deposited films' growth kinetics and morphological evolution. Among these, the paradigms established by Frank–van der Merwe [66,67], Volmer–Weber [68,69], and the Stranski–Krastanov mechanism [70,71] stand out as seminal frameworks.

In the context of the Frank–van der Merwe mechanism, also known as layer-by-layer growth, we observe a unique phenomenon primarily in systems where the film's and substrate's crystalline lattice parameters exhibit minimal discrepancy [72,73]. This leads to the formation of what is colloquially termed a "wetting layer," characterized by the sequential deposition of monatomic layers on the substrate. This mechanism is typically operative when the thermodynamic factors favor minimizing surface and interfacial energies.

Conversely, the Volmer–Weber mechanism is observed in instances with pronounced lattice mismatch between the substrate and the film. Such discrepancies promote the growth of isolated islands on the substrate, negating the formation of an initial wetting layer. The Volmer–Weber mode is driven by the system's need to minimize the strain energy, which leads to a preferential aggregation of atoms into islands rather than a uniform film.

Situated as an intermediate regime between these two is the Stranski–Krastanov mechanism. In this model, the early stages of the film growth process resemble the Frank–van der Merwe mechanism, characterized by establishing an initial wetting layer [74]. However, owing to a slight lattice mismatch, elastic deformations begin to manifest. These deformations induce compressive stresses in the film, ultimately resulting in the nucleation and growth of three-dimensional, rounded islands atop the wetting layer.

Turning our attention to the growth of silicon carbide (SiC) on silicon (Si) substrates, it becomes evident that the porous silicon (por-Si) layer significantly eases the challenges associated with lattice mismatch, as depicted in Figure 9b. Our findings point to the applicability of the Stranski–Krastanov mechanism in this particular heterostructure. This observation substantiates the role of por-Si as a buffer layer, serving as an effective stress-relief intermediary that accommodates the lattice discrepancies between SiC and Si.

Hence, we have presented a facile yet robust methodology for SiC film growth that deftly addresses the perennial technological constraints of lattice mismatch and the compromised crystalline quality of SiC films. Introducing a por-Si buffer layer acts as a catalyst in optimizing the epitaxial growth process of SiC on silicon substrates. The buffer layer's role is pivotal, warranting further exploration for future refinements in the growth methodology. Subsequent research endeavors may be directed towards achieving complete carbide formation in SiC/Si structures and the potential for developing graphene layers on the surface of these heterostructures.

Through this multi-faceted analysis, we aim to contribute to the empirical knowledge pool of SiC film growth and broader efforts aimed at technological advancements in semiconductor applications.

5. Conclusions

The current study meticulously elaborates on an enhanced methodology for synthesizing β-SiC films on silicon substrates by judiciously integrating porous silicon (por-Si) as an intermediary buffer layer. Incorporating this buffer layer culminates in augmented adhesion between the SiC film and the mono-Si substrate. The resultant layered material thus manifests as a hybrid heterostructure of β-SiC/por-Si/mono-Si, with implications for advancing the state of the art in semiconductor technologies.

Examinations of the surface morphology via advanced imaging techniques revealed that the synthesized SiC film predominantly consists of island agglomerates with diametric dimensions ranging from 2 to 6 μm. Further intricate features, such as minuscule pores with sizes varying between 70 and 80 nm, were discerned on the islands' surfaces. Such topological attributes point to an intricate interface interaction and might have ramifications on the mechanical and electronic properties of the resulting heterostructure.

The X-ray Diffraction (XRD) analysis of the synthesized β-SiC/por-Si/mono-Si heterostructure exhibited three distinct peaks at angular positions (2θ) of 28.5°, 35.6°, and 60.1°. These peaks unequivocally correspond to the crystallographic orientations (hkl) of (111) and (220) for silicon (Si) and silicon carbide (SiC) in their cubic symmetry, respectively. This indicates the formation of high-quality crystalline structures. When juxtaposed with the standard peak positions for crystalline silicon, a noticeable shift toward lower angles was observed, particularly in the peak at 2θ = 35.6°. This shift, which represents a deviation from the typical diffraction patterns of bulk silicon, is indicative of quantum size effects. Such effects are consistent with the presence of nanometer-sized objects, as confirmed by Scanning Electron Microscopy (SEM) analyses.

Furthermore, the intensity and uniqueness of the peak at an angular position of 2θ = 35.6°, which aligns with the characteristics of β-SiC in a zinc-blende-type lattice, emphasize the impeccable crystallinity of the SiC layers. The absence of other intense peaks serves as further evidence for this claim, suggesting that the resultant heterostructure retains high structural integrity.

The prospects emerging from our research are expansive. The SiC/por-Si/mono-Si heterostructure promises to revolutionize next-generation electronic devices, given the superior electron mobility and thermal resilience of SiC. In optoelectronics, the heterostructure's unique electronic and optical attributes could spearhead the development of advanced devices such as photodetectors and light-emitting diodes. Furthermore, the chemical robustness of SiC offers avenues for its adoption in advanced sensing technologies, especially in challenging environments. The juxtaposition of SiC's admirable thermal conductivity with por-Si's insulating characteristics paves the path for innovative thermal management solutions for electronic apparatuses. Additionally, the observed quantum size effects signal potential applicability in the burgeoning arena of quantum computing, wherein meticulous control over nanoscale entities is vital.

It is crucial to acknowledge that, while our methodology offers a substantial improvement in overcoming lattice mismatch and adhesion issues that plague the conventional SiC film synthesis on silicon substrates, further research endeavors are requisite. These could include optimizing porosity levels in the buffer layer, a comprehensive assessment of mechanical properties under variable thermal and electrical conditions, and investigating the potential utility of this heterostructure in electronic and optoelectronic applications.

Conclusively, the present study augments the existing body of knowledge by elucidating an efficient, scalable technique for synthesizing high-quality β-SiC films on silicon substrates. The findings portend significant technological benefits and set the stage for subsequent research to refine and expand the application scope of SiC/por-Si/mono-Si heterostructures.

Author Contributions: Conceptualization, Y.S.; methodology, Y.S., S.K. and I.B.; software, Y.S. and S.K.; validation, Y.S., A.L.K., M.V.Z. and A.I.P.; formal analysis, S.K. and I.B.; investigation, Y.S., S.K. and I.B.; resources, Y.S. and A.I.P.; data curation, Y.S., S.K. and A.I.P.; writing—original draft preparation, Y.S., S.K. and A.I.P.; writing—review and editing, Y.S., A.L.K., M.V.Z. and A.I.P.; visualization, Y.S., S.K., A.L.K. and M.V.Z.; supervision, Y.S. and A.I.P.; project administration, Y.S. and A.I.P.; funding acquisition, Y.S. and A.I.P. All authors have read and agreed to the published version of the manuscript.

Funding: The study was supported by the Ministry of Education and Science of Ukraine via Project No. 0122U000129 "The search for optimal conditions for nanostructure synthesis on the surface of A3B5, A2B6 semiconductors and silicon for photonics and solar energy", Project No. 0121U10942 "Theoretical and methodological bases of system fundamentalization of the future nanomaterials experts training for productive professional activity", and Project No. 0123U100110 "System of remote and mixed specialized training of future nanoengineers for the development of new dual-purpose nanomaterials". In addition, the research of A.I.P. and Y.S. was partly supported by COST Action CA20129 "Multiscale irradiation and chemistry driven processes and related technologies" (Multi-Chem). Y.S. was partly supported by COST Action CA20126—Network for research, innovation, and product development on porous semiconductors and oxides (NETPORE). A.I.P., thanks to the Institute of Solid State Physics, University of Latvia, ISSP UL as the Center of Excellence, is supported through the Framework Program for European Universities, Union Horizon 2020, H2020-WIDESPREAD-01–2016–2017-TeamingPhase2, under Grant Agreement No. 739508, CAMART2 project.

Institutional Review Board Statement: Not applicable.

Informed Consent Statement: Not applicable.

Data Availability Statement: Data is contained within the article.

Conflicts of Interest: The authors declare no conflict of interest.

References

1. Chen, D.; Vaqueiro Contreras, M.; Ciesla, A.; Hamer, P.; Hallam, B.; Abbott, M.; Chan, C. Progress in the understanding of light-and elevated temperature-induced degradation in silicon solar cells: A review. *Prog. Photovolt. Res. Appl.* **2021**, *29*, 1180–1201. [CrossRef]
2. Richter, A.; Müller, R.; Benick, J.; Feldmann, F.; Steinhauser, B.; Reichel, C.; Glunz, S.W. Design rules for high-efficiency both-sides-contacted silicon solar cells with balanced charge carrier transport and recombination losses. *Nat. Energy* **2021**, *6*, 429–438. [CrossRef]
3. Andreani, L.C.; Bozzola, A.; Kowalczewski, P.; Liscidini, M.; Redorici, L. Silicon solar cells: Toward the efficiency limits. *Adv. Phys. X* **2019**, *4*, 1548305. [CrossRef]
4. Fell, A.; McIntosh, K.R.; Altermatt, P.P.; Janssen, G.J.; Stangl, R.; Ho-Baillie, A.; Abbott, M.D. Input parameters for the simulation of silicon solar cells in 2014. *IEEE J. Photovolt.* **2015**, *5*, 1250–1263. [CrossRef]
5. Bosio, A.; Pasini, S.; Romeo, N. The History of Photovoltaics with Emphasis on CdTe Solar Cells and Modules. *Coatings* **2020**, *10*, 344. [CrossRef]
6. Çetinkaya, Ç.; Çokduygulular, E.; Kınacı, B.; Güzelçimen, F.; Özen, Y.; Sönmez, N.A.; Özçelik, S. Highly improved light harvesting and photovoltaic performance in CdTe solar cell with functional designed 1D-photonic crystal via light management engineering. *Sci. Rep.* **2022**, *12*, 11245. [CrossRef]
7. Suchikova, Y.; Kovachov, S.; Bohdanov, I. Formation of oxide crystallites on the porous GaAs surface by electrochemical deposition. *Nanomater. Nanotechnol.* **2022**, *12*. [CrossRef]
8. Oshima, R.; Ogura, A.; Shoji, Y.; Makita, K.; Ubukata, A.; Koseki, S.; Imaizumi, M.; Sugaya, T. Ultra-High-Speed Growth of GaAs Solar Cells by Triple-Chamber Hydride Vapor Phase Epitaxy. *Crystals* **2023**, *13*, 370. [CrossRef]
9. Suchikova, Y. Provision of environmental safety through the use of porous semiconductors for solar energy sector. *East.-Eur. J. Enterp. Technol.* **2016**, *6*, 26–33. [CrossRef]
10. Yana, S. Porous indium phosphide: Preparation and properties. In *Handbook of Nanoelectrochemistry: Electrochemical Synthesis Methods, Properties, and Characterization Techniques*, 1st ed.; Springer: Cham, Switzerland, 2015; pp. 283–306. [CrossRef]
11. Suchikova, Y.; Kovachov, S.; Bohdanov, I.; Moskina, A.; Popov, A. Characterization of $Cd_xTe_yO_z$/CdS/ZnO Hetero-structures Synthesized by the SILAR Method. *Coatings* **2023**, *13*, 639. [CrossRef]
12. Suchikova, Y.; Kovachov, S.; Bohdanov, I.; Karipbaev, Z.T.; Pankratov, V.; Popov, A.I. Study of the structural and morphological characteristics of the $Cd_xTe_yO_z$ nanocomposite obtained on the surface of the CdS/ZnO heterostructure by the SILAR method. *Appl. Phys. A* **2023**, *129*, 499. [CrossRef]
13. Shoji, Y.; Tamaki, R.; Okada, Y. Temperature Dependence of Carrier Extraction Processes in GaSb/AlGaAs Quantum Nanostructure Intermediate-Band Solar Cells. *Nanomaterials* **2021**, *11*, 344. [CrossRef] [PubMed]
14. Suchikova, Y.; Kovachov, S.; Bohdanov, I.; Abdikadirova, A.A.; Kenzhina, I.; Popov, A.I. Electrochemical Growth and Structural Study of the $Al_xGa_{1-x}As$ Nanowhisker Layer on the GaAs Surface. *J. Manuf. Mater. Process.* **2023**, *7*, 153. [CrossRef]
15. Sergeyev, D.; Zhanturina, N.; Aizharikov, A.; Popov, A.I. Influence of "Productive" Impurities (Cd, Na, O) on the Properties of the CuZnSnS Absorber of Model Solar Cells. *Latv. J. Phys. Tech. Sci.* **2021**, *58*, 13–23. [CrossRef]

16. Nugroho, H.S.; Refantero, G.; Septiani, N.L.W.; Iqbal, M.; Marno, S.; Abdullah, H.; Prima, E.C.; Nugraha; Yuliarto, B. A progress review on the modification of CZTS(e)-based thin-film solar cells. *J. Ind. Eng. Chem.* **2022**, *105*, 83–110. [CrossRef]
17. Krotkus, A.; Nevinskas, I.; Norkus, R. Semiconductor Characterization by Terahertz Excitation Spectroscopy. *Materials* **2023**, *16*, 2859. [CrossRef]
18. Obaid, W.O.; Hashim, A. Synthesis and Augmented Optical Properties of PC/SiC/TaC Hybrid Nanostructures for Potential and Photonics Fields. *Silicon* **2022**, *14*, 11199–11207. [CrossRef]
19. Vambol, S.O.; Bohdanov, I.T.; Vambol, V.V.; Nestorenko, T.P.; Onyschenko, S.V. Formation of filamentary structures of oxide on the surface of monocrystalline gallium arsenide. *J. Nano-Electron. Phys.* **2017**, *9*, 06016. [CrossRef]
20. Ahmed, H.; Hashim, A. Tuning the spectroscopic and electronic characteristics of ZnS/SiC nanostructures doped organic material for optical and nanoelectronics fields. *Silicon* **2023**, *15*, 2339–2348. [CrossRef]
21. Anjum, S.; Gonzalez, P.E.; Atwater, H.A. Planar and Nanowire InP Thin Solar Cells for Ultralight Space Power Applications. In Proceedings of the 2022 IEEE 49th Photovoltaics Specialists Conference (PVSC), Philadelphia, PA, USA, 5–10 June 2022; pp. 1083–1085. [CrossRef]
22. Yan, G.; Wang, J.L.; Liu, J.; Liu, Y.Y.; Wu, R.; Wang, R. Electroluminescence analysis of VOC degradation of individual subcell in GaInP/GaAs/Ge space solar cells irradiated by 1.0 MeV electrons. *J. Lumin.* **2020**, *219*, 116905. [CrossRef]
23. Köhler, M.; Pomaska, M.; Procel, P.; Santbergen, R.; Zamchiy, A.; Macco, B.; Lambertz, A.; Duan, W.; Cao, P.; Klingebiel, B.; et al. A silicon carbide-based highly transparent passivating contact for crystalline silicon solar cells approaching efficiencies of 24%. *Nat. Energy* **2021**, *6*, 529–537. [CrossRef]
24. Zhou, L.; Xu, Y.; Tan, S.; Liu, M.; Wan, Y. Simulation of Amorphous Silicon Carbide Photonic Crystal Absorption Layer for Solar Cells. *Crystals* **2022**, *12*, 665. [CrossRef]
25. Hsu, C.-H.; Zhang, X.-Y.; Zhao, M.J.; Lin, H.-J.; Zhu, W.-Z.; Lien, S.-Y. Silicon Heterojunction Solar Cells with p-Type Silicon Carbon Window Layer. *Crystals* **2019**, *9*, 402. [CrossRef]
26. Merkininkaite, G.; Gailevicius, D.; Staisiunas, L.; Ezerskyte, E.; Vargalis, R.; Malinauskas, M.; Sakirzanovas, S. Additive Manufacturing of SiOC, SiC, and Si_3N_4 Ceramic 3D Microstructures. *Adv. Eng. Mater.* **2023**, *25*, 2300639. [CrossRef]
27. Hashim, A.; Abbas, M.H.; Al-Aaraji NA, H.; Hadi, A. Controlling the morphological, optical and dielectric characteristics of PS/SiC/CeO2 nanostructures for nanoelectronics and optics fields. *J. Inorg. Organomet. Polym. Mater.* **2023**, *33*, 1–9. [CrossRef]
28. Gaibnazarov, B.B.; Imanova, G.; Khozhiev, S.T.; Kosimov, I.O.; Khudaikulov, I.K.; Kuchkanov, S.K.; Bekpulatov, I.R. Changes in the Structure and Properties of Silicon Carbide under Gamma Irradiation. *Integr. Ferroelectr.* **2023**, *237*, 208–215. [CrossRef]
29. Meteab, M.H.; Hashim, A.; Rabee, B.H. Synthesis and tailoring the morphological, optical, electronic and photodegradation characteristics of PS-PC/MnO2-SiC quaternary nanostructures. *Opt. Quantum Electron.* **2023**, *55*, 187. [CrossRef]
30. Kim, V.V.; Konda, S.R.; Yu, W.; Li, W.; Ganeev, R.A. Harmonics Generation in the Laser-Induced Plasmas of Metal and Semiconductor Carbide Nanoparticles. *Nanomaterials* **2022**, *12*, 4228. [CrossRef]
31. Hashim, A.; Abbas, M.H.; Al-Aaraji NA, H.; Hadi, A. Facile fabrication and developing the structural, optical and electrical properties of SiC/Y2O3 nanostructures doped PMMA for optics and potential nanodevices. *Silicon* **2023**, *15*, 1283–1290. [CrossRef]
32. Scalise, E.; Zimbone, M.; Marzegalli, A. Impact of Inversion Domain Boundaries on the Electronic Properties of 3C-SiC. *Phys. Status Solidi B* **2022**, *259*, 2200093. [CrossRef]
33. Kohn, V.G.; Argunova, T.S. Near-Field Phase-Contrast Imaging of Micropores in Silicon Carbide Crystals with Synchrotron Radiation. *Phys. Status Solidi B* **2022**, *259*, 2100651. [CrossRef]
34. Sameera, J.N.; Islam, M.A.; Islam, S.; Hossain, T.; Sobayel, M.; Akhtaruzzaman; Amin, N.; Rashid, M.J. Cubic Silicon Carbide (3C–SiC) as a buffer layer for high efficiency and highly stable CdTe solar cell. *Opt. Mater.* **2021**, *123*, 111911. [CrossRef]
35. Lebedev, A.S.; Suzdal'tsev, A.V.; Anfilogov, V.N.; Farlenkov, A.S.; Porotnikova, N.M.; Vovkotrub, E.G.; Akashev, L.A. Carbothermal Synthesis, Properties, and Structure of Ultrafine SiC Fibers. *Inorg. Mater.* **2020**, *56*, 20–27. [CrossRef]
36. Capan, I. 4H-SiC Schottky Barrier Diodes as Radiation Detectors: A Review. *Electronics* **2022**, *11*, 532. [CrossRef]
37. Li, G.; Xu, M.; Zou, D.; Cui, Y.; Zhong, Y.; Cui, P.; Cheong, K.Y.; Xia, J.; Nie, H.; Li, S.; et al. Fabrication of Ohmic Contact on N-Type SiC by Laser Annealed Process: A Review. *Crystals* **2023**, *13*, 1106. [CrossRef]
38. Tahani, M.; Postek, E.; Motevalizadeh, L.; Sadowski, T. Effect of Vacancy Defect Content on the Interdiffusion of Cubic and Hexagonal SiC/Al Interfaces: A Molecular Dynamics Study. *Molecules* **2023**, *28*, 744. [CrossRef]
39. Chaturvedi, V.; Haasmann, D.; Moghadam, H.A.; Dimitrijev, S. Electrically Active Defects in SiC Power MOSFETs. *Energies* **2023**, *16*, 1771. [CrossRef]
40. Meli, A.; Muoio, A.; Reitano, R.; Sangregorio, E.; Calcagno, L.; Trotta, A.; Parisi, M.; Meda, L.; La Via, F. Effect of the Oxidation Process on Carrier Lifetime and on SF Defects of 4H SiC Thick Epilayer for Detection Applications. *Micromachines* **2022**, *13*, 1042. [CrossRef]
41. Mukesh, N.; Márkus, B.G.; Jegenyes, N.; Bortel, G.; Bezerra, S.M.; Simon, F.; Beke, D.; Gali, A. Formation of Paramagnetic Defects in the Synthesis of Silicon Carbide. *Micromachines* **2023**, *14*, 1517. [CrossRef]
42. Jorudas, J.; Šimukovič, A.; Dub, M.; Sakowicz, M.; Prystawko, P.; Indrišiūnas, S.; Kovalevskij, V.; Rumyantsev, S.; Knap, W.; Kašalynas, I. AlGaN/GaN on SiC Devices without a GaN Buffer Layer: Electrical and Noise Characteristics. *Micromachines* **2020**, *11*, 1131. [CrossRef]

43. Adamov, R.B.; Pashnev, D.; Shalygin, V.A.; Moldavskaya, M.D.; Vinnichenko, M.Y.; Janonis, V.; Jorudas, J.; Tumėnas, S.; Prystawko, P.; Krysko, M.; et al. Optical Performance of Two Dimensional Electron Gas and GaN:C Buffer Layers in AlGaN/AlN/GaN Heterostructures on SiC Substrate. *Appl. Sci.* **2021**, *11*, 6053. [CrossRef]
44. Triani, R.M.; Neto, J.B.T.D.R.; De Oliveira, P.G.B.; Rêgo, G.C.; Neto, A.L.; Casteletti, L.C. In-Situ Production of Metal Matrix Composites Layers by TIG Surface Alloying to Improve Wear Resistance of Ductile Cast Iron Using a Buffer-Layer and Post Weld Heat Treatment. *Coatings* **2023**, *13*, 1137. [CrossRef]
45. Lenshin, A.; Seredin, P.; Goloshchapov, D.; Radam, A.O.; Mizerov, A. MicroRaman Study of Nanostructured Ultra-Thin AlGaN/GaN Thin Films Grown on Hybrid Compliant SiC/Por-Si Substrates. *Coatings* **2022**, *12*, 626. [CrossRef]
46. Amador-Mendez, N.; Mathieu-Pennober, T.; Vézian, S.; Chauvat, M.P.; Morales, M.; Ruterana, P.; Babichev, A.; Bayle, F.; Julien, F.H.; Bouchoule, S.; et al. Porous nitride light-emitting diodes. *ACS Photonics* **2022**, *9*, 1256–1263. [CrossRef]
47. Faltakh, H.; Bourguiga, R.; Ben Rabha, M.; Bessais, B. Simulation and Optimization of the Performance of Multicrystalline Silicon Solar Cell Using Porous Silicon Antireflection Coating Layer. *Superlattices Microstruct.* **2014**, *72*, 283–295. [CrossRef]
48. Ben Rabha, M.; Mohamed, S.B.; Dimassi, W.; Gaidi, M.; Ezzaouia, H.; Bessais, B. Reduction of Absorption Loss in Multicrystalline Silicon via Combination of Mechanical Grooving and Porous Silicon. *Phys. Status Solidi C* **2011**, *8*, 883–886. [CrossRef]
49. Nafie, N.; Lachiheb, M.A.; Rabha, M.B.; Dimassi, W.; Bouaïcha, M. Effect of the Doping Concentration on the Properties of Silicon Nanowires. *Phys. E Low-Dimens. Syst. Nanostruc.* **2014**, *56*, 427–430. [CrossRef]
50. Lo Nigro, R.; Fiorenza, P.; Greco, G.; Schilirò, E.; Roccaforte, F. Structural and Insulating Behaviour of High-Permittivity Binary Oxide Thin Films for Silicon Carbide and Gallium Nitride Electronic Devices. *Materials* **2022**, *15*, 830. [CrossRef]
51. Fiorenza, P.; Giannazzo, F.; Roccaforte, F. Characterization of SiO_2/4H-SiC Interfaces in 4H-SiC MOSFETs: A Review. *Energies* **2019**, *12*, 2310. [CrossRef]
52. Zhu, S.; Liu, T.; Fan, J.; Maddi, H.L.R.; White, M.H.; Agarwal, A.K. Effects of JFET Region Design and Gate Oxide Thickness on the Static and Dynamic Performance of 650 V SiC Planar Power MOSFETs. *Materials* **2022**, *15*, 5995. [CrossRef]
53. Brzozowski, E.; Kaminski, M.; Taube, A.; Sadowski, O.; Krol, K.; Guziewicz, M. Carrier Trap Density Reduction at SiO_2/4H-Silicon Carbide Interface with Annealing Processes in Phosphoryl Chloride and Nitride Oxide Atmospheres. *Materials* **2023**, *16*, 4381. [CrossRef] [PubMed]
54. Sultan, N.M.; Albarody, T.M.B.; Obodo, K.O.; Baharom, M.B. Effect of Mn^{+2} Doping and Vacancy on the Ferromagnetic Cubic 3C-SiC Structure Using First Principles Calculations. *Crystals* **2023**, *13*, 348. [CrossRef]
55. Kukushkin, S.A.; Osipov, A.V. Anomalous Properties of the Dislocation-Free Interface between Si(111) Substrate and 3C-SiC(111) Epitaxial Layer. *Materials* **2021**, *14*, 78. [CrossRef] [PubMed]
56. Scuderi, V.; Zielinski, M.; La Via, F. Impact of Doping on Cross-Sectional Stress Assessment of 3C-SiC/Si Heteroepitaxy. *Materials* **2023**, *16*, 3824. [CrossRef] [PubMed]
57. Zimbone, M.; Zielinski, M.; Bongiorno, C.; Calabretta, C.; Anzalone, R.; Scalese, S.; Fisicaro, G.; La Magna, A.; Mancarella, F.; La Via, F. 3C-SiC Growth on Inverted Silicon Pyramids Patterned Substrate. *Materials* **2019**, *12*, 3407. [CrossRef]
58. Zhang, R.; Zou, C.; Wei, Z.; Wang, H.; Liu, C. Nano-Phase and SiC–Si Spherical Microstructure in SiC/Al-50Si Composites Solidified under High Pressure. *Materials* **2023**, *16*, 4283. [CrossRef]
59. Suchikova, Y.O.; Kovachov, S.S.; Bardus, I.O.; Lazarenko, A.S.; Bohdanov, I.T. Formation of β-SiC on por-Si/mono-Si Surface According to Stranski—Krastanow Mechanism. *Him. Fiz. Tehnol. Poverhni* **2022**, *13*, 447–454. [CrossRef]
60. Lobanok, M.V.; Mukhammad, A.I.; Gaiduk, P.I. Structural and Optical Properties of SiC/Si Heterostructures Obtained Using Rapid Vacuum-Thermal Carbidization of Silicon. *J. Appl. Spectrosc.* **2022**, *89*, 256–260. [CrossRef]
61. Kulych, V.G.; Fesenko, I.P.; Kovtiukh, M.O.; Tkach, V.M.; Kaidash, O.M.; Kuzmenko, Y.F.; Chasnyk, V.I.; Ivzhenko, V.V. Microstructure and Thermal Conductivity of Reaction-Sintered SiC. *J. Superhard Mater.* **2023**, *45*, 158–160. [CrossRef]
62. Song, Y.; Zhang, G.; Cai, X.; Dou, B.; Wang, Z.; Liu, Y.; Wei, S.H. General Model for Defect Dynamics in Ionizing-Irradiated SiO_2-Si Structures. *Small* **2022**, *18*, 2107516. [CrossRef]
63. Bandarenka, H.; Redko, S.; Nenzi, P.; Balucani, M. Optimization of chemical displacement deposition of copper on porous silicon. *J. Nanosci. Nanotechnol.* **2012**, *12*, 8725–8731. [CrossRef] [PubMed]
64. Dolgyi, A.; Bandarenka, H.; Prischepa, S.; Yanushkevich, K.; Nenzi, P.; Balucani, M.; Bondarenko, V. Electrochemical deposition of Ni into mesoporous silicon. *ECS Trans.* **2012**, *41*, 111. [CrossRef]
65. Artsemyeva, K.; Dolgiy, A.; Bandarenka, H.; Panarin, A.; Khodasevich, I.; Terekhov, S.; Bondarenko, V. Fabrication of SERS-active substrates by electrochemical and electroless deposition of metals in macroporous silicon. *ECS Trans.* **2013**, *53*, 85. [CrossRef]
66. Wang, H.; Yao, Z.; Jung, G.S.; Song, Q.; Hempel, M.; Palacios, T.; Kong, J. Frank-van der Merwe Growth in Bilayer Graphene. *Matter* **2021**, *4*, 3339–3353. [CrossRef]
67. Zhai, P.; Kou, S.; Peng, Y.; Shi, Y.; Jiang, H.; Li, G. Change Volmer–Weber to Frank–van der Merwe Growth Model of Epitaxial $BiVO_4$ Film. *J. Phys. D Appl. Phys.* **2022**, *55*, 324004. [CrossRef]
68. Gastellóu, E.; García, R.; Herrera, A.M.; Ramos, A.; García, G.; Hirata, G.A.; Ramírez, Y.D. Optical and Structural Analysis of GaN Microneedle Crystals Obtained via GaAs Substrates Decomposition and their Possible Growth Model Using the Volmer–Weber Mechanism. *Phys. Status Solidi B* **2023**, *260*, 2200201. [CrossRef]
69. Qiu, X.P.; Liu, X.; Jiang, S.M. Growth Mechanism for Zinc Coatings Deposited by Vacuum Thermal Evaporation. *J. Iron Steel Res. Int.* **2021**, *28*, 1047–1053. [CrossRef]

70. Dirko, V.V.; Lozovoy, K.A.; Kokhanenko, A.P.; Kukenov, O.I.; Korotaev, A.G.; Voitsekhovskii, A.V. Peculiarities of the 7×7 to 5×5 Superstructure Transition during Epitaxial Growth of Germanium on Silicon (111) Surface. *Nanomaterials* **2023**, *13*, 231. [CrossRef]
71. Penha, F.M.; Andrade, F.R.D.; Lanzotti, A.S.; Moreira Junior, P.F.; Zago, G.P.; Seckler, M.M. In Situ Observation of Epitaxial Growth during Evaporative Simultaneous Crystallization from Aqueous Electrolytes in Droplets. *Crystals* **2021**, *11*, 1122. [CrossRef]
72. Sun, X.W.; Huang, H.C.; Kwok, H.S. On the initial growth of indium tin oxide on glass. *Appl. Phys. Lett.* **1996**, *68*, 2663–2665. [CrossRef]
73. Floro, J.A.; Hearne, S.J.; Hunter, J.A.; Seel, S.C.; Thompson, C.V. The dynamic competition between stress generation and relaxation mechanisms during coalescence of Volmer-Weber thin films. *J. Appl. Phys.* **2001**, *89*, 4886–4897. [CrossRef]
74. Pan, L.; Lew, K.-K.; Redwing, J.M.; Dickey, E.C. Stranski-Krastanow growth of germanium on silicon nanowires. *Nano Lett.* **2005**, *5*, 1081–1085. [CrossRef] [PubMed]

Disclaimer/Publisher's Note: The statements, opinions and data contained in all publications are solely those of the individual author(s) and contributor(s) and not of MDPI and/or the editor(s). MDPI and/or the editor(s) disclaim responsibility for any injury to people or property resulting from any ideas, methods, instructions or products referred to in the content.

Article

Electrical Discharge Machining of Alumina Using Cu-Ag and Cu Mono- and Multi-Layer Coatings and ZnO Powder-Mixed Water Medium

Anna A. Okunkova [1,*], Marina A. Volosova [1], Khaled Hamdy [1,2] and Khasan I. Gkhashim [1]

[1] Department of High-Efficiency Processing Technologies, Moscow State University of Technology STANKIN, Vadkovsky per. 1, 127994 Moscow, Russia
[2] Production Engineering and Mechanical Design Department, Faculty of Engineering, Minia University, Minia 61519, Egypt
* Correspondence: a.okunkova@stankin.ru; Tel.: +7-909-913-12-07

Abstract: The paper aims to extend the current knowledge on electrical discharge machining of insulating materials, such as cutting ceramics used to produce cutting inserts to machine nickel-based alloys in the aviation and aerospace industries. Aluminum-based ceramics such as Al_2O_3, AlN, and SiAlON are in the most demand in the industry but present a scientific and technical problem in obtaining sophisticated shapes. One of the existing solutions is electrical discharge machining using assisting techniques. Using assisting Cu-Ag and Cu mono- and multi-layer coatings of 40–120 μm and ZnO powder-mixed deionized water-based medium was proposed for the first time. The developed coatings were subjected to tempering and testing. It was noticed that Ag-adhesive reduced the performance when tempering had a slight effect. The unveiled relationship between the material removal rate, powder concentration, and pulse frequency showed that performance was significantly improved by adding assisting powder up to 0.0032–0.0053 mm^3/s for a concentration of 14 g/L and pulse frequency of 2–7 kHz. Further increase in concentration leads to the opposite trend. The most remarkable results corresponded to the pulse duration of 1 μs. The obtained data enlarged the knowledge of texturing insulating cutting ceramics using various powder-mixed deionized water-based mediums.

Keywords: alumina; assisting coating; brass wire; electrical discharge machining; sublimation; tempering

1. Introduction

The issues of micro-texturing of working surfaces of critical machine-building products made of difficult-to-machine materials, including those aluminum-based ceramics, have been the subject of special attention in recent years and the object of theoretical and experimental research by leading scientific groups [1–6]. Such cutting ceramics (Al_2O_3, AlN, SiAlON) exhibit exceptional thermomechanical and tribological properties making them unreplaceable materials for machining nickel-based alloys for aircraft and aerospace industries [7–11]. It should be noted that manufacturing cutting ceramic inserts involves processing a large number of microstructures and often requests additional coating [12–14]. Surface micromachining includes micro-profiling to create a three-dimensional specific relief on the surface, the dimensions and roughness of which are determined based on the operation characteristics and the textured material's physicochemical properties. These microtextures' functional purpose is to reduce the intensity of friction and adhesive setting between mating surfaces and increase the wear resistance and service life of the cutting tool many times under a wide range of operational loads [15–17]. The formed microrelief provides a significant reduction in the actual contact area. Additionally, it serves as micro-reservoirs for grease, microencapsulated lubricants, anti-friction materials, and liquids capable of forming and retaining anti-friction films between the contact surfaces for a long time [18,19].

Traditionally, diamond tools are used to process cutting ceramics. However, laser, micro-abrasive, water-jet machining, and chemical processing are well-known technologies of directed action on the surface layer [20–23]. Electrical discharge machining of dielectrics using assisting techniques is one of the promising ways of texturing the surface of products [24,25]. The advantages of this method include high reproducibility, accuracy, the ability to process complex geometry surfaces, high locality, and the ability to process superhard materials for which traditional mechanical methods are ineffective. The electrical discharge machining makes it possible to avoid time-consuming and expensive post-processing operations while maintaining the high quality of the workpiece. Due to the removal of material by electrical erosion that includes thermal and chemical dissociation of the material at a small distance between electrodes (the interelectrode gap is about 0.02 mm), the absence of physical contact between the tool and the workpiece is achieved. Consequently, the productivity of electrical discharge machining does not depend on the mechanical properties such as hardness, strength, and brittleness of the material to be processed but on its electrical properties. In addition, it is possible to increase the productivity of the process by reducing the processing time of complex geometry textures due to the possibility of using electrodes of various configurations or the technological mobility of the tool electrode with an accuracy of 80–100 nm.

It is well known that any material with a specific electrical conductivity above the threshold value of 10^5 $\mu\Omega\cdot$cm can be subjected to electrical erosion. However, in 1986, Soviet scientists [26] invented and patented a method for electrical discharge machining of dielectrics. The method was further developed by leading scientists worldwide, developing its application concerning tool ceramics for 20 years [27–31].

The assisting electrode coating, which has proven itself in the works of many authors, is adhesive copper tape [32–35]. Insulating copper (II) oxide is unstable in the presence of hydrogen and reduces to conductive metallic copper [36–40]. At the same time, oxides of another copper-group metal, silver, such as Ag_2O and Ag_2O_2, dissociate at temperatures above 280 °C and 100 °C, respectively, and exhibit conductive properties (specific electrical conductivity γ of 60.0–$62.5\cdot10^6$ S·m^{-1}) exceeding values for copper (γ of 58.0–$59.5\cdot10^6$ S·m^{-1}), making them even more attractive for developing technology of electrical discharge machining of insulating materials.

ZnO material was chosen for assisting powder due to the following reasons:

- It is commercially available;
- Stable under fire exposure conditions and is not reactive to water;
- Refers to the materials that require considerable preheating, under all ambient temperature conditions, before ignition and combustion can occur;
- Is a widely used n-type semiconductor;
- The band gap E_g = 3.30–3.36 eV at room temperature [41–43]; the less the band gap, the more conductive properties the material exhibits [44,45]);
- Exhibits a chemical affinity for aluminum and copper [46–48].

It should be noted that ZnO powder is hazardous to aquatic organisms according to GHS Hazard pictograms; it is toxic and causes foundry fever if the dust is inhaled. It is coded by NFPA: Standard System for the Identification of the Hazards of Materials for Emergency Response as follows:

- Code 2: Intense or continued but not chronic exposure could cause temporary incapacitation or possible residual injury, for health;
- Code 1: Materials that require considerable preheating, under all ambient temperature conditions, before ignition and combustion can occur, for flammability;
- Code 0: Normally stable, even under fire exposure conditions, and is not reactive with water, for instability–reactivity.

However, it is relatively safe for the personnel working with micro-sized particles in suspension and actively used as additive of toothpaste and cement in therapeutic dentistry and cosmetic sunscreens.

Thus, developing electrical discharge machining techniques for increasing the productivity of electrical discharge machining of aluminum-based cutting ceramics in the case of study alumina using an assisting electrode coating and powder-mixed dielectric medium is relevant and in demand among modern tool production.

The object of research is electrical discharge machining of aluminum-based ceramics in the case of studying alumina using assisting techniques such as Cu-Ag and Cu mono- and multi-layer coating and ZnO powder-mixed deionized water medium. The subject of the study is the performance of electrical discharge machining of insulating alumina using a combined assisting electrode technique.

The study aims to find an alternative approach to electrical discharge machining of alumina and increase the productivity of insulating cutting ceramics machining using the assisting electrode technique by optimizing machining factors such as powder concentration, pulse frequency, and duration.

2. Materials and Methods

2.1. Sintering of the Samples

The corundum α-Al_2O_3 A16SG (Alcoa, New York, NY, USA) was used for producing samples. The average particle diameter was 0.53 µm. A detailed description of powder characterization [49], preparing suspensions [50], drying, machining graphite molds of MPG-6 grade cold-pressed blanks using a carbide tool with a multi-layer combined PVD-coating [51–53] and diagnostic system [54–56], consequent pouring powder mixture and sintering [7,8,57–59] on a spark plasma sintering machine KCE FCT-H HP D-25 SD (FCT Systeme GmbH, Rauenstein, Germany) are presented in the previously published works. The sintered discs were 65.5 mm in diameter and 10 mm in thickness. Optical control was carried out on an Olympus BX51M instrument (Ryf AG, Grenchen, Switzerland). Scanning electron microscopy of the obtained kerfs and their chemical analyses were conducted on a VEGA3 instrument (Tescan, Brno, Czech Republic). For each kerf, at least five spectra scannings were produced.

2.2. Electrical Discharge Machining

A two-axis wire electrical discharge machine ARTA 123 Pro (NPK "Delta-Test", Fryazino, Russia) was used for experiments (Table 1). The open tank system of the machine allows using any deionized water- or oil-based dielectric medium out of the filtration system [60].

Table 1. Main characteristics of wire electrical discharge machine ARTA 123 Pro.

Parameters	Value and Description
Max axis motions $X \times Y \times Z$, mm	$125 \times 200 \times 80$
Tool positioning accuracy, µm	±1
Average surface roughness parameter R_a, µm	0.6
Dielectric medium	Any
Max power consumption, kW	<6

A brass wire-electrode of 0.25 mm in diameter made of CuZn35 brass provided texture formation on alumina without taking into account the spark gap (path offset). The preliminary testing in a deionized water medium reduced the range of the machining factors. The range of the chosen factors is presented in Table 2. The choice of the factors is based on [60]. However, the range of factors was intentionally extended for exploratory research. The previously conducted work with alumina showed that the optimal value of the operational current is 0.3–0.4 A [49]. The main correlation between current, operational voltage, and character of the obtained wells on the surface or material removal rate is as follows [61]:

$$\sum F_{imp} = I \cdot U_o, \tag{1}$$

where ΣF_{imp} is the summarized force of working impulses in the system's action, N; I is current, A; U_0 is operational voltage, V.

Table 2. Range of electrical discharge machining factors.

Factor	Measuring Units	Value
Operational voltage, U_0	V	108; 72; 60; 48; 36
Pulse frequency, f	kHz	2; 5; 8; 11; 15; 17; 20; 25; 30
Pulse duration, D	µs	0.5; 1; 1.5; 1.75; 2; 2.35; 2.5; 2.68; 2.7
Rewinding speed, v_W	m/min	3; 3.4; 7; 10
Feed rate, v_F	mm/min	0.1; 0.3; 0.4; 0.5; 1
Wire tension F_T	N	0.05; 0.1; 0.25; 0.3; 0.4

Machining is carried out according to the control program of the translational movement of the wire-electrode along the X-axis from the zero position to a depth of 0.98 mm (taking into account the spark gap) [62–64]. The development of factors was carried out following the results of pre-implemented experiments for each type of the developed assisting coating. Each failed experiment minimized the range of the factors and assisted in determining the optimal values in terms of the material removal rate (productivity) [65].

Adaptive control based on voltage control in the gap (control of factors) is necessary to avoid contact between the tool electrode and the conductive assisting coating that leads to a short circuit. When electric shortcuts are registered in the gap, the wire tool is retracted automatically to establish the required interelectrode gap and ensure effective machining when the number of working pulses (aimed at destroying the workpiece or assisting coating) exceeds the number of idle pulses (aimed at destroying erosion products, debris). The effective ratio of working pulses to the number of impulses is 0.7–0.9 [66–68]. It should be noted that voltage in the gap and concentration of conductive debris strongly influence surface roughness (density of formed wells) when the current influences their overall size [69–73]. Further improvement in roughness is not observed when the concentration exceeds a certain level. Moreover, with increased concentration, the more frequent appearance of shorts can be observed until the wire tool is blocked in the kerf, and further machining is not possible. At the same time, the concentration of non-conductive particles reduces the process productivity and leads as well to the blockage of the wire tool.

The coated blank was fixed on the machine table during experiments (Figure 1). The basing was carried out by wire tool approach along the X- and Y- axes; the surface of the coated workpiece was taken as zero. The wire tool was adjusted vertically along the Z-axis by a spark. The position +2–+3 mm from the assisting coating was taken as zero of the wire tool. The upper nozzle was placed at +2–+3 of the coated workpiece to ensure adequate flushing by turbulent flows [74,75]. Electrical discharge machining was carried out with immersion of the workpiece in the ZnO-powder-mixed deionized water medium. The dielectric fluid level was established 1–2 mm above the workpiece. The workpiece was held for 8–10 min in a dielectric before machining to avoid the influence of thermal fluctuations. The obtained sample was wiped with a rag over [76]. At least 5 kerfs were produced for each parameters' set.

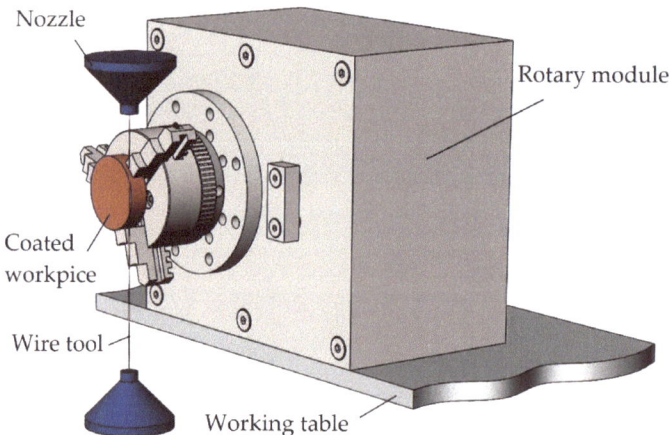

Figure 1. Scheme of coated workpiece microtexturing by wire electrical discharge machining.

Formed textures (kerfs) were controlled optically. Here and below, the optical measurement error was calculated by the formula [77–81]:

$$\delta_l = \pm 3 + \frac{L}{30} + \frac{g \cdot L}{4000}, \quad (2)$$

$$\delta_t = \pm 3 + \frac{L}{50} + \frac{g \cdot L}{2500}, \quad (3)$$

where δ_l is the longitudinal measurement error, μm; δ_t is the transversal measurement error, μm; L is the measured length, mm; g is the product height above microscope table glass (taken equal to zero), mm.

The material removal rate (MRR) was calculated as follows [82]:

$$MRR = \frac{V}{t}, \quad (4)$$

where the volume of removed material is calculated using the formula:

$$V = S \cdot l, \quad (5)$$

where S is the kerf area in the plan, mm²; l is the length of the kerf, μm. A detailed description of calculating S is provided in [50]. The processing time t is calculated from the feed rate v_F of the electrode-tool and workpiece height (kerf length) h:

$$t(s) = \frac{h(mm)}{v_F\left(\frac{mm}{s}\right)} \quad (6)$$

2.3. Assisting ZnO-Powder-Mixed Deionized Water Medium

Deionized water (LLC "Atlant", pos. Marusino, Lyubertsy district, Moscow region, The Russian Federation) following ASTM D-5127-90 with specific electrical resistivity up to 18.0 MΩ·cm was chosen as a suspension basis to avoid the formation of insulating Al_4C_3 or $Al_2(C_2)_3$) [83,84].

ZnO-powder-mixed deionized water medium was tested at concentrations of 7, 14, 21, 35, 50, and 100 g/L to improve the performance of the developed system. A total of 76 × 5 experiments were carried out. The zinc oxide ZnO of "Ch" grade, 99% of purity (LLC "Unihim", Saint Petersburg, The Russian Federation), following GOST 10262-73, bulk density of 5.61 g/cm^3, was used for producing suspension. The chemical composition is presented in Table 3.

Table 3. Chemical composition of ZnO powder ("Ch" grade, 99% of purity).

Chemical Substances	Chemical Formula	wt.%
Zinc oxide	ZnO	Balance
Manganese	Mn	≤0.0005
Arsenic	As	≤0.0002
Cadmium	Cd	not standardized
Potassium permanganate	KMnO$_4$	≤0.01
Potassium	K	≤0.005
Calcium	Ca	≤0.01
Substances insoluble in hydrochloric acid	-	≤0.01
Sulfates	SO$_4$	≤0.01
Phosphates	PO$_4$RR'R''	not standardized
Chlorides	Cl$_x$R (x = 1–5)	≤0.004
Iron	Fe	≤0.001
Sodium	Na	not standardized
Copper	Cu	≤0.001
Lead	Pb	≤0.01

ZnO is a colorless crystalline powder, insoluble in water, turning yellow when heated, and subliming at 1800 °C [85–88]. Zinc oxide is a direct-gap semiconductor with a band gap E_g of 3.30–3.36 eV [41–43]. Natural doping with oxygen makes it an *n*-type semiconductor. When heated, the substance changes color: white at room temperature, and zinc oxide becomes yellow. This is explained by a decrease in the band gap and a shift of the edge in the absorption spectrum from the UV region to blue.

The powder was subjected to granulometric analysis and optical microscopy. An EL104 laboratory balance (Mettler Toledo, Columbus, OH, USA) with a measurement range of 0.0001–120 g weighed powder with an error of 0.0001 g. ZnO powder was sifted using an analytical sieving machine AS200 basic (Retsch, Dusseldorf, Germany) with a test sieve (10 µm by ISO 3310-1). The previously conducted studies showed that the smallest size of suspended particles led to the highest productivity [69–73].

The prepared suspension was constantly stirred during experiments. For the higher powder concentration, electrical discharge machining was intensified by ultrasonic vibrations using an ultrasonic unit IL100-6/1 (LLC "Ultrasonic Technology—INLAB", Saint Petersburg, Russia) to avoid powder conglomerations in the discharge gap at the higher particle concentration (100 g/L) at frequency 22 kHz [89]. It should be noted that higher than 30 kHz and up to 1 MHz could be harmful to the biological process in the human body since arising cavitation with bubble formation with a diameter of less than 1 µm (ultrasound surgery). The emitting tip was placed at the tank to provide ultrasonic vibrations in volume. As known, the speed of propagation of vibrations in an elastic body is much higher than in a liquid or gaseous medium. Thus, the suspension is subjected to bulk ultrasonic vibrations.

After processing, the samples were cleaned with alkali.

2.4. Assisting Electrode

A few types of copper-based assisting coatings were developed (Table 4, Figure 2a). A HomaFix 404 20 m/10 mm copper tape (JSC Electroma, Lipetsk, The Russian Federation) was used as a basis for the developed coatings (Figure 2b). The main properties of the tape are provided in Table 5. A conductive polymer-based silver-containing adhesive (synthetic resins) Kontaktol (Keller, Yekaterinburg, The Russian Federation), with an electrical resistance of 10^{-6} Ω·m (γ of 10^6 S·m^{-1}) was used in the sandwich-type of assisting coating electrode forming a complete continuous uniform coating of variable thickness of 20–200 µm on the sample. The ratio of silver powder to acetone is (120–140)/(40–60). The recommended operating temperature is up to +160 °C. The adhesive coating was deposed using a brush. Half of the samples were additionally kept in an oven (drying cabinet) at temperatures of +160, +200, and +240 °C for 60, 90, 120, and 180 min to ensure the drying of the polymer base of the tape, removing organic soluble media and reducing stress.

Table 4. Developed assisting coatings.

Assisting Coating	Adhesive Type	Thickness, mm	Electrical Conductivity γ [1], $S \cdot cm^{-1}$	Specific Electrical Resistivity [2], $\Omega \cdot mm^2 \cdot m^{-1}$
Silver Adhesive	Polymer-based + Silver powder	0.100–0.110	0.009486 ± 0.00001	1.0542×10^{-6}
Copper tape, 1 layer	Resin-based	0.040		
Copper tape, 2 layers	Resin-based	2×0.040	0.580046 ± 0.00001	0.01724×10^{-6}
Copper tape, 3 layers	Resin-based	3×0.040		
Sandwich "Copper tape + Silver Adhesive", 1 layer	Polymer-based + Silver powder	0.150		
Sandwich "Copper tape + Silver Adhesive", 2 layers	Polymer-based + Silver powder	2×0.150	0.584112 ± 0.00001	0.01712×10^{-6}
Sandwich "Copper tape + Silver Adhesive", 3 layers	Polymer-based + Silver powder	3×0.150		
Graphite [3]	-	-	-	8.00
Distilled water [3]	-	-	-	10^3–10^4

[1] experimental values; [2] calculated values; [3] for reference.

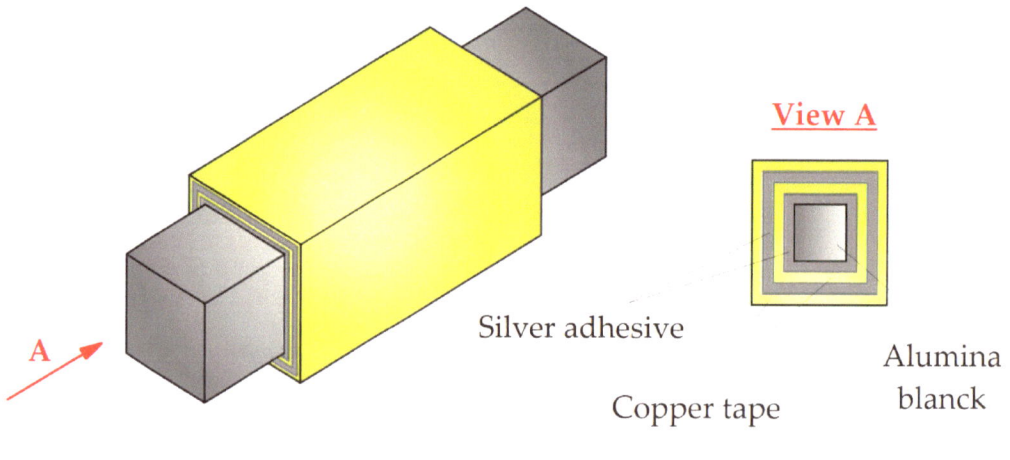

(a)

Figure 2. Cont.

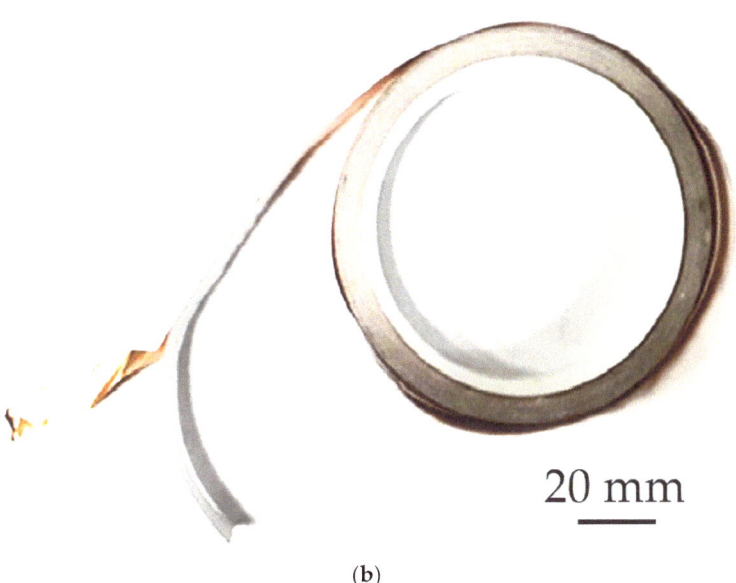

Figure 2. (**a**) Developed sandwich-type coating (schematic presentation of Sandwich "Copper tape + Silver Adhesive", 2 layers); (**b**) Copper tape.

Table 5. Parameters of copper tape.

Parameter	Value
Thickness of copper basis, mm	0.035 ± 0.0002
Tensile strength, N/cm	115
Elongation (Extension ratio), %	<2
Specific electrical resistivity, $\Omega \cdot mm^2 \cdot m^{-1}$	0.016–0.017
Operating temperature, °C	From -40 to $+110 \pm 5$
Tape width, mm	10

The specific electrical resistance of the formed assisting coatings (Table 3) was controlled by a Fischer Sigmascope SMP10 device (Helmut Fischer GmbH, Sindelfingen, Germany). The thickness of the first layer of the developed coatings was controlled with a Calowear instrument (CSM Instruments, Needham, MA, USA). It was developed to carry out wear tests on a small scale using the spherical notch method. In other words, it forms a recess by erasing the sample material during the rotation of a ball of a certain diameter (20 mm) coated with an abrasive medium (spherical microabrasion method). The method has proven to be fast for analyzing the thickness of any coating (mono- or multi-layer) and determining the wear coefficient of massive materials and coatings [90–92].

Before coating, the ceramic samples were placed in an ultrasonic tank and cleaned using a soap solution at a temperature of 60 °C for 20 min and alcohol for 5 min [93–95]. Approbation of the coated samples was conducted in a deionized water medium. Coating removal was conducted by a complex method: washing in an ultrasonic tank and mechanical cleaning.

3. Results

3.1. Characterization of ZnO Powder

Granulomorphometric analysis and optical microscopy (Figure 3) of the ZnO powder showed that the powder sample had an average inner diameter of 63.79 μm and 33.46 μm for 50% of the particles, while an average area diameter of 80.06 μm and 47.19 for 50% of the particles. The zinc oxide particles have a larger diameter with a high percentage of particles. The average circularity of zinc oxide powder is about 0.597 μm and about 0.625 μm for 50% of the particles. The powder was sieved before further processing.

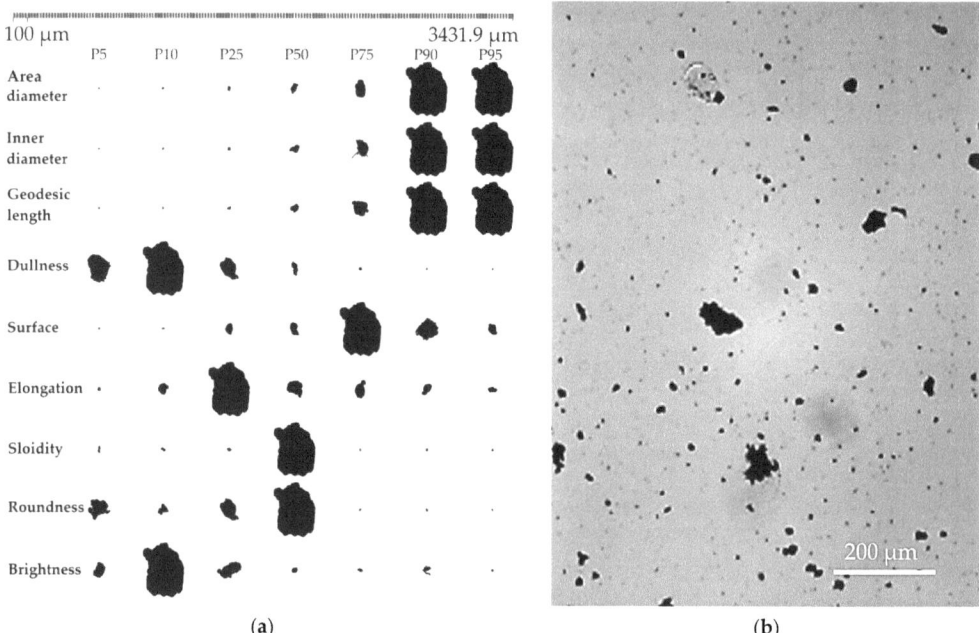

Figure 3. Granulomorphometric analysis of the composition of ZnO powder (a) and morphology (b).

3.2. Preliminary Testing of the Developed Coatings in Deionized Water Medium

Figure 4a shows the optical microscopy of kerf after electrical discharge machining the samples coated with silver adhesive. The traces of the coating erosion are observed. However, the traces of ceramic workpiece sublimation were not remarked (absence of the erosion of ceramics under discharge pulses). In this context, the term "sublimation" is used for the electrical erosion of ceramics since:

The alumina's boiling point is about 2980 °C [96], and the temperature in the discharge spark is about 10,000 °C [97–99]. With such a difference between the boiling point and surrounding temperature in a pulse period of 1–100 μs under conditions of continuously during the pulse expanding rarefied low-temperature gas-plasma bubble (region of low pressure) (Figure 4b), the material cannot pass the stages from solid to liquid, vapor, and plasma steadily, and direct sublimation of the material occurs from solid to vapor and ion plasma state [100].

It should be noted that a detailed phase diagram of the state of aluminum oxide at elevated and reduced pressure requires additional research and is still not presented in the literature, as well as for many other substances. An analysis of the preliminary testing results for sandwich types assisting coatings showed that the holding time in the oven does not significantly affect the adhesion of the copper tape to ceramics and material removal rate for the whole range of the developed coatings (Figure 4c). Differences in adhesion between two and three-layer coatings (80 and 120 μm in thickness) were also not observed. Approbation of samples of the "silver + copper" sandwich coatings showed similar results with variations with low tempering modes and multi-layer structure. The difference (Figures 5 and 6) does not have a fundamental nature to the purposes of the current study.

Figure 4. (a) The preliminary testing results of the electrical discharge machining of samples coated with a silver adhesive (optical microscopy), 10×; (b) phase transitions and states of oxide ceramics; (c) the preliminary testing results of the electrical discharge machining of alumina using Cu-Ag and Cu mono- and multi-layer coatings and various coating tempering modes.

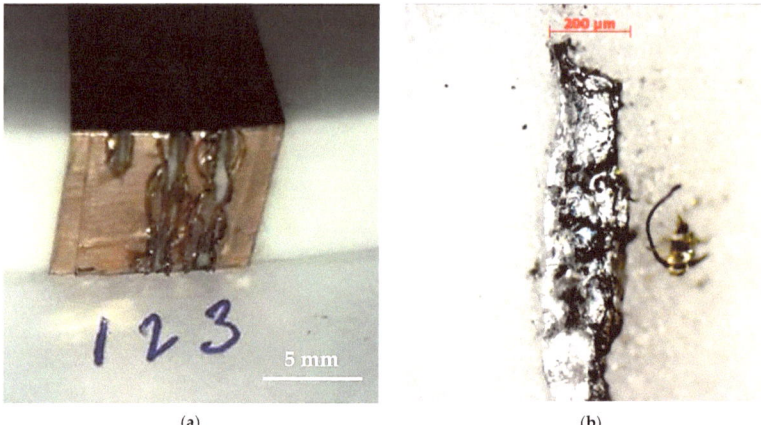

Figure 5. The preliminary testing results of the electrical discharge machining of samples coated with self-adhesive copper tape: (**a**) a sample after preliminary testing; (**b**) optical microscopy of the formed texture, 10×.

Figure 6. The preliminary testing results of the electrical discharge machining of samples with a sandwich-type mono-layer coating "silver + copper" after an hour of tempering at a temperature of 160 °C: (**a**) a sample after preliminary testing; (**b**) optical microscopy of the formed texture before coating removal, 10×; (**c**) optical microscopy of the formed texture, 10×.

The samples after the preliminary testing were cleaned. The deposition of metallic copper and its adhesion to the ceramic base is observed after cleaning. The use of a sandwich-type coating visually improved the adhesion of the coating (Figure 6) but significantly reduced the effect of the coating when electrical discharge machining compared with the sample with self-adhesive copper tape (Figure 5). The reduction in effect was consistent with the expected decline in the bulk conductivity of the coating due to the use of synthetic resin-based adhesives. Further experiments were conducted with the double-layer copper tape coating without silver adhesive (copper tape, 2 layers). The range of the machining factors was reduced (Table 6).

Table 6. Range of electrical discharge machining factors.

Factor	Measuring Units	Value
Operational voltage, U_0	V	108
Pulse frequency, f	kHz	2; 5; 7; 10; 15; 20; 25; 30
Pulse duration, D	µs	0.5; 1.0; 1.5; 2.0; 2.5; 2.64
Rewinding speed, v_W	m/min	7
Feed rate, v_F	mm/min	0.3
Wire tension F_T	N	0.25

3.3. Electrical Discharge Machining in ZnO Powder-Mixed Deionized Water Medium

The general view of the kerfs and results of experiments obtained by optical microscopy for the various powder concentration, pulse frequencies, and duration are shown in Figure 7. During conducting experiments with the powder-mixed deionized water medium at a concentration of 7 g/L, the wire tool was interrupted at a frequency of 30 kHz, pulse duration from 1.0 to 2.64 µs, except for the experiment with a pulse duration of 2.5 µs (Figure 7a). There were little erosion marks at a frequency of 5, 10, 15, and 30 kHz and a pulse duration of 2.5 µs. The non-stable erosion results were remarked for pulse frequency of 5–25 kHz and pulse duration of 1.0, 1.5, 2.0 µs and pulse frequency of 7 and 25 kHz and pulse duration of 2.5 µs.

At a concentration of 14 g/L, the effect of powder adding was visually reduced: the most pronounced result was achieved at a frequency of 7 kHz and a pulse duration of 1.0 µs (Figure 7b). Non-stable erosion marks were noticed for a pulse frequency of 5, 10 kHz and a pulse duration of 1.0, 1.5 µs. Little erosion marks were observed at a pulse frequency of 7, 10 kHz and a pulse duration of 1.5, 2.0, 2.5 µs.

At a concentration of 21 g/L, the effect of powder addition was also visually reduced. The most pronounced result was achieved at a pulse frequency of 5 and 7 kHz and a pulse duration of 0.5 and 1.0 µs, respectively (Figure 7c). Non-stable erosion marks were noticed for:
- A pulse frequency of 7 and 10 kHz and a pulse duration of 0.5 µs;
- A pulse frequency of 5 and 10 kHz and a pulse duration of 1.0 µs.

Little erosion marks were observed at a pulse frequency of 5 and 10 kHz and a pulse duration of 1.5 µs. It was noticed that visually electrical discharge machining is more stable at a pulse duration of 1.0 µs.

At a concentration of 35 g/L, the most pronounced result was achieved at a pulse frequency of 5 and 10 kHz and a pulse duration of 1.0 µs (Figure 7d). Little erosion marks were observed at a pulse frequency of 7 kHz and a pulse duration of 1.0 µs.

At a concentration of 50 g/L, the most pronounced result was achieved at a pulse frequency of 2 and 10 kHz and a pulse duration of 1.0 µs (Figure 7e). Little erosion marks were observed at a pulse frequency of 5, 7, and 15 kHz and a pulse duration of 1.0 µs.

At a concentration of 100 g/L, the most pronounced result was achieved at a pulse frequency of 2, 7, and 10 kHz and a pulse duration of 1.0 µs (Figure 7f). Non-stable erosion marks were observed at a pulse frequency of 5 kHz and a pulse duration of 1.0 µs.

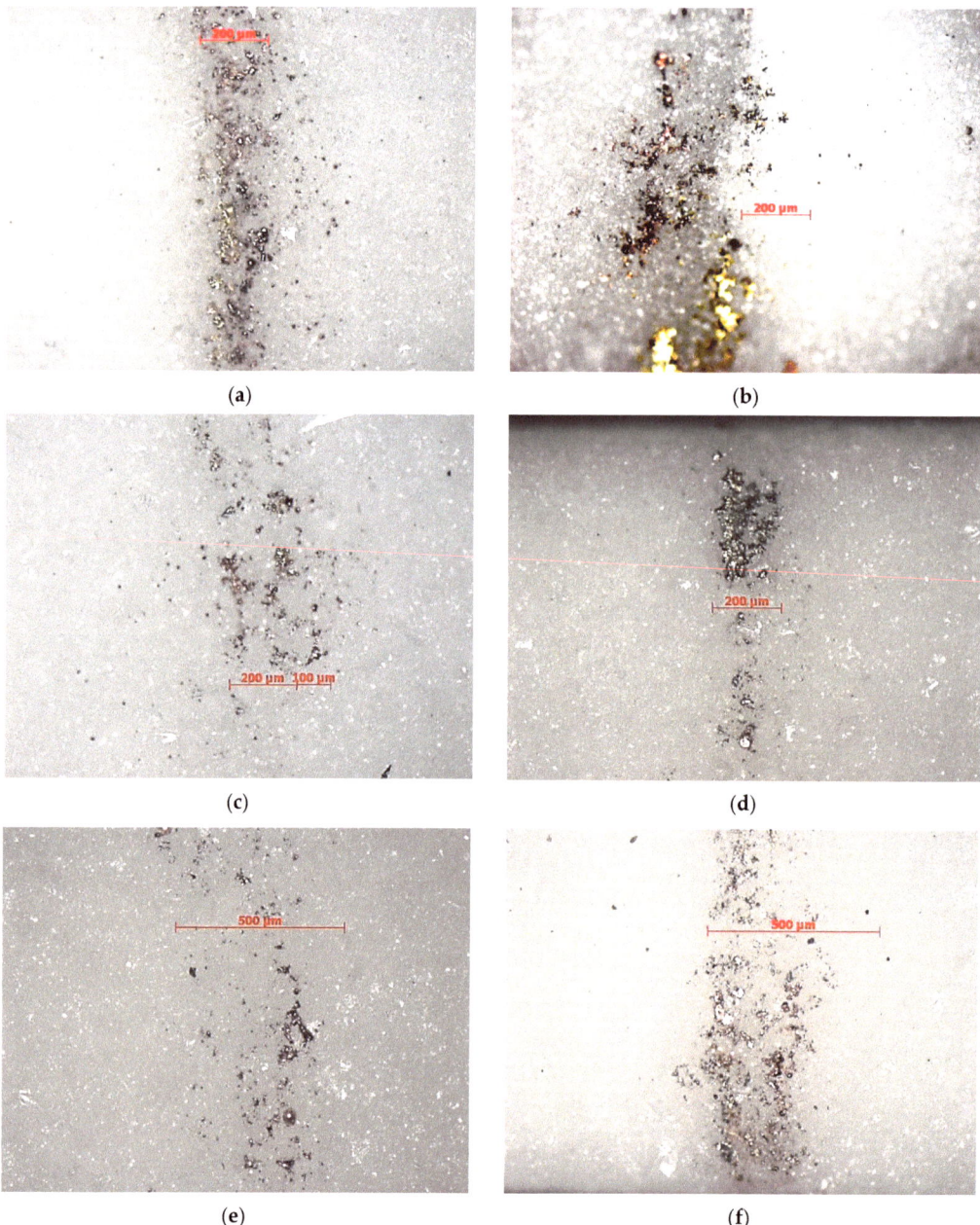

Figure 7. Optical microscopy of the obtained kerf in Al_2O_3 samples coated with a double-layer copper tape coating of 80 μm (general view): (**a**) concentration of 7 g/L, pulse frequency f = 5 kHz, pulse duration D = 1.0 μs; (**b**) concentration of 14 g/L, pulse frequency f = 5 kHz, pulse duration D = 1.0 μs; (**c**) concentration of 21 g/L, pulse frequency f = 5 kHz, pulse duration D = 0.5 μs; (**d**) concentration of 35 g/L, pulse frequency f = 5 kHz, pulse duration D = 1.0 μs; (**e**) concentration of 50 g/L, pulse frequency f = 5 kHz, pulse duration D = 1.0 μs; (**f**) concentration of 100 g/L, pulse frequency f = 7 kHz, pulse duration D = 1.0 μs.

As can be seen, the effect of adding powder is noticeable at powder concentrations of 7 and 14 g/L (Figure 7a,b). The effect is visually reduced at a powder concentration of 21–50 g/L (Figure 7c–e) and appears again at a powder concentration of 100 g/L (Figure 7f). The technological gaps are presented in Table 7.

Table 7. Technological gaps of electrical discharge machining alumina in ZnO powder-mixed deionized water medium.

Powder Concentration, g/L	Pulse Frequency, kHz	Pulse Duration, µs					
		0.5	1.0	1.5	2.0	2.5	2.64
7	2	x	x	x	x	x	x
	5	x	1	1	1	0	x
	7	x	1	1	1	1	x
	10	x	1	1	1	0	x
	15	x	1	1	1	0	x
	20	x	1	1	1	x	x
	25	x	1	1	1	1	x
	30	x	x	x	x	0	x
14	2	x	x	x	x	x	x
	5	x	1	1	x	x	x
	7	x	2	1	0	0	x
	10	x	1	0	0	0	x
	15	x	x	x	x	x	x
	20	x	x	x	x	x	x
	25	x	x	x	x	x	x
	30	x	x	x	x	x	x
21	2	x	x	x	x	x	x
	5	2	1	0	x	x	x
	7	1	2	x	x	x	x
	10	1	1	0	x	x	x
	15	x	x	x	x	x	x
	20	x	x	x	x	x	x
	25	x	x	x	x	x	x
	30	x	x	x	x	x	x
35	2	x	0	x	x	x	x
	5	0	2	0	x	x	x
	7	x	1	x	x	x	x
	10	0	2	0	x	x	x
	15	x	x	x	x	x	x
	20	x	x	x	x	x	x
	25	x	x	x	x	x	x
	30	x	x	x	x	x	x
50	2	0	2	0	x	x	x
	5	x	1	x	x	x	x
	7	x	1	x	x	x	x
	10	0	2	0	x	x	x
	15	x	1	x	x	x	x
	20	x	x	x	x	x	x
	25	x	x	x	x	x	x
	30	x	x	x	x	x	x
100	2	x	2	0	x	x	x
	5	x	1	0	x	x	x
	7	0	2	0	x	x	x
	10	0	2	x	x	x	x
	15	x	0	x	x	x	x
	20	x	x	x	x	x	x
	25	x	x	x	x	x	x
	30	x	x	x	x	x	x

NB: x is the absence of the traces, 0 is little erosion marks, 1 is non-stable erosion marks, 2 is the most pronounced result.

3.4. Scanning Electron Microscopy and Chemical Analyses

The results of scanning electron microscopy (SEM) in secondary electrons and qualitative and quantitative analysis of the obtained kerfs after electrical discharge machining of alumina using a copper assisting coating are shown in Figure 8. SEM-image (Figure 8a) shows the workpiece's microstructure with the deposited material and the traces of the material's sublimation (thermal dissociation). Figure 8b demonstrates a uniform distribution of aluminum and oxygen and localization of the deposited copper in the kerf. The energy-dispersive spectroscopy of the deposited material (Figure 8c) demonstrates the prevalence of copper in the spectra. The carbon corresponds to the normal atmospheric contamination of the samples. The presence of the formed oxides is partly related to the normal contamination of the samples and partly related to the material of the insulating ceramic (bound with aluminum). The presence of aluminum is minor. A quantitative analysis of the five spectra is presented in Table 8.

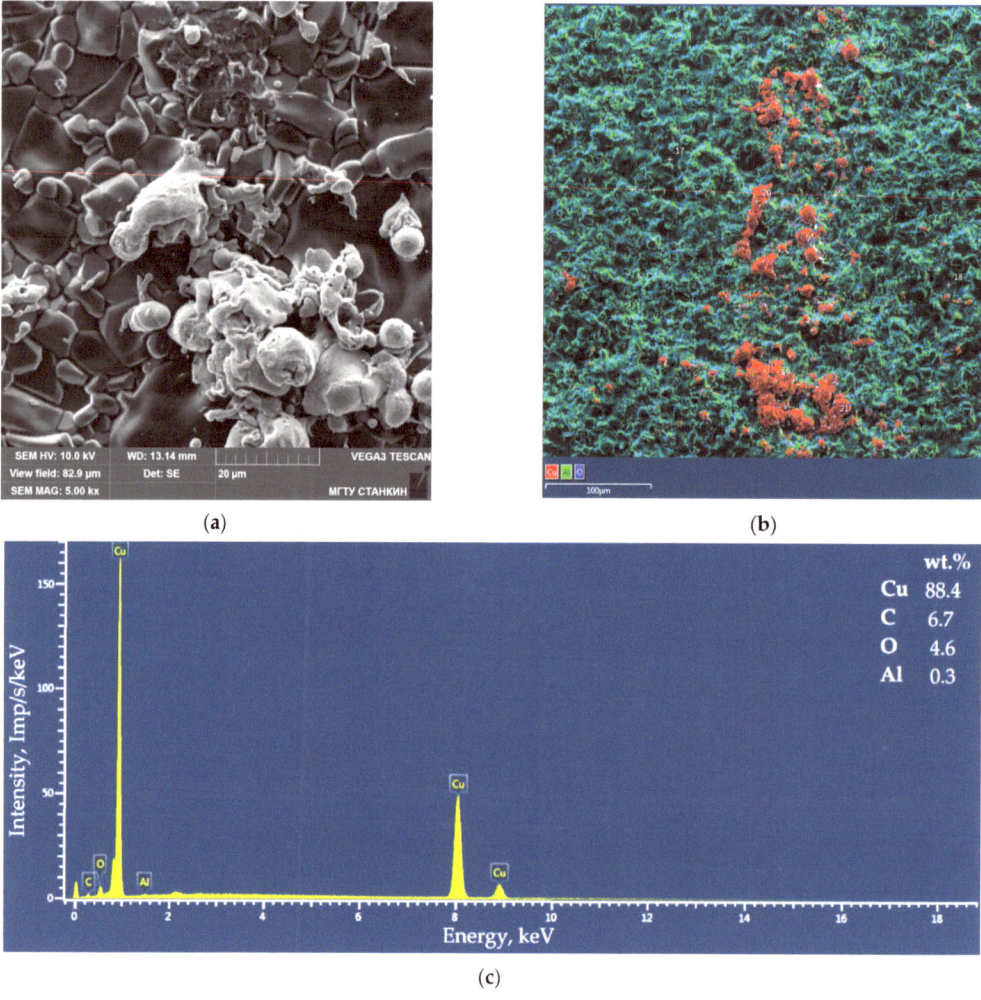

Figure 8. SEM-analyses of the machined kerf in the Al_2O_3 sample: (**a**) image in secondary electrons, 5.00 k×; (**b**) chemical mapping, 1.00 k×; (**c**) energy dispersive spectroscopy of the deposed material.

Table 8. Chemical analysis of the kerf.

Spectrum	Weight Ratio of O, wt.%	Weight Ratio of Al, wt.%	Weight Ratio of Cu, wt.%	Weight Ratio of C, wt.%
1	54.19	45.81	-	-
2	54.95	45.05	-	-
3	2.43	-	86.68	10.88
4	3.41	1.48	84.08	11.03
5	4.64	0.3	88.37	6.69

4. Discussion

A graphical presentation of the calculated material removal rate, which shows the relationship between material removal rate, pulse frequency, and concentration of ZnO-suspension for a pulse duration of 1 µs, is shown in Figure 9. The optimized values of the factors of electrical discharge machining alumina using a double-layer copper tape assisting coating of 80 µm in thickness with ZnO powder-mixed water medium of 14 g/L were: pulse frequency of 5 and 7 kHz, pulse duration of 1 µs.

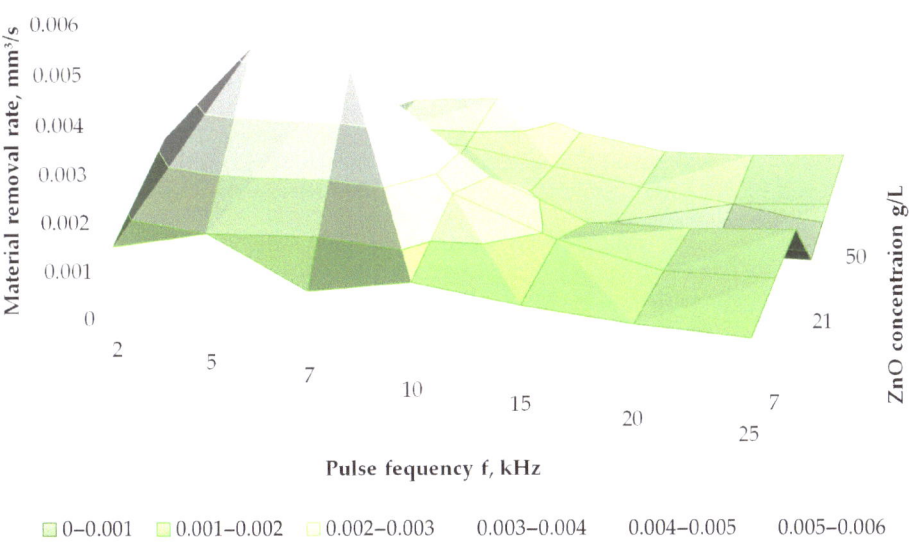

Figure 9. The electrical discharge machining alumina performance using a double-layer copper tape assisting coating of 80 µm in thickness with ZnO powder-mixed deionized water-based dielectric medium for a pulse duration of 1 µs.

In comparison with the previously published data, the achieved performance does not extend analogs [33,49,50,69], where the maximum achievable performance was 0.0084 mm^3/s [50] for deionized water-based medium and 0.0213 mm^3/s for hydrocarbons [69] but significantly enlarged the current knowledge on electrical discharge machining of insulating ceramics as follows:

- Even with the lower concentration of the suspension powder, it is possible to achieve higher values of material removal rate (7 g/L for ZnO in the current study comparing 150 g/L for TiO$_2$ [50]) in combination with a copper coating that can be related to the electrical properties of the assisting powder such as band gap more than to specific electrical resistance of the coating;

- Complex multi-layer complex coating can reduce performance due to a decrease in the electrical conductivity: there is a noticeable difference between mono- and multi-layer Cu-Ag and Cu sandwich coatings and a mono- and double-layer coatings, but no effect was observed between double and triple copper coating;
- Temperature and holding time during coating tempering do not demonstrate any noticeable effect. However, a slight improvement was observed between tempered and not tempered samples that do not have a principle character but significantly enlarge the labor intensity of the work.

5. Conclusions

The paper solves a scientific and technical problem of electrical discharge machining of insulating alumina using assisting mono- and multi-layer Cu-Ag and Cu coatings of 40–120 μm and ZnO powder-mixed deionized water-based medium. Using mono- and multi-layer Cu-Ag and Cu coatings and ZnO powder-mixed water medium was proposed for the first time.

The research showed that tempering temperature and holding time have an insignificant effect on the electrical discharge machining of alumina performance, while multi-layer coating visually improves the performance compared to mono-layer coating. At the same time, there is no visible difference between double- and triple-layer coatings and using silver adhesive reduces the effect of the assisting coating.

The conducted work established a relationship between the material removal rate, powder concentration, and pulse frequency. Using ZnO-powder significantly improves performance up to noticeable values of 0.0015–0.0020 mm^3/s for a concentration of 7 g/L and pulse frequency of 2–5 kHz and 0.0032–0.0053 mm^3/s for a concentration of 14 g/L and pulse frequency of 2–7 kHz. Further concentration increase leads to the opposite trend: at a concentration of 100 g/L, a slight increase in performance is observed (0.0023–0.0025 g/L) for a pulse frequency of 5–7 kHz. The most remarkable results corresponded to the pulse duration of 1 μs.

The obtained data enlarge knowledge on texturing insulating cutting ceramics using various types of powder-mixed deionized water-based mediums and can be used for producing a new class of cutting inserts for machining nickel-based alloys.

Author Contributions: Conceptualization, M.A.V.; methodology, A.A.O.; software, K.I.G.; validation, K.H.; formal analysis, K.H.; investigation, A.A.O.; resources, K.H.; data curation, K.I.G.; writing—original draft preparation, A.A.O.; writing—review and editing, A.A.O.; visualization, K.H. and K.I.G.; supervision, M.A.V.; project administration, M.A.V.; funding acquisition, M.A.V. All authors have read and agreed to the published version of the manuscript.

Funding: This work was supported financially by the Ministry of Science and Higher Education of the Russian Federation (project No FSFS-2021-0006).

Institutional Review Board Statement: Not applicable.

Informed Consent Statement: Not applicable.

Data Availability Statement: Data are available in a publicly accessible repository.

Acknowledgments: The study was carried out on the equipment of the Center of collective use of MSUT "STANKIN" supported by the Ministry of Higher Education of the Russian Federation (project No. 075-15-2021-695 from 26.07.2021, unique identifier RF-2296.61321X0013).

Conflicts of Interest: The authors declare no conflict of interest.

References

1. Oppong Boakye, G.; Geambazu, L.E.; Ormsdottir, A.M.; Gunnarsson, B.G.; Csaki, I.; Fanicchia, F.; Kovalov, D.; Karlsdottir, S.N. Microstructural Properties and Wear Resistance of Fe-Cr-Co-Ni-Mo-Based High Entropy Alloy Coatings Deposited with Different Coating Techniques. *Appl. Sci.* **2022**, *12*, 3156. [CrossRef]
2. Mahesh, K.; Philip, J.T.; Joshi, S.N.; Kuriachen, B. Machinability of Inconel 718: A critical review on the impact of cutting temperatures. *Mater. Manuf. Process.* **2021**, *36*, 753–791. [CrossRef]
3. Padalko, A.G.; Ellert, O.G.; Efimov, N.N.; Novotortsev, V.M.; Talanova, G.V.; Zubarev, G.I.; Fedotov, V.T.; Suchkov, A.N.; Solntsev, K.A. High-pressure phase transformations, microstructure, and magnetic properties of the hypereutectic alloy 10Ni-90Al. *Inorg. Mater.* **2013**, *49*, 1098–1105. [CrossRef]
4. Stepanov, N.D.; Shaysultanov, D.G.; Tikhonovsky, M.A.; Zherebtsov, S.V. Structure and high temperature mechanical properties of novel non-equiatomic Fe-(Co, Mn)-Cr-Ni-Al-(Ti) high entropy alloys. *Intermetallics* **2018**, *102*, 140–151. [CrossRef]
5. Zhong, Q.; Wei, K.; Yue, X.; Zhou, R.; Zeng, X. Powder densification behavior and microstructure formation mechanism of W-Ni alloy processed by selective laser melting. *J. Alloys Compd.* **2022**, *908*, 164609. [CrossRef]
6. Pingale, A.D.; Belgamwar, S.U.; Rathore, J.S. A novel approach for facile synthesis of Cu-Ni/GNPs composites with excellent mechanical and tribological properties. *Mater. Sci. Eng. B Solid-State Mater. Adv. Technol.* **2020**, *260*, 114643. [CrossRef]
7. Grigoriev, S.; Peretyagin, P.; Smirnov, A.; Solis, W.; Diaz, L.A.; Fernandez, A.; Torrecillas, R. Effect of graphene addition on the mechanical and electrical properties of Al_2O_3-SiCw ceramics. *J. Eur. Ceram. Soc.* **2017**, *37*, 2473–2479. [CrossRef]
8. Grigoriev, S.N.; Volosova, M.A.; Peretyagin, P.Y.; Seleznev, A.E.; Okunkova, A.A.; Smirnov, A. The Effect of TiC Additive on Mechanical and Electrical Properties of Al_2O_3 Ceramic. *Appl. Sci.* **2018**, *8*, 2385. [CrossRef]
9. Díaz, L.A.; Montes-Morán, M.A.; Peretyagin, P.Y.; Vladimirov, Y.G.; Okunkova, A.; Moya, J.S.; Torrecillas, R. Zirconia-alumina-nanodiamond composites with gemological properties. *J. Nanoparticle Res.* **2014**, *16*, 2257. [CrossRef]
10. Grigoriev, S.N.; Volosova, M.A.; Fedorov, S.V.; Okunkova, A.A.; Pivkin, P.M.; Peretyagin, P.Y.; Ershov, A. Development of DLC-Coated Solid SiAlON/TiN Ceramic End Mills for Nickel Alloy Machining: Problems and Prospects. *Coatings* **2021**, *11*, 532. [CrossRef]
11. Rashid, A.; Bilal, A.; Liu, C.; Jahan, M.P.; Talamona, D.; Perveen, A. Effect of Conductive Coatings on Micro-Electro-Discharge Machinability of Aluminum Nitride Ceramic Using on-Machine-Fabricated Microelectrodes. *Materials* **2019**, *12*, 3316. [CrossRef] [PubMed]
12. Zawada-Michałowska, M.; Pieśko, P.; Józwik, J. Tribological Aspects of Cutting Tool Wear during the Turning of Stainless Steels. *Materials* **2020**, *13*, 123. [CrossRef] [PubMed]
13. Shabani, M.; Sacramento, J.; Oliveira, F.J.; Silva, R.F. Multilayer CVD Diamond Coatings in the Machining of an Al6061-15 Vol % Al_2O_3 Composite. *Coatings* **2017**, *7*, 165. [CrossRef]
14. Staszuk, M.; Pakuła, D.; Pancielejko, M.; Tański, T.; Dobrzański, L.A. Investigations on wear mechanisms of PVD coatings on carbides and sialons. *Arch. Metall. Mater.* **2017**, *62*, 2095–2100. [CrossRef]
15. Yang, Q.; Wu, T.; Wang, L.; Chen, Y.; Chen, L.; Lou, D.; Cheng, J.; Liu, D. Laser-induced gradient microstrcutres on Si3N4 ceramics and their wettability analysis. *Mater. Chem. Phys.* **2021**, *270*, 124749. [CrossRef]
16. Kang, Z.; Fu, Y.; Ji, J.; Tian, L. Numerical Investigation of Microtexture Cutting Tool on Hydrodynamic Lubrication. *J. Tribol.* **2017**, *139*, 054502. [CrossRef]
17. Kuzin, V.; Grigoriev, S.N.; Volosova, M.A. The role of the thermal factor in the wear mechanism of ceramic tools: Part 1. Macrolevel. *J. Frict. Wear* **2014**, *35*, 505–510. [CrossRef]
18. Stepanov, N.D.; Shaysultanov, D.G.; Chernichenko, R.S.; Tikhonovsky, M.A.; Zherebtsov, S.V. Effect of Al on structure and mechanical properties of Fe-Mn-Cr-Ni-Al non-equiatomic high entropy alloys with high Fe content. *J. Alloys Compd.* **2019**, *770*, 194–203. [CrossRef]
19. Mohsan, A.U.H.; Liu, Z.; Padhy, G.K. A review on the progress towards improvement in surface integrity of Inconel 718 under high pressure and flood cooling conditions. *Int. J. Adv. Manuf. Technol.* **2017**, *91*, 107–125. [CrossRef]
20. Filippov, A.V.; Tarasov, S.Y.; Podgornyh, O.A.; Shamarin, N.N.; Filippova, E.O. Oriented microtexturing on the surface of high-speed steel cutting tool. *AIP Conf. Proc.* **2016**, *1783*, 020057.
21. Wu, Z.; Deng, J.; Su, C.; Luo, C.; Xia, D. Performance of the micro-texture self-lubricating and pulsating heat pipe self-cooling tools in dry cutting process. *Int. J. Refract. Met. Hard Mater.* **2014**, *45*, 238–248. [CrossRef]
22. Pakuła, D.; Staszuk, M.; Dziekońska, M.; Kožmín, P.; Čermák, A. Laser Micro-Texturing of Sintered Tool Materials Surface. *Materials* **2019**, *12*, 3152. [CrossRef] [PubMed]
23. Niketh, S.; Samuel, G.L. Surface textured drill tools-an effective approach for minimizing chip evacuation force and burr formation during high aspect ratio machining of titanium alloy. *J. Manuf. Sci. Eng. Trans. ASME* **2021**, *143*, 041005. [CrossRef]
24. Lazarenko, B.R.; Lazarenko, N.I. Electric spark machining of metals in water and electrolytes. *Surf. Eng. Appl. Electrochem.* **1980**, *1*, 5–8.
25. Lazarenko, B.R.; Duradzhi, V.N.; Bryantsev, I.V. Effect of Incorporating an additional inductance on the characteristics of anode and cathode processes. *Surf. Eng. Appl. Electrochem.* **1979**, *5*, 8–13.
26. Lukashenko, S.V.; Kovtun, A.V.; Dashuk, P.N.; Sokolov, B.N. The Method of Electrical Discharge Machining of Dielectrics. Patent 1,542,715, 10 December 1986.
27. Kumar, M.; Vaishya, R.O.; Suri, N.M.; Manna, A. An Experimental Investigation of Surface Characterization for Zirconia Ceramic Using Electrochemical Discharge Machining Process. *Arab. J. Sci. Eng.* **2021**, *46*, 2269–2281. [CrossRef]

28. Raju, P.; Babasaheb, S. Study on analysis of plasma resistance variation in WEDM of insulating zirconia. *Mater. Manuf. Process.* **2021**, *36*, 59–72. [CrossRef]
29. Guo, Y.; Hou, P.; Shao, D.; Li, Z.; Wang, L.; Tang, L. High-Speed Wire Electrical Discharge Machining of Insulating Zirconia with a Novel Assisting Electrode. *Mater. Manuf. Process.* **2014**, *29*, 526–531. [CrossRef]
30. Shinde, B.; Pawade, R. Study on analysis of kerf width variation in WEDM of insulating zirconia. *Mater. Manuf. Process.* **2021**, *36*, 1010–1018. [CrossRef]
31. Hou, P.; Guo, Y.; Shao, D.; Li, Z.; Wureli, Y.; Tang, L. Influence of open-circuit voltage on high-speed wire electrical discharge machining of insulating Zirconia. *Int. J. Adv. Manuf. Technol.* **2014**, *73*, 229–239. [CrossRef]
32. Ji, R.; Liu, Y.; Zhang, Y.; Wang, F.; Cai, B.; Fu, X. Single discharge machining insulating Al_2O_3 stantaneous pulse energy in kerosene. *Mater. Manuf. Process.* **2012**, *27*, 676–682. [CrossRef]
33. Moudood, M.A.; Sabur, A.; Ali, M.Y.; Jaafar, I.H. Effect of Peak Current on Material Removal Rate for Electrical Discharge Machining of Non-Conductive Al_2O_3 Ceramic. *Adv. Mater. Res.* **2014**, *845*, 730–734. [CrossRef]
34. Liu, Y.H.; Li, X.P.; Ji, R.J.; Yu, L.L.; Zhang, H.F.; Li, Q.Y. Effect of technological parameter on the process performance for electric discharge milling of insulating Al_2O_3 ceramic. *J. Mater. Process. Technol.* **2008**, *208*, 245–250. [CrossRef]
35. Lin, Y.-J.; Lin, Y.-C.; Wang, A.-C.; Wang, D.A.; Chow, H.M. Machining characteristics of EDM for non-conductive ceramics using adherent copper foils. In *Advanced Materials Research, Materials Processing Technologies, Proceedings of the International Conference on Advances in Materials and Manufacturing Processes, Shenzhen, China, 6–8 November 2010*; Jiang, Z.Y., Liu, X.H., Bu, J.L., Eds.; Trans Tech Publications Ltd.: Durnten-Zurich, Switzerland, 2010. [CrossRef]
36. Panova, T.V.; Kovivchak, V.S. Formation of Oxide Layers on the Surface of Copper and its Alloys Modified by a High-Power Ion Beam. *J. Surf. Investig.* **2019**, *13*, 1098–1102. [CrossRef]
37. Murzin, S.P.; Kryuchkov, A.N. Formation of ZnO/CuO heterostructure caused by laser-induced vibration action. *Procedia Eng.* **2017**, *176*, 546–551. [CrossRef]
38. Silva, N.; Ramírez, S.; Díaz, I.; Garcia, A.; Hassan, N. Easy, Quick, and Reproducible Sonochemical Synthesis of CuO Nanoparticles. *Materials* **2019**, *12*, 804. [CrossRef] [PubMed]
39. Zheng, W.; Chen, Y.; Peng, X.; Zhong, K.; Lin, Y.; Huang, Z. The Phase Evolution and Physical Properties of Binary Copper Oxide Thin Films Prepared by Reactive Magnetron Sputtering. *Materials* **2018**, *11*, 1253. [CrossRef] [PubMed]
40. Ikim, M.I.; Spiridonova, E.Y.; Belysheva, T.V.; Gromov, V.F.; Gerasimov, G.N.; Trakhtenberg, L.I. Structural properties of metal oxide nanocomposites: Effect of preparation method. *Russ. J. Phys. Chem. B* **2016**, *10*, 543–546. [CrossRef]
41. Ming, F. Effects of low melting point oxides addition on sintering characteristics of ZnO-glass varistors. *Rare Met. Mater. Eng.* **2002**, *31*, 221–224.
42. Brzezińska, M.; García-Muñoz, P.; Ruppert, A.M.; Keller, N. Photoactive ZnO Materials for Solar Light-Induced Cu_xO-ZnO Catalyst Preparation. *Materials* **2018**, *11*, 2260. [CrossRef]
43. Lee, H.; Zhang, X.; Hwang, J.; Park, J. Morphological Influence of Solution-Processed Zinc Oxide Films on Electrical Characteristics of Thin-Film Transistors. *Materials* **2016**, *9*, 851. [CrossRef]
44. Sahu, A.K.; Chatterjee, S.; Nayak, P.K.; Mahapatra, S.S. Study on effect of tool electrodes on surface finish during electrical discharge machining of Nitinol. *IOP Conf. Ser. Mate. Sci. Eng.* **2018**, *338*, 012033. [CrossRef]
45. Movchan, B.A.; Grechanyuk, N.I. Microhardness and microbrittleness of condensates of the TiC-Al_2O_3 system. *Inorg. Mater.* **1981**, *17*, 972–973.
46. Zhikhareva, I.G.; Zhikharev, A.I. Modeling of Electrodeposited Precipitation Structure. *Izv. Vyss. Uchebnykh Zaved. Khimiya Khimicheskaya Tekhnologiya* **1993**, *36*, 52–58.
47. Kubaschewski, O. Experimental Thermochemistry of Alloys. *Thermochim. Acta* **1988**, *129*, 11–27. [CrossRef]
48. Volosova, M.A.; Okunkova, A.A.; Fedorov, S.V.; Hamdy, K.; Mikhailova, M.A. Electrical Discharge Machining Non-Conductive Ceramics: Combination of Materials. *Technologies* **2020**, *8*, 32. [CrossRef]
49. Okunkova, A.A.; Volosova, M.A.; Kropotkina, E.Y.; Hamdy, K.; Grigoriev, S.N. Electrical Discharge Machining of Alumina Using Ni-Cr Coating and SnO Powder-Mixed Dielectric Medium. *Metals* **2022**, *12*, 1749. [CrossRef]
50. Grigoriev, S.N.; Okunkova, A.A.; Volosova, M.A.; Hamdy, K.; Metel, A.S. Electrical Discharge Machining of Al2O3 Using Copper Tape and TiO2 Powder-Mixed Water Medium. *Technologies* **2022**, *10*, 116. [CrossRef]
51. Grigoriev, S.; Volosova, M.; Fyodorov, S.; Lyakhovetskiy, M.; Seleznev, A. DLC-coating application to improve the durability of ceramic tools. *J. Mater. Eng. Perform.* **2019**, *28*, 4415–4426. [CrossRef]
52. Grigor'ev, S.N.; Fedorov, S.V.; Pavlov, M.D.; Okun'kova, A.A.; So, Y.M. Complex surface modification of carbide tool by Nb + Hf + Ti alloying followed by hardfacing (Ti + Al)N. *J. Frict. Wear* **2013**, *34*, 14–18. [CrossRef]
53. Volosova, M.; Grigoriev, S.; Metel, A.; Shein, A. The role of thin-film vacuum-plasma coatings and their influence on the efficiency of ceramic cutting inserts. *Coatings* **2018**, *8*, 287. [CrossRef]
54. Grigoriev, S.N.; Gurin, V.D.; Volosova, M.A.; Cherkasova, N.Y. Development of residual cutting tool life prediction algorithm by processing on CNC machine tool. *Mater. Werkst.* **2013**, *44*, 790–796. [CrossRef]
55. Grigoriev, S.N.; Kozochkin, M.P.; Sabirov, F.S.; Kutin, A.A. Diagnostic systems as basis for technological improvement. *Proc. CIRP* **2012**, *1*, 599–604. [CrossRef]
56. Grigoriev, S.N.; Sinopalnikov, V.A.; Tereshin, M.V.; Gurin, V.D. Control of parameters of the cutting process on the basis of diagnostics of the machine tool and workpiece. *Meas. Tech.* **2012**, *55*, 555–558. [CrossRef]

57. Viswanathan, V.; Laha, T.; Balani, K.; Agarwal, A.; Seal, S. Challenges and advances in nanocomposite processing techniques. *Mater. Sci. Eng. R* **2006**, *54*, 121–285. [CrossRef]
58. Wang, L.; Zhang, J.; Jiang, W. Recent development in reactive synthesis of nanostructured bulk materials by spark plasma sintering. *Int. J. Refract. Met. Hard Mater.* **2013**, *39*, 103–112. [CrossRef]
59. Zhang, Y.F.; Wang, L.J.; Jiang, W.; Chen, L.-D. Microstructure and properties of Al2O3-TiC composites fabricated by combination of high-energy ball milling and spark plasma sintering (SPS). *J. Inorg. Mater.* **2005**, *20*, 1445.
60. Grigoriev, S.N.; Volosova, M.A.; Okunkova, A.A.; Fedorov, S.V.; Hamdy, K.; Podrabinnik, P.A.; Pivkin, P.M.; Kozochkin, M.P.; Porvatov, A.N. Electrical Discharge Machining of Oxide Nanocomposite: Nanomodification of Surface and Subsurface Layers. *J. Manuf. Mater. Process.* **2020**, *4*, 96. [CrossRef]
61. Grigoriev, S.N.; Volosova, M.A.; Okunkova, A.A.; Fedorov, S.V.; Hamdy, K.; Podrabinnik, P.A.; Pivkin, P.M.; Kozochkin, M.P.; Porvatov, A.N. Wire Tool Electrode Behavior and Wear under Discharge Pulses. *Technologies* **2020**, *8*, 49. [CrossRef]
62. Grigoriev, S.N.; Kozochkin, M.P.; Porvatov, A.N.; Volosova, M.A.; Okunkova, A.A. Electrical discharge machining of ceramic nanocomposites: Sublimation phenomena and adaptive control. *Heliyon* **2019**, *5*, e02629. [CrossRef]
63. Ay, M.; Etyemez, A. Optimization of the effects of wire EDM parameters on tolerances. *Emerg. Mater. Res.* **2020**, *9*, 527–531. [CrossRef]
64. Markopoulos, A.P.; Papazoglou, E.-L.; Karmiris-Obratański, P. Experimental Study on the Influence of Machining Conditions on the Quality of Electrical Discharge Machined Surfaces of aluminum alloy Al5052. *Machines* **2020**, *8*, 12. [CrossRef]
65. Moghaddam, M.A.; Kolahan, F. An optimised back propagation neural network approach and simulated annealing algorithm towards optimisation of EDM process parameters. *Int. J. Manuf. Res.* **2015**, *10*, 215–236. [CrossRef]
66. Melnik, Y.A.; Kozochkin, M.P.; Porvatov, A.N.; Okunkova, A.A. On adaptive control for electrical discharge machining using vibroacoustic emission. *Technologies* **2018**, *6*, 96. [CrossRef]
67. Grigoriev, S.N.; Kozochkin, M.P.; Kropotkina, E.Y.; Okunkova, A.A. Study of wire tool-electrode behavior during electrical discharge machining by vibroacoustic monitoring. *Mech. Ind.* **2016**, *17*, 717. [CrossRef]
68. Grigor'ev, S.N.; Kozochkin, M.P.; Fedorov, S.V.; Porvatov, A.N.; Okun'kova, A.A. Study of Electroerosion Processing by Vibroacoustic Diagnostic Methods. *Meas. Tech.* **2015**, *58*, 878–884. [CrossRef]
69. Kucukturk, G.; Cogun, C. A New Method for Machining of Electrically Nonconductive Workpieces Using Electric Discharge Machining Technique. *Mach. Sci. Technol.* **2010**, *14*, 189–207. [CrossRef]
70. Kumar, A.; Mandal, A.; Dixit, A.R.; Das, A.K. Performance evaluation of Al_2O_3 nano powder mixed dielectric for electric discharge machining of Inconel 825. *Mater. Manuf. Process.* **2018**, *33*, 986–995. [CrossRef]
71. Tzeng, Y.F.; Lee, C.Y. Effects of powder characteristics on electrodischarge machining efficiency. *Int. J. Adv. Manuf. Technol.* **2001**, *17*, 586–592. [CrossRef]
72. Fukuzawa, Y.; Tani, T.; Mohri, N. Machining characteristics of insulated Si_3N_4 ceramics by electrical discharge method—Use of powder-mixed machining fluid. *J. Ceram. Soc. Jpn.* **2000**, *108*, 184–190. [CrossRef]
73. Zhang, W.; Li, L.; Wang, N.; Teng, Y.L. Machining of 7Cr13Mo steel by US-PMEDM process. *Mater. Manuf. Process.* **2021**, *9*, 1060–1066. [CrossRef]
74. Gavrin, V.N.; Kozlova, Y.P.; Veretenkin, E.P.; Logachev, A.V.; Logacheva, A.I.; Lednev, I.S.; Okunkova, A.A. Reactor target from metal chromium for "pure" high-intensive artificial neutrino source. *Phys. Part. Nucl. Lett.* **2016**, *13*, 267–273. [CrossRef]
75. Volosova, M.A.; Okunkova, A.A.; Povolotskiy, D.; Podrabinnik, P.A. Study of electrical discharge machining for the parts of nuclear industry usage. *Mech. Ind.* **2015**, *7*, 706. [CrossRef]
76. Grigoriev, S.N.; Volosova, M.A.; Okunkova, A.A.; Fedorov, S.V.; Hamdy, K.; Podrabinnik, P.A. Elemental and Thermochemical Analyses of Materials after Electrical Discharge Machining in Water: Focus on Ni and Zn. *Materials* **2021**, *14*, 3189. [CrossRef]
77. Grigoriev, S.N.; Teleshevskii, V.I. Measurement Problems in Technological Shaping Processes. *Meas. Tech.* **2011**, *54*, 744–749. [CrossRef]
78. Li, Y.; Zheng, G.; Cheng, X.; Yang, X.; Xu, R.; Zhang, H. Cutting performance evaluation of the coated tools in high-speed milling of AISI 4340 steel. *Materials* **2019**, *12*, 3266. [CrossRef]
79. Zakharov, O.V.; Brzhozovskii, B.M. Accuracy of centering during measurement by roundness gauges. *Meas. Tech.* **2006**, *49*, 1094–1097. [CrossRef]
80. Rezchikov, A.F.; Kochetkov, A.V.; Zakharov, O.V. Mathematical models for estimating the degree of influence of major factors on performance and accuracy of coordinate measuring machines. *MATEC Web Conf.* **2017**, *129*, 01054. [CrossRef]
81. Zakharov, O.V.; Balaev, A.F.; Kochetkov, A.V. Modeling Optimal Path of Touch Sensor of Coordinate Measuring Machine Based on Traveling Salesman Problem Solution. *Procedia Eng.* **2017**, *206*, 1458–1463. [CrossRef]
82. Grigoriev, S.N.; Volosova, M.A.; Okunkova, A.A.; Fedorov, S.V.; Hamdy, K.; Podrabinnik, P.A. Sub-Microstructure of Surface and Subsurface Layers after Electrical Discharge Machining Structural Materials in Water. *Metals* **2021**, *11*, 1040. [CrossRef]
83. Vozniakovskii, A.A.; Kidalov, S.V.; Kol'tsova, T.S. Development of composite material aluminum-carbon nanotubes with high hardness and controlled thermal conductivity. *J. Compos. Mater.* **2019**, *53*, 2959–2965. [CrossRef]
84. Yuan, H.; Zhu, F.; Yang, B.; Xu, B.; Dai, Y. Latest progress in aluminum production by alumina carbothermic reduction in vacuum. *J. Vac. Sci. Technol.* **2011**, *31*, 765–774.
85. Slepchenkov, M.M.; Kolosov, D.A.; Glukhova, O.E. Novel Van Der Waals Heterostructures Based on Borophene, Graphene-like GaN and ZnO for Nanoelectronics: A First Principles Study. *Materials* **2022**, *15*, 4084. [CrossRef]

86. Xin, M. Crystal Structure and Optical Properties of ZnO:Ce Nano Film. *Molecules* **2022**, *27*, 5308. [CrossRef] [PubMed]
87. Platonov, V.; Nasriddinov, A.; Rumyantseva, M. Electrospun ZnO/Pd Nanofibers as Extremely Sensitive Material for Hydrogen Detection in Oxygen Free Gas Phase. *Polymers* **2022**, *14*, 3481. [CrossRef]
88. Li, J.; Mushtaq, N.; Arshad, N.; Shah, M.A.K.Y.; Irshad, M.S.; Yan, R.; Yan, S.; Lu, Y. Proton-Ion Conductivity in Hexagonal Wurtzite-Nanostructured ZnO Particles When Exposed to a Reducing Atmosphere. *Crystals* **2022**, *12*, 1519. [CrossRef]
89. Metel, A.S.; Grigoriev, S.N.; Tarasova, T.V.; Filatova, A.A.; Sundukov, S.K.; Volosova, M.A.; Okunkova, A.A.; Melnik, Y.A.; Podrabinnik, P.A. Influence of Postprocessing on Wear Resistance of Aerospace Steel Parts Produced by Laser Powder Bed Fusion. *Technologies* **2020**, *8*, 73. [CrossRef]
90. Vereschaka, A.S.; Grigoriev, S.N.; Sotova, E.S.; Vereschaka, A.A. Improving the efficiency of the cutting tools made of mixed ceramics by applying modifying nano-scale multilayered coatings. *Adv. Mater. Res.* **2013**, *712–715*, 391–394. [CrossRef]
91. Vereschaka, A.A.; Grigoriev, S.N.; Volosova, M.A.; Batako, A.; Vereschaka, A.S.; Sitnikov, N.N.; Seleznev, A.E. Nano-scale multi-layered coatings for improved efficiency of ceramic cutting tools. *Int. J. Adv. Manuf. Technol.* **2017**, *90*, 27–43. [CrossRef]
92. Grigoriev, S.N.; Vereschaka, A.A.; Fyodorov, S.V.; Sitnikov, N.N.; Batako, A.D. Comparative analysis of cutting properties and nature of wear of carbide cutting tools with multi-layered nano-structured and gradient coatings produced by using of various deposition methods. *Int. J. Adv. Manuf. Technol.* **2017**, *90*, 3421–3435. [CrossRef]
93. Grigoriev, S.; Volosova, M.; Vereschaka, A.; Sitnikov, N.; Milovich, F.; Bublikov, J.; Fyodorov, S.; Seleznev, A. Properties of (Cr, Al, Si)N-(DLC-Si) composite coatings deposited on a cutting ceramic substrate. *Ceram. Int.* **2020**, *46*, 18241–18255. [CrossRef]
94. Grigoriev, S.; Pristinskiy, Y.; Volosova, M.; Fedorov, S.; Okunkova, A.; Peretyagin, P.; Smirnov, A. Wire electrical discharge machining, mechanical and tribological performance of TiN reinforced multiscale SiAlON ceramic composites fabricated by spark plasma sintering. *Appl. Sci.* **2021**, *11*, 657. [CrossRef]
95. Fedorov, S.V.; Pavlov, M.D.; Okunkova, A.A. Effect of structural and phase transformations in alloyed subsurface layer of hard-alloy tools on their wear resistance during cutting of high-temperature alloys. *J. Frict. Wear* **2013**, *34*, 190–198. [CrossRef]
96. Kondratev, V.V.; Ershov, V.A.; Shakhray, S.G.; Ivanov, N.A. Preliminary heating of calcined anode. *Tsvetnye Met.* **2015**, *1*, 54–56.
97. Hill, J.D. Basic Principles of EDM. *Cut. Tool Eng.* **1978**, *30*, 8–10.
98. Pamfilov, E.A. Features of wear and ways to improve the wear resistance of dies and wood-cutting tools. *Trenie I Iznos* **1997**, *18*, 321–330.
99. Reibakh, S.Y.; Segida, A.P. Thermal deformation of electrical-discharge blanking machines. *Sov. Eng. Res.* **1989**, *9*, 113–114.
100. Derevyanko, M.S.; Kondrat'ev, A.V. Phase transformations and thermodynamic properties of oxide systems. *Izv. Ferr. Metall.* **2022**, *65*, 188–189. [CrossRef]

Disclaimer/Publisher's Note: The statements, opinions and data contained in all publications are solely those of the individual author(s) and contributor(s) and not of MDPI and/or the editor(s). MDPI and/or the editor(s) disclaim responsibility for any injury to people or property resulting from any ideas, methods, instructions or products referred to in the content.

Article

Regenerating Iron-Based Adsorptive Media Used for Removing Arsenic from Water

Ilaria Ceccarelli [1], Luca Filoni [1], Massimiliano Poli [1], Ciro Apollonio [2], and Andrea Petroselli [3,*]

1. GAJARDA SRL, Via Fosso Meneghina snc, 01100 Viterbo, Italy; i.ceccarelli@gajarda.com (I.C.); l.filoni@gajarda.com (L.F.); m.poli@gajarda.com (M.P.)
2. Department of Agriculture and Forest Sciences (DAFNE), Tuscia University, 01100 Viterbo, Italy; ciro.apollonio@unitus.it
3. Department of Economics, Engineering, Society and Business (DEIM), Tuscia University, 01100 Viterbo, Italy
* Correspondence: petro@unitus.it

Abstract: Of all the substances that can be present in water intended for human consumption, arsenic (As) is one of the most toxic. Many treatment technologies can be used for removing As from water, for instance, adsorption onto iron media, where commercially available adsorbents are removed and replaced with new media when they are exhausted. Since this is an expensive operation, in this work, a novel and portable plant for regenerating iron media has been developed and tested in four real case studies in Central Italy. The obtained results highlight the good efficiency of the system, which was able, from 2019 to 2023, to regenerate the iron media and to restore its capability to adsorb the As from water almost entirely. Indeed, when the legal threshold value of 10 µg/L is exceeded, the regeneration process is performed and, after that, the As concentration in the water effluent is at the minimum level in all the investigated case studies.

Keywords: arsenic; drinking water; iron-based adsorptive media; regeneration

Citation: Ceccarelli, I.; Filoni, L.; Poli, M.; Apollonio, C.; Petroselli, A. Regenerating Iron-Based Adsorptive Media Used for Removing Arsenic from Water. *Technologies* **2023**, *11*, 94. https://doi.org/10.3390/technologies11040094

Academic Editor: Nam-Trung Nguyen

Received: 8 May 2023
Revised: 18 June 2023
Accepted: 10 July 2023
Published: 12 July 2023

Copyright: © 2023 by the authors. Licensee MDPI, Basel, Switzerland. This article is an open access article distributed under the terms and conditions of the Creative Commons Attribution (CC BY) license (https://creativecommons.org/licenses/by/4.0/).

1. Introduction

According to Directive 2020/2184 of the European Parliament and the Council on the quality of water intended for human consumption, "Water intended for human consumption" is treated or untreated water intended for drinking, for preparing food, drink, or other domestic uses, and water used for the manufacture, treatment, and storage of substances intended for human consumption [1]. Water intended for human consumption, particularly drinking water, must be healthy and clean, and must not contain microorganisms, parasites, or other substances in quantities or concentrations representing a potential danger to human health.

Arsenic (As) is a chemical element with the atomic number 33. It is a semimetal that occurs in three different allotropic forms: yellow, black, and gray. It is found in rocks, water, and animal and vegetable organisms and its content in the rocks of the Earth's crust has been estimated at 1.5 g per ton of rock. As is a common contaminant in water. The most dangerous problem in this context is that in many areas of the world, drinking water is contaminated with As. Due to its high toxicity, especially with regard to arsenite, drinking water regulations allow only very low concentrations of arsenic. Of all the substances that can be present in water, As is one of the most toxic, potentially resulting in skin cancer or other cancers [2]. Indeed, arsenic is easily found in nature, from the atmosphere to soils and rocks, but also in natural waters and organisms [3], being the twentieth most abundant element in all-natural metalloids [4]. Since millions of people are nowadays being exposed to excessive As through the consumption of contaminated drinking water [5], it is mandatory to remove this substance from the water intended for human consumption, respecting the maximum level for As in drinking water fixed at 10 µg/L by many Agencies such as the World Health Organization (WHO) or the US Environmental Protection Agency [6].

In Italy, the data on the presence of As in water comprise a rapidly evolving set of information; following new abstractions, this element can be found in significant concentrations even in aquifers whose hydrogeological conditions would exclude its presence. However, it is evident that this element constitutes a strong criticality in the water use system, especially for water intended for human consumption.

Arsenic is a water contaminant widespread in many areas of the Earth, not only in Italy, of course. Some of the territories where arsenic is found in water in high concentrations are western India, Alaska, Mexico, Chile, and Argentina.

In general terms, it can be stated that in Italy, the supply of qualitatively suitable water is pursued with rigorous and consolidated practices of management of the water system, is controlled through a tested surveillance system, and is regulated on a legislative level, with regard to frequency, typology, and methods of control. Without prejudice to certain circumstances, generally limited in terms of time and territory, for which non-compliance may occur due to the presence of non-standard chemical or microbiological parameters following which limitations on the use of water may also be ordered with adequate information actions on the populations concerned, the water distributed is fit for human consumption and can be consumed in safe conditions from a health point of view, so there is no need for its treatment downstream of the "point of delivery".

In any case, the As presence in water poses the problem of its removal before the water's distribution to the population, so many treatment technologies can be used for removing As from water, including the following [7]:

(1) Oxidation (oxidation and filtration; photochemical oxidation; photocatalytic oxidation; biological oxidation; in situ oxidation);
(2) Membrane technologies (microfiltration; ultrafiltration; nanofiltration; reverse osmosis);
(3) Coagulation/flocculation;
(4) Ion exchange;
(5) Adsorption onto solid media (AM) (activated alumina; iron-based sorbents; zero-valent iron; indigenous filters; miscellaneous sorbents; metal–organic frameworks).

A review of such technologies has been made by a number of authors [8–12]. In many cases, these technologies have been proven as expensive or complex, with the exception of AM, which has been accepted as a suitable removal technology, particularly for developing regions, because of its simple operation, potential for regeneration, and little toxic sludge generation [13].

In AM technology, As and other anions are adsorbed onto a packed bed of media. The employed removal mechanism is usually an exchange of anions for surface hydroxides of the media [14]. When the As concentration of the effluent from an adsorption system reaches the regulatory limit of 10 µg/L, the media are usually removed and replaced with new media.

In the literature, a lot of research can be found dealing with the adsorbent materials employed in AM technology, i.e., with metal oxide/hydroxides, including iron, aluminum, zirconium, and titanium, constituting the majority of the commercially available adsorbents [8,11,15]. In particular, many iron media products have been introduced in the drinking water treatment market in recent years since iron-based adsorbents generally have been found to have higher arsenic adsorptive capacity and efficiency [16,17]. As aforementioned, when the adsorptive iron media no longer has the ability to reduce the As in water effluent to values lower than the maximum contaminant level of 10 µg/L, it is removed and replaced with the media, and the exhausted media can be disposed of in a sanitary landfill [18], but this is an expensive operation. The price of the adsorptive iron media on the market remains quite high and there does not seem to be any sign at the moment of a sensible decrease. In addition, the presence of competitive and/or inhibiting species in the treated waters, perhaps underestimated during the design phase, has led to a reduction in the estimated durations of activity of these products before their exhaustion, thus aggravating the cost of water purification. Indeed, the substitution of the adsorptive iron media accounts on average for 80% of all the operation and maintenance costs of the system, including media replacement, chemicals, electricity, and labor [16].

In order to diminish the operation and maintenance costs, one option to exhaust iron media substitution is its regeneration and reuse. The regeneration process is based on the mechanism of adsorption of As on ferric hydroxides, a chemical exchange that by its nature is reversible. The reactivation process brings the adsorbent material back to its initial state by extracting and bringing back into solution the species that are bound to it during the "work" phase of the masses, leading to the achievement of the desired result: the restoration of the absorption capacity of the As [17].

In the literature, few studies have been conducted on the regeneration and reuse of adsorptive iron media [19,20], while some authors have even suggested that it is not possible to regenerate such media because the process will cause particle degradation [21], although it is true that regeneration offers a potential option to reduce the cost of the system. Starting from these premises, the aim of the present manuscript is the introduction of a novel technological plant that can be used for removing As from water intended for human consumption, employing adsorptive iron media and pursuing its regeneration and reuse. The novelty of this work consists particularly of the presentation of a technological system characterized by its portability. Indeed, it is worth noting that the majority of existing systems for regenerating iron media are non-portable.

The present manuscript is organized as follows. In Section 2, the iron media formation and its use in removing As from water are briefly presented. In Section 3, the proposed portable technological plant is introduced. In Section 4, the selected case studies that have been analyzed are presented. Section 5 describes the performed analysis, while the results are described and discussed in Section 6. Finally, the conclusions are summarized in Section 7.

2. Iron Media Formation and Its Use for Removing As from Water

The process of formation of iron oxide/hydroxide, like the type used in the proposed technological plant, mainly consists of the following three macro-phases: a chemical reaction, a dehydration phase, and a grinding/granulating phase.

In the first macro-phase, i.e., the chemical reaction, iron oxide/hydroxide can be obtained by combining a positive trivalent iron ion with a hydroxide ion. For instance, the following formula shows the chemical reaction between ferric chloride ($FeCl_3$) and sodium hydroxide (NaOH):

$$3Na(OH) + FeCl_3 \rightarrow 3NaCl + Fe(OH)_3 \qquad (1)$$

Based on the reaction of the two selected reagents (sometimes, instead of ferric chloride, ferric sulfate can be used), the necessary time passes for the ferric hydroxide to flocculate, and then it is pressed in a filter press.

The sludge resulting from the first phase contains a significant percentage of water; therefore, a dehydration process is required, i.e., the second macro-phase. Such a process occurs in two ways. (1) Sludge freezing allows further separation from the sludge related to the portion of water that is not mechanically separated; the product obtained has a percentage of residual water of about 50% and is translucent and granular in appearance. (2) Sludge drying by a thermal process such as a rotary drum or belt dryers can reduce the water content to less than 20% in the media.

The third macro-phase, i.e., the grinding/granulating phase, consists of the grinding of the dehydrated sludge (e.g., ferric hydroxide), in order to obtain the required granulometry.

The obtained iron media can be used for removing As from water based on the mechanism of adsorption of arsenic on ferric hydroxides, a chemical exchange that by its nature is reversible. Various studies have found that a classic three-step regeneration process of (1) backwashing the iron media, (2) caustic regeneration, and (3) acid neutralization conditioning has been shown to be effective in removing As and other contaminants absorbed by the iron media [22,23]. Hence, the study of the reactivation process and the subsequent development of the design of the novel technological plant presented here were based on experiences conducted by the international scientific community, in particular by the US Environmental Protection Agency, both as laboratory tests and small pilot plants on real-scale interventions.

3. The Proposed Portable Plant and the Regeneration Process

Based on the aforementioned, the novel technological plant presented here has been designed to be portable and work on site and is constituted by the following elements.

- Two centrifugal pumps for the reagents' (acid and base) recirculation, characterized by a flow rate of 16 m^3/h and a pressure head of 20 m;
- A static mixer, made of plastic material, with a maximum flow rate of 30 m^3/h;
- A piston pump for ferric hydroxide dosing, in 30% solution, in plastic material, with a flow rate of 1500 L/h;
- A piston pump for 50% sulfuric acid solution dosing, in plastic material, with a flow rate range of 100–300 L/h;
- Polyvinyl chloride (PVC), with a flow rate range of 0–30 m^3/h, diameter (D) 50 mm; various valves (1″ PVC ball and check valves);
- Pipes in plastic material;
- Electromagnetic flowmeter, with a range of 0–30 m^3/h, D 40 mm.

Additionally, the following on-site accessories are required for the completion of the regeneration:

- A polyethylene tank, the volume of which must be defined based on that of the specific filter, for depositing the eluates;
- Chemical reagent tanks.

The novel technological plant is shown in Figures 1–3.

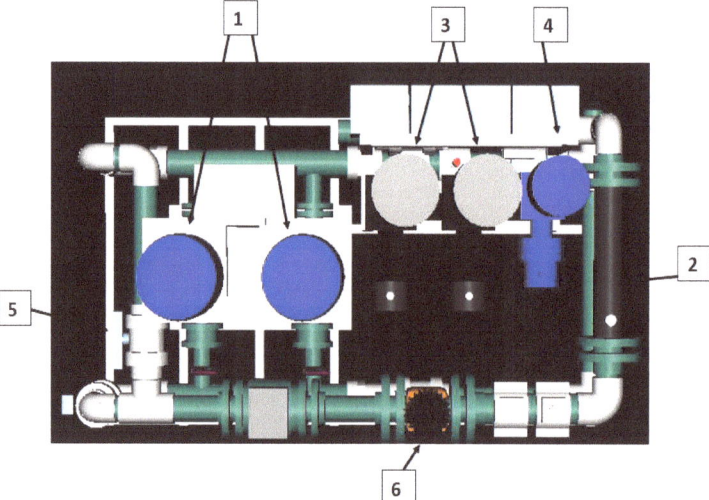

Figure 1. Technological plant aerial view (dimensions 1.7 m × 1 m). (1) Recirculation pumps. (2) Static mixer. (3) Sodium hydroxide dosing pumps. (4) Sulfuric acid dosing pump. (5) Manual regulation valve. (6) Electromagnetic flowmeter.

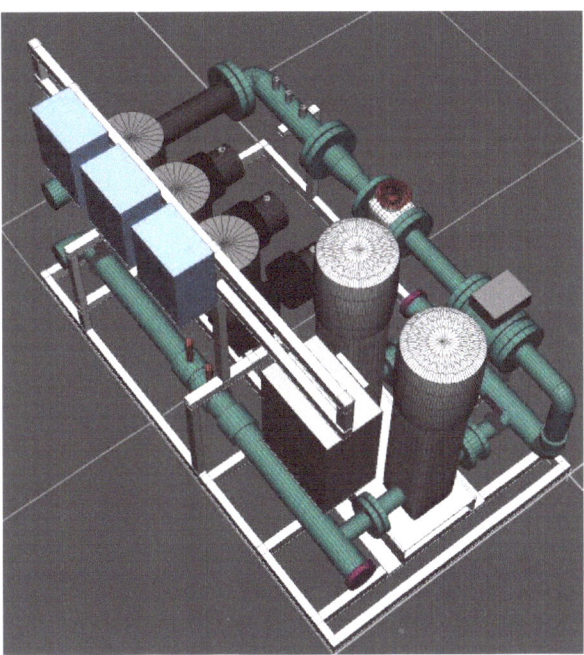

Figure 2. Technological plant isometric view.

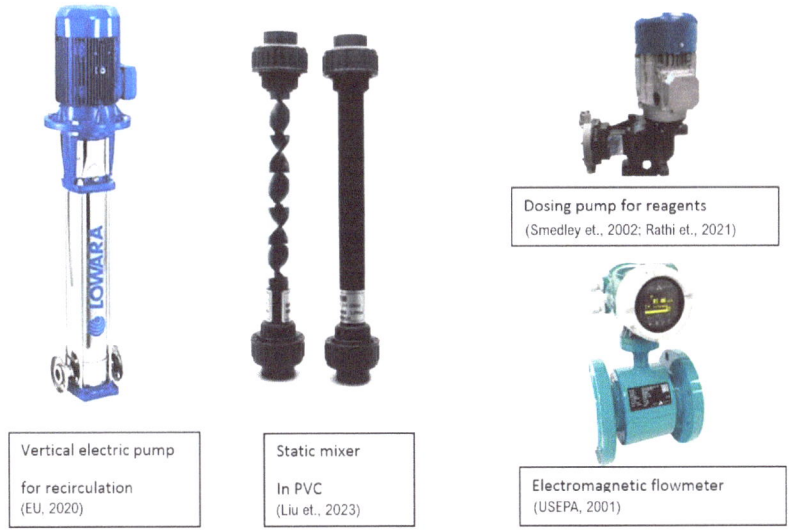

Figure 3. Technological plant details. Numbers refer to Figure 1 ([1–4,6]).

The filter filled with iron media is used as a reactor for the reactivation process while the remaining part of the equipment necessary for the process (hereafter called the skid) is brought directly on site, as well as a tank of adequate capacity where the process eluates will flow. Hoses and hydraulic fittings complete the system equipment. Once the inspection has been carried out and the data for the process are collected, a specific procedure is created in order to plan all the operations that will be carried out during the intervention itself.

From a descriptive point of view, the regeneration process presents a preliminary phase in which the filter containing the oxides to be reactivated is isolated, partially or totally, based on the characteristics of the plant, from the rest of the system. At the same time, the skid and the eluate collection tank are positioned. Once the various equipment has been hydraulically connected via flexible pipes, the system is ready to be started.

The first phase of the regeneration process consists of an alkaline washing of the iron media masses with an aqueous solution of sodium hydroxide that is recirculated through the filter. The aim of alkaline solution recirculation is to increase the contact time with the media and bring a favorable environment to the ion exchange process.

At the end of the alkaline phase of the process, the eluates produced are progressively segregated in the collection tank and replaced with water. This operation continues until a conspicuous lowering of the saline concentration of the water leaving the filter is obtained.

During this phase of "displacement" of the alkaline solution, the eluates are neutralized before reaching the collection tank. At this point, almost all of the species absorbed in the mass were transferred to the alkaline solution and segregated in the storage tank. It is therefore possible to proceed to the second phase of the process in order to restore the alkalinity of the system. The water still present in the filter is recirculated, proceeding at the same time with the controlled injection of sulfuric acid, which acts as a neutralizer. The neutralization operation will continue until a neutral pH is obtained. The process ends with the restoration of the original hydraulic connections and with the backwashing of the filter in order to ensure uniform hydraulic conductivity through the filtering masses.

From an operational point of view, the operational phases regarding the regeneration process are as follows:

- Before starting the regeneration process, a backwashing of the filter media is performed;
- The loading manhole (located at the top of the filter) is opened, and water is allowed to flow about 10–15 cm above the filter media bed;
- The outlet of the pilot system is connected to the previously opened loading manhole, and the return of the system is connected to the lower part of the filter;
- The pumps of the pilot system are activated, and water is circulated in a closed loop;
- The dosing pumps for soda are activated, and soda is circulated inside the filter;
- After the soda has been completely introduced, it continues to circulate for about an hour;
- Subsequently, the eluate is discharged into an existing tank;
- The upper loading manhole is closed, and a backwash with clean water is performed. It is then reopened when the water level is 10–15 cm above the filter media bed. The dosing pump for acid is activated, and the mass is neutralized;
- The last step is to close the upper loading manhole and wash the filter before putting it back into operation.

It should be noted that the procedure is well-suited for regenerating a conventional pressure filter with nozzles (perforated plate with nozzles) for use in series or parallel. The filter must have vents and drains for emptying, as well as a loading/unloading manhole for the filter media.

On the other hand, during the installation of a new filter, the stages of filling the filter media are as follows:

- Placement of a layer of quartz sand support (particle size 2.00–3.15 mm) to cover the nozzles;
- G-OX (filter media bed height between 0.8 and 1.6 m);
- Partial filling with water to protect the nozzles during the sand-filling process;
- Introduction of quartz sand (DIN EN 12904 grade) as a support layer according to the supplier's instructions; leveling and rinsing the layer;
- Introduction of ferric hydroxide through the upper loading manhole;
- Free space, approximately 50% of the G-OX bed, for expansion during backwashing.

4. The Selected Case Studies

The novel technological plant is a potential solution to address the issue of As contamination in water sources and was tested in four different case studies to examine its performance in real-world applications. The four case studies were located in the Viterbo province, in Central Italy, situated 80 km north of Rome.

Italy is generally characterized by high As concentrations in water. High As concentrations have been historically observed (since 1999) in the Central Italian Alps [24], where water is collected from both cold springs and thermal springs. Also, Central Italy has been known to have high levels of As in its water sources. Zuzolo et al. [25], in their comprehensive study aimed at evaluating the occurrence, distribution, and potential health impacts of As on a national scale in Italy, found that significant As concentrations in tap water and soil (up to 27.20 µg/L and 62.20 mg/kg, respectively) are mainly governed by geological features, and that in the central parts of Italy, where alkaline volcanic materials and consequently high levels of As occur, there can be health issues for residents. Baiocchi et al. [26] and Cinti et al. [27] analyzed many water samples from springs and wells in the Sabatini and Vicano-Cimino Volcanic Districts (Central Italy), determining high As concentrations and highlighting risks to the population due to the fact that water mostly sourced from shallow and cold aquifers hosted within volcanic rocks represents the main public drinking water supply. The aforementioned studies highlight that high carcinogenic and non-carcinogenic risk is associated with water ingestion for those living in Northern Italy and Central and Southern Italy (including the capital Rome), also pointed out by [28]. This is true, although it is well known in the literature that As can be found in different oxidation states (As(III) and As(V), i.e., arsenites and arsenates) which have different effects on different organisms, from moderate (skin diseases) to severe (cancer) [29].

The European Union (EU) allows a maximum concentration of 10 µg/L for As in drinking water. However, in the Viterbo area, the concentration levels far exceed the limit, with the highest value recorded at 75 µg/L.

Being a volcanic area, the Viterbo province has a natural composition that includes arsenic. As a result, the water in the region is considered undrinkable by law and must be purified before it can be consumed. The installation of As removal plants is common in the region, and in 2014, 50 million euros were invested in dedicated purifiers to help mitigate the situation. Despite this, many municipalities have reported difficulties in managing the purifiers due to their high operational costs.

The novel technological plant presents a promising solution to improve the situation in the Viterbo province while keeping operational costs low. The plant has been designed to remove As from water sources efficiently, and its effectiveness has been demonstrated in the four case studies mentioned earlier. By implementing the new plant, the region can improve the quality of its water sources and ensure that the concentration of As in drinking water remains within the limits set by the EU.

The four case studies were selected with the primary goal of thoroughly testing the system under different circumstances in order to ensure its effectiveness and durability. Such a decision was driven by the need to ensure that the system is robust enough to handle various environmental factors and external influences that may affect its functionality. Testing the same portable plant multiple times provided valuable insight into how the system performs in the real world and allowed us to identify any potential issues or weaknesses that needed to be addressed. Through the testing process, the system's design and functionality have been optimized, ensuring that it met the required performance standards. The testing of the plant also served as a validation of the system's reliability, which is critical for ensuring the success of any large-scale technological project.

The novel portable technological plant has been employed in different dearsenification plants in the Viterbo province of Central Italy, named WT01, WT02, WT03, and WT04 (Figure 4).

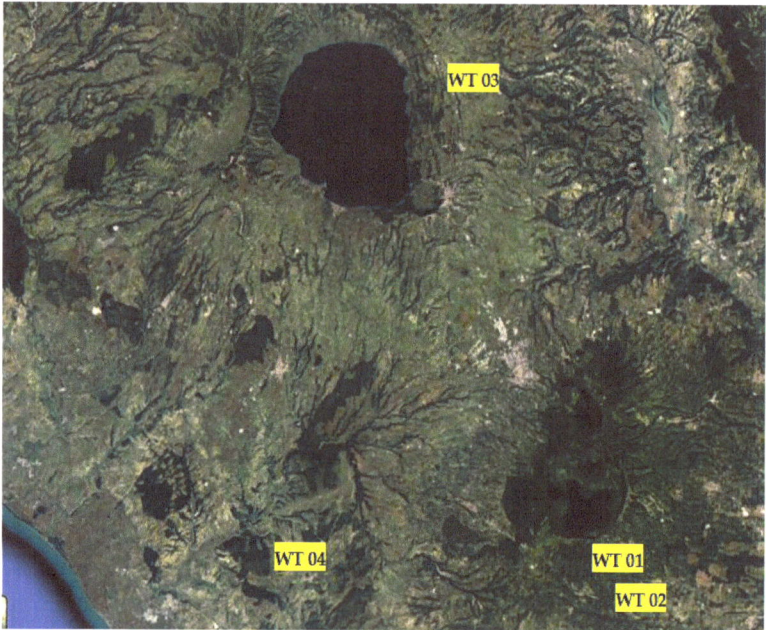

Figure 4. Map of the 4 technological plants in the province of Viterbo, Central Italy.

The WT01 dearsenification plant has two filters in series, each one containing 1800 L of ferric hydroxides. The nominal plant flow rate is 12 L/s and the input As concentration is 34 µg/L.

The WT02 dearsenification plant has two filters in series, each one containing 1000 L of ferric hydroxides. The input water is particularly critical in terms of concentration not only for As (58 µg/L) but also vanadium (12 µg/L) and silica (64 µg/L). The nominal plant flow rate is 4.5 L/s.

The WT03 dearsenification plant has two filters in series, each one containing 3200 L of ferric hydroxides. The nominal plant flow rate is 17 L/s.

Finally, the WT04 dearsenification plant has one filter (named F1) of 4700 L of ferric hydroxides and two filters (named F2-F3) each with 5700 L of ferric hydroxides, connected in parallel. The nominal plant flow rate is 26 L/s.

5. The Performed Analysis

For the four investigated case studies, the As concentrations were periodically monitored in the water effluent, before and after the iron media regeneration, employing national and international guidelines [30–32]. Each analysis was performed two times and then the results were averaged. Other than As, pH and hydraulic conductivity were measured (continuously), but for the sake of brevity, only As and pH results are presented in detail in the following section. It is noteworthy that pH could have an effect on the regeneration process and its efficiency, and this could be a subject of future research, as specified in the conclusion section.

6. Results and Discussion

Figures 5–9 show the As time series monitored in the water effluent in the four investigated case studies.

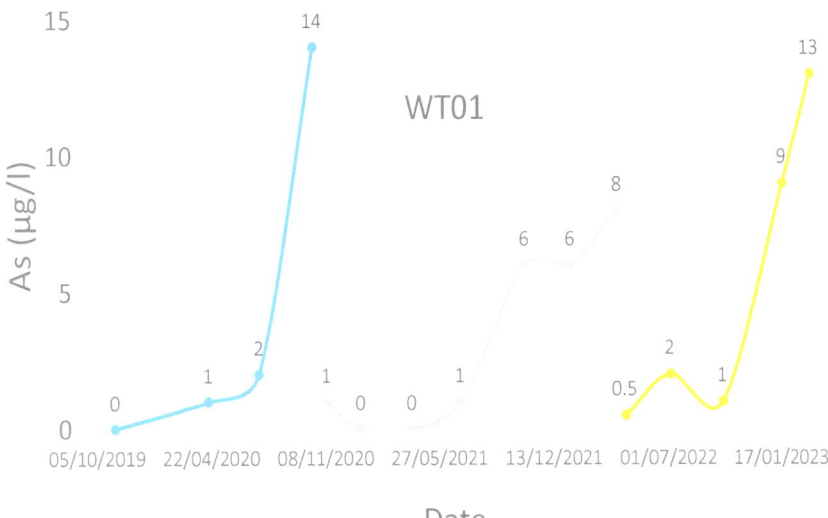

Figure 5. WT01 technological plant water effluent's As time series.

Figure 6. WT02 technological plant water effluent's As time series.

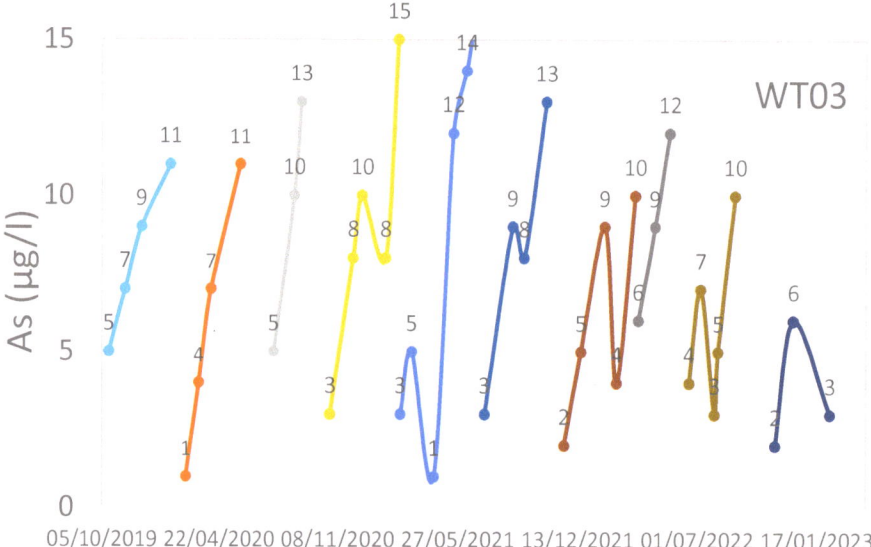

Figure 7. WT03 technological plant water effluent's As time series.

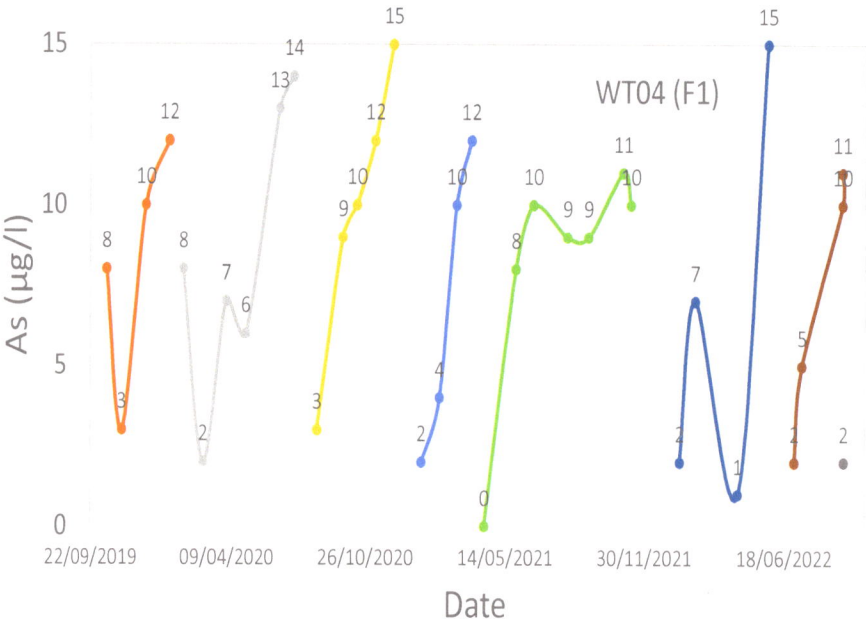

Figure 8. WT04 (F1) technological plant water effluent's As time series.

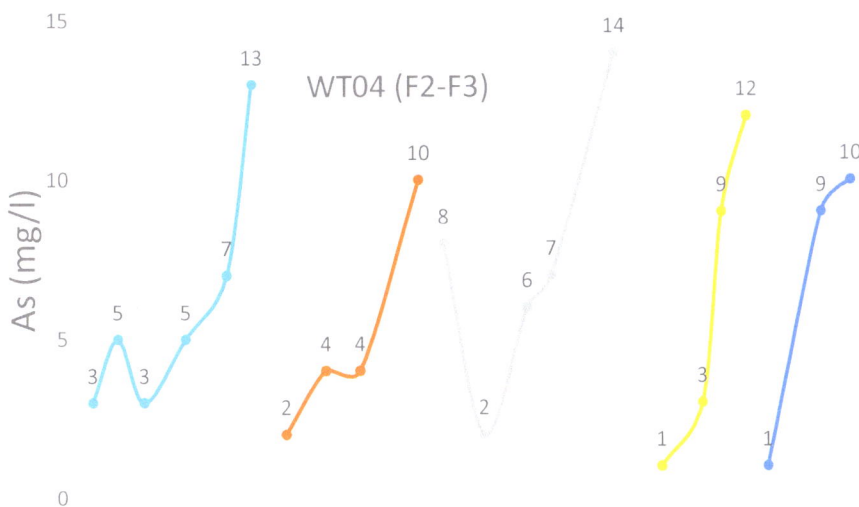

Figure 9. WT04 (F2-F3) technological plant water effluent As time series.

Iron media regenerations were performed for WT01 in October 2019, September 2020, March 2022, and March 2023.

Iron media regenerations were performed for WT02 in March 2019, January 2020, March 2021, July 2021, March 2022, June 2022, and November 2022.

For WT03, iron media regenerations were performed in September 2019, February 2020, July 2020, October 2020, February 2021, July 2021, November 2021, March 2022, June 2022, and October 2022.

Iron media regenerations were performed for WT04 (F1) in October 2019, February 2020, July 2020, December 2020, March 2021, December 2021, June 2022, and September 2022.

For WT04 (F2-F3), iron media regenerations were performed in August 2019, March 2020, July 2020, December 2020, March 2021, July 2021, and December 2021.

Regarding WT01, the iron media kept the As level below the legal threshold of 10 µg/L until the passage of about 67,000 bed volumes (BV), equal to about 240,000 m³ of treated water. Then, it was decided to regenerate only one filter. The regeneration was performed not at the exact moment when the threshold was reached but sometime later, when the system had reached 74,000 BV (approximately 267,000 m³).

The new As legal threshold value of 10 µg/L achievement occurred at around 104,000 BV. It is interesting to note how the duration of the regeneration (104,000 − 74,000 = 30,000 BV) is aligned with the expected value, i.e., just under half of the previous duration, equal to 67,000/2 = 33,500 BV. In view of this second exceeding of the legal limit, it was decided to proceed with a new regeneration of the iron media.

In particular, the decision was again to reactivate only one filter, the one at the top of the series. Given the excellent response that brought the plant over 28,000 BV, the second filter was reactivated the following year. The response in terms of duration was also higher than expected, with approximately 31,000 BV.

The excellent results meant that the regeneration of the iron media can be carried out on both filters also not at the same moment, but one after the other. At present, the iron media continue to absorb arsenic and produce water with pollutant levels well below the legal limits.

Figure 5 shows the As concentration from 2019 to today, and it can be seen that periodically, the value of As increases (close to or above the limit for water intended for human consumption of 10 µg/L) and then decreases.

This happens in correspondence with the regenerations carried out at this plant in October 2019, September 2020, March 2022, and March 2023. It is noteworthy that after the iron media regeneration, the As concentration in the effluent is close to 0 µg/L, testifying to the effectiveness of the process.

Regarding WT02, it was decided to proceed with the regeneration of the iron media, as their duration did not allow for the treatment of more than 50,000 m^3 before the performance decreased.

Since the first reactivation, carried out in March 2019, the iron media have been regenerated other five times (January 2020, March 2021, July 2021, November 2022, and July 2022) and, presumably, this operation will also be carried out in the future. Figure 6 shows the As concentration from 2019 to today: also for this case study, it can be seen how periodically, the value of As increases (close to or above the limit for water intended for human consumption of 10 µg/L) and then decreases. This happens in correspondence with the regenerations carried out at this plant. Unlike the previous case study, the As concentration in the effluent after the iron media regeneration is not close to 0 µg/L, having values close to 1 or 2 µg/L. This circumstance could be due to the input water that is particularly critical in terms of concentration not only for As but also for other elements. In any case, the water effluent As concentration is much lower than the legal threshold equal to 10 µg/L.

Regarding WT03 and WT04, as can be seen from Figures 7–9, the trend is rather variable, but there are evident drops in the value of As near the regenerations, which took place in September 2019, February 2020, July 2020, October 2020, February 2021, July 2021, November 2021, March 2022, and October 2022 for WT03, while for WT04, the regenerations were performed at different moments based on the filter to reactivate (F1 or F2-F3).

Regarding the pH analysis, a general trend was observed in all four investigated case studies. The monitored pH was high (values greater than 12) during the desorption of As and other adsorbed elements, low (values not greater than 3–4) during the reacidification process, and neutral (values approximately equal to 7) during the last phases of the regeneration process.

In any case, the investigated case studies testify to the effectiveness of the proposed system related to the regeneration of the iron media, with the following fundamental advantages:

- Environmental sustainability: Iron media are no longer considered as waste to be disposed of in landfills (according to the European Waste Catalogue, 19 September 2001, "Solid waste produced by primary filtration and screening processes") but indeed can be reused effectively according to the principles of a circular economy.
- Cost saving: Considering as an example a standard filter with a diameter of 2 m and a filter bed height of 1.1 m, the volume to be considered for the unique purchase of the filling material is approximately 3–4 m^3, which corresponds to an expense of EUR 15,000–20,000 for the replacement of the masses. Moreover, other costs that should be taken into account are related to the landfill disposal of the exhausted iron media, which can be approximately quantified as EUR 2500, and the disposal of eluates, which can be approximately quantified as EUR 3500 (in the same example of the aforementioned standard filter). Of course, we also have to consider the cost of the reagents used in the regeneration process. Their exact quantity depends on the initial characteristics of the filter media, but in the same example of the aforementioned standard filter, we can roughly estimate an approximate cost of EUR 1000–2000.
- Using the regenerated material significantly reduces the production of the "new" media, with considerable savings in raw materials and convenience in terms of energy and economics.
- Low environmental impact, due to the reuse of materials and savings on disposal costs.
- Less dependence on producers (that at the moment are no more than 2–3 in Europe).

Moreover, the advantages deriving from regeneration with a mobile plant, directly on site, like in the here-investigated case studies, are as follows:
- Brief shutdown of the plant.
- Washing of the iron media on site, avoiding the emptying/filling of the iron media.
- Low environmental impact.

It is noteworthy that oxidation is known as a suitable method for the removal of As, but it is also true that oxidation alone can only transform As(III) to As(V), which is less toxic, and that the sorption processes of arsenite and arsenates on iron oxides are quite different and complex, as shown by [33]. In our opinion, this circumstance does not diminish the effectiveness and relevance of the presented technological plant and the related technical and economic implications.

7. Conclusions

Adsorptive media technology is a commonly employed approach for removing arsenic (As) from water, especially for drinking purposes. However, this method usually requires iron-based media, which are disposed of after one use. Since the replacement of iron media accounts for approximately 80% of the total operational cost, it would be more cost-effective to regenerate and reuse the media. In light of this issue, a novel and portable technological plant has been designed and tested in four real case studies to regenerate iron media, proving its feasibility. The following conclusions can be drawn:

(1) The obtained results highlight the good efficiency of the system, which is able to regenerate the iron media and restore its capability to adsorb As from water almost entirely. When the legal threshold value of 10 μg/L is exceeded, the regeneration process is performed, and after that, the As concentration in the water effluent is at the minimum level in all the investigated case studies.

(2) Multiple regenerations occurring from 2019 to 2023 seem to validate the system's reliability, which is critical for ensuring the success of any large-scale technological project.

(3) The main advantages of the proposed portable technological plant that emerged from the results are the renewal of the filter bed with the restoration of its original adsorption capacity; no solid waste to dispose of, thereby eliminating disposal costs; low environmental impact; a reduction in the production of "new" media, saving raw materials and improving energy efficiency; minimized system downtime; and limited equipment movement.

Further research is needed to test various types of iron media to investigate the possibility of extending the time between regenerations. The system's good efficiency indicates that it has the potential to be used on a larger scale, which would help minimize the amount of iron media being thrown away, making the technology more sustainable and cost-effective. As aforementioned, the pH of the water to be treated could have an effect on the regeneration process and its efficiency, so this could also be a subject of future research.

The innovative approach described here has the potential to significantly reduce the operational cost of As removal from water. This is particularly important in developing countries where the cost of As removal is often prohibitive, leading to the consumption of As-contaminated water, which can have serious health implications. By using this technology, we can provide clean and safe drinking water to the affected population, improving their overall health and well-being.

In conclusion, the development of this novel and portable system for regenerating iron media is a promising step forward in As removal technology. Its potential for large-scale implementation could make it an attractive solution for developing countries where As contamination is a significant problem. Future research will continue to explore and optimize this technology, making it more effective, sustainable, and accessible.

Author Contributions: Conceptualization, I.C., L.F. and M.P.; methodology, I.C., L.F. and M.P.; software, I.C., L.F. and M.P.; validation, I.C., L.F. and M.P.; formal analysis, I.C., L.F. and M.P.; investigation, I.C., L.F. and M.P.; resources, I.C., L.F. and M.P.; data curation, I.C., L.F. and M.P.; writing—original draft preparation, I.C., L.F., M.P., C.A. and A.P.; writing—review and editing, I.C., L.F., M.P., C.A. and A.P.; project administration, I.C., L.F. and M.P.; funding acquisition, I.C., L.F. and M.P. All authors have read and agreed to the published version of the manuscript.

Funding: This research received no external funding.

Institutional Review Board Statement: Not applicable.

Informed Consent Statement: Not applicable.

Data Availability Statement: Data are available from the authors upon request.

Acknowledgments: We thank the company Gajarda Srl for its technical and operational support (materials for experiments and highly specialized technicians).

Conflicts of Interest: The authors declare no conflict of interest.

References

1. EU. Directive 2020/2184 of the European Parliament and of the Council of 16 December 2020 on the quality of water intended for human consumption (recast). *Off. J. EU.* **2020**, *435*, 1–62.
2. Liu, Y.; Chen, Z.; Yin, X.; Chen, Y.; Liu, Y.; Yang, W. Selective and efficient removal of As(V) and As(III) from water by resin-based hydrated iron oxide. *J. Mol. Struct.* **2023**, *1273*, 134361. [CrossRef]
3. Smedley, P.L.; Kinniburgh, D.G. A review of the source, behaviour and distribution of arsenic in natural waters. *Appl. Geochem.* **2002**, *17*, 517–568. [CrossRef]
4. Rathi, B.S.; Kumar, P.S. A review on sources, identification and treatment strategies for the removal of toxic arsenic from water system. *J. Hazard. Mater.* **2021**, *418*, 126299. [CrossRef]
5. He, Z.L.; Tian, S.L.; Ning, P. Adsorption of arsenate and arsenite from aqueous solutions by cerium-loaded cation exchange resin. *J. Rare Earths* **2012**, *30*, 563–572. [CrossRef]
6. USEPA. National Primary Drinking Water Regulations; Arsenic and Clarification to Compliance and New Source Contaminant Monitoring. *Final Rule Fed. Reg.* **2001**, *66*, 6076.
7. Nicomel, N.R.; Leus, K.; Folens, K.; Van Der Voort, P.; Du Laing, G. Technologies for arsenic removal from water: Current status and future perspectives. *Int. J. Environ. Res. Public Health* **2015**, *13*, 62. [CrossRef]
8. Choong, T.S.Y.; Chuah, T.G.; Robiah, Y.; Gregory, F.L.; Koay, G.; Azni, I. Arsenic toxicity, health hazards and removal techniques from water: An overview. *Desalination* **2007**, *217*, 139–166. [CrossRef]
9. Cundy, A.B.; Hopkinson, L.; Whitby, L.D. Use of iron-based technologies in contaminated land and groundwater remediation: A review. *Sci. Total Environ.* **2008**, *400*, 42–51. [CrossRef] [PubMed]
10. Giles, D.E.; Mohapatra, M.; Issa, T.B.; Anand, S.; Singh, P. Iron and aluminium based adsorption strategies for removing arsenic from water. *J. Environ. Manag.* **2011**, *92*, 3011–3022. [CrossRef] [PubMed]
11. Jain, C.K.; Singh, R.D. Technological options for the removal of arsenic with special reference to South East Asia. *J. Environ. Manag.* **2012**, *107*, 1–18. [CrossRef]
12. Mondal, P.; Bhowmick, S.; Chatterjee, D.; Figoli, A.; Wan der Bruggen, B. Remediation of inorganic arsenic in groundwater for safe water supply: A critical assessment of technological solutions. *Chemosphere* **2013**, *92*, 157–170. [CrossRef] [PubMed]
13. Asere, T.G.; Stevens, C.V.; Du Laing, G. Use of (modified) natural adsorbents for arsenic remediation: A review. *Sci. Total Environ.* **2019**, *676*, 706–720. [CrossRef] [PubMed]
14. Clifford, D.A.; Sorg, T.J.; Ghurye, G.L. Ion exchange and adsorption of inorganics contaminants. In *Water Quality and Treatment*; A Handbook on Drinking Water; AWWA: Denver, CO, USA, 2011.
15. Lakshmanan, D.; Clifford, D.; Samanta, G. Arsenic removal by coagulation with aluminum, iron, titanium and zirconium. *J. Am. Water Work. Assoc.* **2008**, *100*, 76–88. [CrossRef]
16. Wang, L.; Chen, A.S.C. *Costs of Arsenic Removal Technologies for Small Water Systems: U.S. EPA Arsenic Removal Technology Demonstration Program*; EPA/600/R-04/210; USEPA, Office of Research and Development: Cincinnati, OH, USA, 2011.
17. Chen, A.S.C.; Sorg, T.J.; Wang, L. Regeneration of iron-based adsorptive media used for removing arsenic from groundwater. *Water Res.* **2015**, *77*, 85–97. [CrossRef]
18. Cornwell, D.A.; Roth, D.K. Water treatment plant residuals management. In *Water Quality and Treatment*; A Handbook on Drinking Water; AWWA: Denver, CO, USA, 2011.
19. Mamindy-Pajany, Y.; Hure, C.; Marmier, N.; Romoe, M. Arsenic (V) adsorption from aqueous solution onto goethite, hematite, magnetite and zero-based iron: Effects of pH, concentration and reversibility. *Desalination* **2011**, *281*, 93–99. [CrossRef]
20. Roy, P.; Mondal, N.K.; Das, K. Modeling of the adsorptive removal of arsenic: A statistical approach. *J. Environ. Chem. Eng.* **2013**, *189*, 1–13. [CrossRef]

21. Mohan, D.; Pittman, C.U., Jr. Arsenic removal from water/wastewater using adsorbents e a critical review. *J. Hazard. Mater.* **2007**, *142*, 1–53. [CrossRef]
22. Sorg, T.J.; Chen, A.S.C.; Wang, L.; Kolisz, R. Regenerating an arsenic removal iron-based adsorptive media system, part 1: The regeneration process. *J. Am. Water Work. Assoc.* **2017**, *109*, 13–24. [CrossRef] [PubMed]
23. Sorg, T.J.; Chen, A.S.C.; Wang, L.; Kolisz, R. Regenerating an arsenic removal iron-based adsorptive media system, part 2: Performance and cost. *J. Am. Water Work. Assoc.* **2017**, *109*, E122–E128. [CrossRef]
24. Peña Reyes, F.A.; Crosta, G.B.; Frattini, P.; Basiricò, S.; Della Pergola, R. Hydrogeochemical overview and natural arsenic occurrence in groundwater from alpine springs (upper Valtellina, Northern Italy). *J. Hydrol.* **2015**, *529*, 1530–1549. [CrossRef]
25. Zuzolo, D.; Cicchella, D.; Demetriades, A.; Birke, M.; Albanese, S.; Dinelli, E.; Lima, A.; Valera, P.; De Vivo, B. Arsenic: Geochemical distribution and age-related health risk in Italy. *Environ. Res.* **2020**, *182*, 109076. [CrossRef] [PubMed]
26. Baiocchi, A.; Lotti, F.; Piscopo, V.; Chiocchini, U.; Madonna, S.; Manna, F. Hydraulic interactions between aquifers in Viterbo area (Central Italy). In *Urban Groundwater Meeting the Challenge*; Howard, K.W.F., Ed.; Taylor & Francis Group: London, UK, 2007; Volume 8, pp. 223–238.
27. Cinti, D.; Vaselli, O.; Poncia, P.P.; Brusca, L.; Grassa, F.; Procesi, M.; Tassi, F. Anomalous concentrations of arsenic, fluoride and radon in volcanic sedimentary aquifers from central Italy: Quality indexes for management of the water resource. *Environ. Pollut.* **2019**, *253*, 525–537. [CrossRef]
28. Flora, S.J.S. *Handbook of Arsenic Toxicology*; Academic Press: Cambridge, MA, USA, 2023; p. 954.
29. Ozturk, M.; Metin, M.; Altay, V.; Ahmad Bhat, R.; Ejaz, M.; Gul, A.; Turkyilmaz Unal, B.; Hasanuzzaman, M.; Nibir, L.; Nahar, K.; et al. Arsenic and Human Health: Genotoxicity, Epigenomic Effects, and Cancer Signaling. *Biol. Trace Elem. Res.* **2022**, *200*, 988–1001. [CrossRef] [PubMed]
30. American Public Health Association. *Standard Methods for the Examination of Water and Wastewater*, 21st ed.; APHA: Washington, DC, USA, 2005.
31. APAT/IRSA-CNR. *Metodi Analitici per le Acque*; Alluminio: Roma, Italy, 2003.
32. World Health Organization. *Guidelines for Drinking-Water Quality*; WHO: Geneva, Switzerland, 2006.
33. Dixit, S.; Hering, J.G. Comparison of Arsenic(V) and Arsenic(III) Sorption onto Iron Oxide Minerals: Implications for Arsenic Mobility. *Environ. Sci. Technol.* **2003**, *37*, 4182–4189. [CrossRef] [PubMed]

Disclaimer/Publisher's Note: The statements, opinions and data contained in all publications are solely those of the individual author(s) and contributor(s) and not of MDPI and/or the editor(s). MDPI and/or the editor(s) disclaim responsibility for any injury to people or property resulting from any ideas, methods, instructions or products referred to in the content.

Article

Comparative Effect of the Type of a Pulsed Discharge on the Ionic Speciation of Plasma-Activated Water

Victor Panarin [1], Eduard Sosnin [1,2], Andrey Ryabov [3], Victor Skakun [1], Sergey Kudryashov [3] and Dmitry Sorokin [1,4,*]

1. Optical Radiation Laboratory, Institute of High Current Electronics SB RAS, 2/3 Akademichesky Ave., 634055 Tomsk, Russia; panarin@loi.hcei.tsc.ru (V.P.); badik@loi.hcei.tsc.ru (E.S.); skakun54@bk.ru (V.S.)
2. Faculty of Innovative Technologies, Tomsk State University, 36 Lenina Ave., 634050 Tomsk, Russia
3. Institute of Petroleum Chemistry, 4 Akademichesky Ave., 634055 Tomsk, Russia; andrey@ipc.tsc.ru (A.R.); ks@ipc.tsc.ru (S.K.)
4. Faculty of Physics, Tomsk State University, 36 Lenina Ave., 634050 Tomsk, Russia
* Correspondence: sdma-70@loi.hcei.tsc.ru

Abstract: The comparison of ion concentrations, pH index, and conductivity in distilled and ground water after exposure to low-temperature plasma formed by barrier and bubble discharges is performed. It has been found that in the case of groundwater, the best performance for the production of NO_3^- anions is provided by the discharge inside the gas bubbles. For distilled water, the barrier discharge in air, followed by saturation of water with plasma products, is the most suitable from this point of view. In both treatments, the maximum energy input into the stock solution is ensured. After 10 min treatment of ground water, the pH index increases and then it decreases. The obtained numerical indicators make it possible to understand in which tasks the indicated treatment modes should be used, their comparative advantages, and disadvantages. From the point of view of energy consumption for obtaining approximately equal (in order of magnitude) amounts of NO_3^- anions, both types of discharge treatment are suitable. The research results point to a fairly simple way to convert salts (calcium carbonates) from an insoluble form to soluble one. Namely, when interacting with NO_3^- anions, insoluble carbonates pass into soluble nitrates.

Keywords: plasma-activated water; barrier discharge; bubble discharge; physicochemical properties; process conditions

Citation: Panarin, V.; Sosnin, E.; Ryabov, A.; Skakun, V.; Kudryashov, S.; Sorokin, D. Comparative Effect of the Type of a Pulsed Discharge on the Ionic Speciation of Plasma-Activated Water. *Technologies* **2023**, *11*, 41. https://doi.org/10.3390/technologies11020041

Academic Editors: Sergey N. Grigoriev, Marina A. Volosova and Anna A. Okunkova

Received: 15 February 2023
Revised: 13 March 2023
Accepted: 13 March 2023
Published: 14 March 2023

Copyright: © 2023 by the authors. Licensee MDPI, Basel, Switzerland. This article is an open access article distributed under the terms and conditions of the Creative Commons Attribution (CC BY) license (https://creativecommons.org/licenses/by/4.0/).

1. Introduction

Interdisciplinary scientific research on the effect of non-equilibrium low-temperature plasma of electric discharges in the air on water and aqueous solutions is being carried out very intensively [1–6]. All research in this direction can be conditionally divided into three groups: (1) Study of the speciation of treated water and the kinetics of reactions in the discharge plasma and aqueous solutions; (2) development and testing of discharge reactors; (3) revealing the useful properties of laboratory-produced solutions, or determination of the degree of purification of such solutions from pollutants of various natures. These lines of research complement each other.

A special place is occupied by the direction associated with the development of equipment for the production of plasma-activated water (PAW). PAW is a result of plasma action on water or aqueous solutions (e.g., phosphate-buffered saline, etc.) in the presence of oxygen O_2 or a mixture of O_2 and nitrogen N_2, at atmospheric pressure [7]. In biomedicine, the benefits of PAW have been demonstrated in the tasks of biofilm removal, wound healing, and bacterial inactivation [8–10]. In agriculture, PAW is proposed to be used to increase the rate of seed germination and subsequently accelerate the growth of seedlings and plants, to inactivate plant-associated pathogens and rescue fungus-infected plants, and to preserve crops [11–15]. In particular, in studies [16,17] it has been shown that treating water with a discharge and then irrigating with this water significantly increases the growth rate of plants such as spinach (*Spinacia oleracea*), radish (*Raphanus sativus* var. *sativus*), strawberry (*Fragaria ananassa*), and Chinese cabbage (*Brassica campestris*). An increase in the concentration of nitrogen in the leaves was also observed.

Currently, it is believed that this PAW activity is due to the action of the following chemicals: reactive oxygen and nitrogen species (RONS), as well as relatively short-lived radicals (•OH, NO•), superoxide (O_2^-), peroxynitrate ($OONO_2^-$), and peroxynitrite ($ONOO^-$). The above effect of improving plant growth with PAW is attributed to the action of aqueous nitrates, nitrites, and ammonium ions, as well as hydrogen peroxide. In addition, the ability of PAW to produce fungicidal and antimicrobial effects is traditionally associated with short-lived reactive oxygen species [7,14,18–20]. PAW is considered a sustainable and promising solution for biotechnological applications due to the transient nature of its biochemical activity and the potential economic and environmental benefits of using ambient air rather than scarcely available or expensive chemicals as the raw material. This approach can potentially reduce the cost of treatment technology.

Methods and devices for water treatment with an electric discharge are of decisive importance for future technology. Authors from different scientific teams classify these methods in different ways, but it is better to give a classification that is most consistent with the processes of obtaining plasma-activated water. These are the following methods of discharge treatment of aqueous solutions: electrohydraulic, heterophase, bubble discharge, as well as some types of remote discharge reactors. Each method has its own advantages, but it also has disadvantages (see, for example [21]), an overview of which, should be briefly discussed.

Electrohydraulic discharge reactors [22] are characterized by the fact that the discharge occurs directly in an aqueous solution; for example, between two pointed or specially shaped electrodes. Or, in a narrow channel between them; for example, a capillary one. Another option may be a barrier configuration, i.e., when the high-voltage electrode is covered with a dielectric (barrier) layer, e.g., polyethylene or ceramic. Most often, pulsed arc and corona discharges are used. The advantage of this approach to the formation of a discharge, is that the plasma is formed directly in contact with water, which accelerates the PAW production. However, when overloaded, this can also lead to the decay of useful chemicals. In addition, the formation of active oxygen and nitrogen species in this case is limited by the low concentration of nitrogen and oxygen dissolved in water. Therefore, in order for hydrodynamic installations for the PAW production to be productive, water must be constantly saturated with the indicated gases or air.

From this point of view, a much more common way to produce PAW is the heterophase method. It consists in the fact that the discharge is formed in a gaseous medium (air and other plasma-forming gases) from an electrode with a small radius of curvature (pin or wire), and closes to the liquid phase (solution). The type of discharge formed in such installations is glow, corona or spark, and supply voltage can be constant, alternative or pulsed. The type of discharge in air and its polarity affect the products of plasma decay and the resulting content of chemicals in an aqueous solution. For example, the positive corona discharge in air saturates an aqueous solution mainly with ozone, while the glow discharge leads to the formation of aqueous nitrates and nitrites, which can be used to stimulate plant development [23].

Another promising method for the PAW production is based on the fact that the discharge occurs directly in the volume and on the surface of air bubbles. There are many options for the execution of a nozzle and its orientation, which ultimately determine the shape and number of bubbles, which in turn affect the distribution of an electric field in the gap. Like heterophase methods, bubble discharges make it easier to initiate a discharge compared to electrohydraulic installations. In addition, the use of bubbles allows better mixing of the resulting PAW and provides better uniformity in the treatment of the liquid. It is also noted that bubbles contribute to a significant increase in the concentration of reactive species, both in plasma and in liquid. In this case, pulse voltage is applied, and the type of particles and their concentration are determined by the type of plasma-forming gas [24]. Equally important is the matching of the discharge supply parameters with the bubble formation process [25].

Devices with discharge treatment of water droplets, or a thin water film to produce PAW, have been studied little. But from the data available in the literature, it can be assumed that their productivity measured in "liters per hour" will apparently be the lowest [2].

The idea of remote discharge reactors is based on the fact that gaseous plasma products are formed by a discharge in a separate zone, and then they are injected into an aqueous solution. Many ozonating plants operate on this principle. If the plasma-forming gas is air (or synthetic air, or humid air), then gaseous plasma products can include not only ozone, but also other chemically-active particles, such as •OH, NO• radicals, H_2O_2 peroxide, and recombination products [26].

Apparently, the simplest way for obtaining PAW is a device that consists of a tube placed in a solution, through which a plasma-forming gas is injected. In this case, a rod electrode is placed inside the tube, to which voltage pulses are applied. Under these conditions, a single-barrier discharge is implemented between the inner wall of the tube and the electrode, and the formed plasma products immediately enter the solution without any losses for transportation. This scheme, as well as options for its implementation, were described in detail in the review article [27].

Despite the abundance of experimental data, when developing a technology, it is necessary to take into account factors that may be out of sight of scientists. One or another method of treatment of aqueous solutions can be effective, in comparison with others, in terms of the yield of chemical products; but at the same time, can be inferior in terms of time and energy costs. Therefore, to create a discharge water treatment technology, comparative studies of various treatment modes are required, one of which will be presented in this article.

The purpose of this study is to compare the ionic composition of water treated with plasma produced with pulsed barrier and bubble discharges. This knowledge will be in demand in the development of specific technologies for the conscious use of certain modes of discharge water treatment. The choice of these types of electric discharges was due, firstly, to their prevalence in applied research [5,12,14]. Secondly, both types of discharge are, quite simply, constructively implemented, which can further facilitate the creation of technological installations. The third reason is their potentially high performance (production of active particles) in the case of using high-voltage pulses, which, however, had to be verified. Based on the results of ongoing research, recommendations will be made regarding the use of these discharges.

2. Experimental Setup and Techniques

During the experiments, distilled water (initial conductivity is 0.2 μS/cm; LLC Oils and Lubricants, Russia), as well as ground water (spring water; initial conductivity is 280 μS/cm; its composition will be discussed below) were treated. The studies were carried out on an experimental setup, a block diagram of which is shown in Figure 1a. Power supply 1 produced voltage pulses with an amplitude of 10 kV, a pulse duration of 1.3 μs, and a rise time of 1000 ns. Voltage pulses following with a pulse repetition rate of 54 kHz were applied to electrode 2 (high-voltage (HV) electrode). It was a metallic rod covered with a polytetrafluoroethylene (PTFE; fluoroplast) shell with a thickness of 0.5 mm. The HV electrode was placed in glass tube 3 connected to membrane pump 5. Ambient air injected through the tube 3 entered water 4, which filled quartz vessel 6. Foil electrode 7 (with an area of 127 cm^2) placed on the outer surface of the quartz vessel and connected with capacitor 8 (capacity is C_0 = 10 pF; charging voltage is U_0 = 20 kV) provided capacitive decoupling between electrode 2 and ground. An inner diameter and a height of the vessel were 2 and 30 cm, respectively. The inner and outer diameters of the glass tube were 5.5 and 7.5 mm, respectively. The glass tube was tapered towards the bottom. The inner diameter of this part was 0.8 mm. In a single experiment, a volume of water treated with plasma was 78 mL.

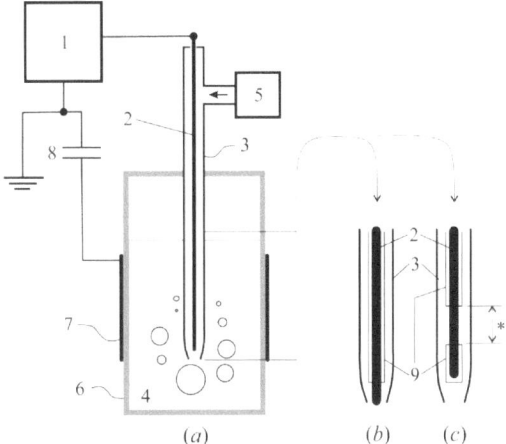

Figure 1. Block diagram of the experimental setup: general view (*a*) and zoomed sketches (indicated by arrows) of two design options of the high-voltage part (*b*,*c*). 1—power supply; 2—high-voltage electrode; 3—air feeding tube; 4—aqueous solution; 5—membrane air pump; 6—quartz vessel; 7—foil electrode; 8—capacitor; 9—dielectric shell; (*)—10-cm-length section freed from the dielectric shell.

Two designs of the electrode assembly were used. In the first version presented in Figure 1b, the metal high-voltage rod was almost completely covered with fluoroplast 9. The part not covered by the dielectric was located in the tapering part of the glass tube. The discharge was ignited at the tube outlet—between the end of the HV rod and the wall of an air bubble formed there. In this case, during the bubble formation, several breakdowns can occur. In other words, the mode of transfer of discharge energy into the bubbles is important here. This mode was chosen in accordance with the data obtained in [26]. With this type of excitation, chemically-active particles (reactive species) are produced both directly in the discharge in air and at the air–water interface. We will call this mode bubble discharge. A characteristic oscillogram of the voltage on the HV electrode for the bubble discharge is shown in Figure 2.

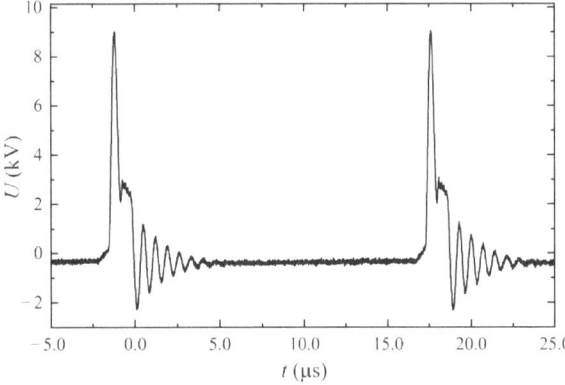

Figure 2. Waveform of the voltage at the HV electrode in the case of the bubble discharge.

In the second version of the electrode assembly design (Figure 1c), plasma enriched with reactive species was formed in a barrier discharge. To do this, the HV electrode had a 10-cm-length section denoted as (*) in Figure 1c freed from the dielectric. The end of the electrode 2 was also covered with PTFE tube. When applying HV pulses, the barrier discharge in the air was ignited in this region, and due to pumping, the chemically-active particles of the discharge plasma were transported into water bulk, saturating it and changing its characteristics. During the entire cycle of studies, the air pumping rate (200 mL/min), water volume, as well as the repetition rate of voltage pulses were fixed. This made it possible to compare the performance and composition of the obtained products for different assemblies.

Since the breakdown in bubbles can occur in different phases of voltage change (during the rise time, at the peak, or during the fall time of the pulse), which at this stage leads to a scatter and ambiguity in the electrical characteristics, the thermodynamic approach was used to estimate the input energy. This approach, in our opinion, makes it possible to quite correctly estimate the energy input for both types of excitation. Therefore, the energy input in this case was estimated by calculating the thermal energy released in the discharge. For this, the heating of the volume of the aqueous solution in the vessel was determined for a fixed time period. The temperature was measured using a temperature sensor built into a tester pH HI98108 (Hanna Instruments Ltd., Nușfalău, Romania). The temperature measurement accuracy was ±0.5°. Measurements for a single sample were carried out three times, and then the obtained values were averaged.

The ionic composition of the aqueous solution after plasma treatment was determined using a "Kapel-105/105M" (Lumeks Co., Ltd., Saint Petersburg, Russia) capillary electrophoresis system for the detection of nitrate NO_3^-, calcium Ca^{++} magnesium Mg^{++} ions. This device operates on the principle of spectrophotometric detection. The dispersing element is a diffraction monochromator with an operating spectral range from 190 to 380 nm. The accompanying methodological support makes it possible to analyze various anions and cations with a detection limit of 0.5 µg/cm^3. The hydrogen index (pH) was measured with an "Ionomer I-160MI" (LLC Izmeritelnaya Technika, Moscow, Russia) pH meter using calibration buffer solutions. The electrical conductivity of water was determined using an "ANION-4120" (Infraspark-Analit NPP, Yekaterinburg, Russia) laboratory conductometer. For each sample of water obtained after discharge treatment, measurements of the concentration of NO_3^-, Ca^{++}, and Mg^{++} ions, and electrical conductivity were carried out three times and then averaged. Thus, in what follows, all the data presented are averaged over three measurements.

3. Results and Discussion

Figure 3 demonstrates the change in the concentration of NO_3^- nitrate ions over time for various modes of formation of a chemically-active plasma. It is believed that the formation of NO_3^- anions in an aqueous solution is ensured by the conversion of nitrogen and oxygen molecules present in the air. Their activation and conversion directly depend on the performance of the discharge. It is seen that the discharge in bubbles provides a noticeably higher performance for these anions in ground water; while in distilled water, the best performance is provided by the barrier discharge.

It was found out how the data presented in Figure 3 correlate with the energy release in an aqueous solution. Typically, this characteristic for gas discharge devices is determined by calculating active power based on current and voltage waveforms. However, in our case, this approach was not applicable for two reasons. First, during the formation of a bubble at the end of the glass tube 3 (Figure 1b), several breakdowns of the gap between the rod tip and the inner surface of the bubble occur. Their number can vary from a few to hundreds, depending on the pulse repetition rate and the bubble formation time. Therefore, the energy input for each of these breakdowns will be different, which complicates statistical accounting for individual waveforms. Secondly, in both variants of the treatment, the conductivity and temperature of the solution change. This also entails changes in the energy release. Therefore, a lower estimate of the amount of heat imparted to the liquid during 10-min discharge treatment was made. At the same time, the estimate does not take into account the fact that this heat is also dissipated on the quartz walls of the vessel. In addition, the fact that not all of the energy deposited to water was further turned to heat (a fraction of the energy spent on forming chemical compounds), was not taken into account.

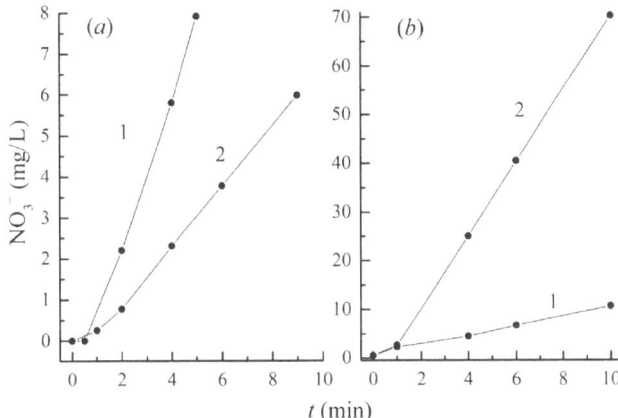

Figure 3. Time behavior of a specific amount of nitrate ions in distilled (*a*) and ground (*b*) water: barrier discharge (1); bubble discharge (2).

Table 1 shows the results of calculating the thermal power for the four cases shown in Figure 3. From the presented data, it can be seen that the higher the thermalized power, the higher the discharge performance in relation to the production of NO_3^- nitrate anions. At the same time, in distilled water, the barrier discharge shows the best performance, and in ground water—the bubble one. From the point of view of electrolytic physics [28], it can be assumed that ground water differs from the distilled one by a significant difference in the concentrations of ions in initial solution (mineralization of water samples). For ground water, a higher concentration of ions in the solution causes its greater conductivity and, consequently, a higher dissipation of the discharge energy in water.

Table 1. The value of the discharge power thermalized in an aqueous solution depending on the discharge type and the type of treated water.

Type of Treated Water	Thermalized Power, W	
	Bubble Discharge	Barrier Discharge
Distilled water	3.3 ± 0.5	4 ± 0.3
Ground water	11.1 ± 0.3	5.7 ± 0.4

When distilled water was treated with both types of discharges, the acidity decreases with time from about 7.5 to 3.5, while the conductivity increases from several to tens of µS/cm. The situation is different for ground water treatment. This is shown in Figure 4. It is seen that when ground water is exposed to both bubble and barrier discharges, the pH index first increases and then reaches a plateau. In addition, during the treatment with a barrier discharge, an anomalous "jump" in conductivity is observed during the first ~6 min. Then the conductivity returns to the starting level. The validity of this anomaly is confirmed by the fact that the conductivity measurements were carried out by two different methods. A similar behavior of the conductivity took place in both cases. When treated with the bubble discharge, no such behavior of the conductivity of the aqueous solution was observed.

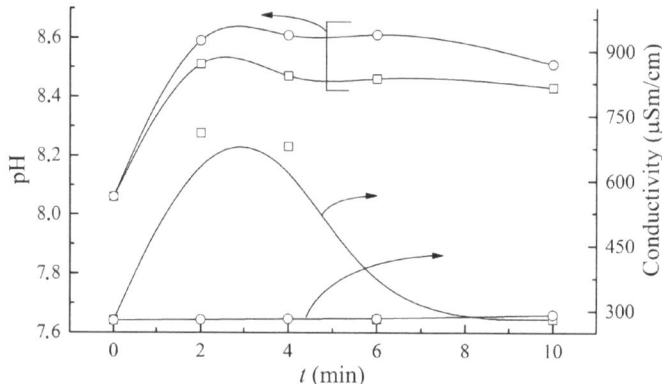

Figure 4. Time behavior of the pH-index and conductivity in ground water: barrier discharge (□); bubble discharge (○).

Table 2 shows the concentrations of magnesium and calcium ions before and after 10-min treatment of ground water with different types of discharge. It should be noted that the calcium concentration reached its maximum value after 2 min of treatment and then remained at the same level. It can be seen that plasma treatment leads to a significant increase in the concentration of Ca^{++} Mg^{++} in ground water. It is possible to assume that initially these elements are present in the solution in the form of $CaCO_3$ and $MgCO_3$ carbonates. This may be due to the fact that $Mg(OH)^+$ and $Ca(OH)^+$ hydroxide ions are easily converted into carbonates in ground water in the presence of air. Insoluble salts in the form of carbonates in the solution do not precipitate due to the small size of (fine) particles in suspension, and determine the constant hardness of water. Further, under the discharge action, both hydroxide ions and nitric acid (i.e., H^+ and NO_3^- ions) can be formed, as a result of which the formation of ions is possible, including as a result of the reaction $Me(OH)^+ + H^+ \rightarrow Me^{++} + H_2O$. The ion formation process must be accompanied by the consumption of hydrogen ions H^+, since an increase in the hydrogen index from 8.1 to 8.5 for the barrier discharge and to 8.6 for the bubble discharge is observed (Figure 4). It should be noted that the question of the formation of calcium and magnesium ions under these conditions requires a separate, more detailed study. Nevertheless, it can be assumed that the pH index growth is due mainly to the increase in the concentration of calcium and magnesium ions during the first 2 min of treatment.

Table 2. The concentration (mg/L) of calcium and magnesium ions in ground water before and after plasma treatment.

Sample	Ion	Bubble Discharge	Barrier Discharge
Reference	Mg^{++}	0.0526	0.0526
	Ca^{++}	1.1395	1.1395
10 min treatment	Mg^{++}	9.694	9.792
	Ca^{++}	66.43	42.71

In addition to magnesium and calcium ions, potassium and sodium cations were also monitored in water samples. But, as measurements showed, treatment with discharges did not significantly affect their concentration in water; the content of potassium and sodium cations does not change during 10 min of exposure, and amounts to 1.065 and 9.395 mg/L, respectively. Potassium and sodium carbonates are soluble salts and determine the initial conductivity of an aqueous sample at 280 μS/cm. The appearance of additional NO_3^- and

NO_2^- anions, as a result of the action of the discharge, leads to the formation of potassium and sodium nitrates, which are also soluble salts.

The low concentration of Ca^{++} Mg^{++} in the reference sample (Table 2), and the slightly alkaline environment of ground water, imply the presence of these ions in the solution in the form of insoluble salts. For example, calcium and magnesium carbonates. It is known [5] that when a discharge is ignited in atmospheric air in the presence of water, it leads to the formation of various active particles in the plasma, including NO_3^- and NO_2^- anions, which can interact with water-insoluble carbonates [5,16]. This, in turn, leads to the formation of water-soluble salts—calcium and magnesium nitrates or nitrites, which contributes to an increase in the concentration of their ions in water. However, the alkaline reaction of the solution, which persists throughout the entire time of water treatment with the barrier discharge, indicates that the amount of NO_3^- and NO_2^- anions formed is insufficient to shift the equilibrium to a neutral or acidic reaction.

A further increase in the treatment time led to the fact that the concentration of these anions in the solution increased and the pH began to decrease.

The results described above are of interest from the point of view of access to technological installations for the PAW production, since data on the discharge treatment of distilled and deionized water are usually presented in the scientific literature. Of course, on an industrial scale, the use of such water is impractical. Our data show that, in the case of ground water, we can also obtain high concentrations of NO_3^- anions.

Importantly, ground water treatment also opens the way to the study of relatively easy methods for converting salts (calcium carbonates) from an insoluble form to a soluble one. Namely, when interacting with NO_3^- anions, insoluble carbonates pass into soluble nitrates.

4. Conclusions

As a result of the research, the set goal was achieved—the ionic composition of two types (distilled and ground) of water treated with low-temperature plasma formed by two types of pulsed discharge (barrier and in bubbles) in atmospheric pressure air was revealed. It has shown that the bubble discharge in ground water gives the maximum performance for the NO_3^- anions. In this case, the energy thermalized in the solution is maximum. At the same time, the complex compounds that affect the hardness of water, the most Ca^{++} ions, are released into the solution. These features of the process of water treatment using pulsed discharges should be taken into account when designing installations for large-scale PAW production.

These results are important for an industry such as hydroponic plant growing technologies (see, e.g., [16,17]), where an aqueous solution enriched with NO_3^- anions is required.

The data obtained are of interest due to the fact that data on the discharge treatment of distilled and deionized water are usually presented in the scientific literature. Of course, on an industrial scale, the use of such water is impractical. Therefore, research on discharge treatment of groundwater is needed. Our research starts this process. However, to prove the beneficial properties of plasma-activated water (PAW) produced in a bubble discharge in ground water, laboratory and field studies on the effects of such water on economically-valuable plants are needed.

Author Contributions: V.P., E.S. and V.S. performed the experiments. D.S. helped initiate the research. A.R. and S.K. conducted a physicochemical analysis. E.S., D.S. and S.K. wrote the manuscript. All authors have read and agreed to the published version of the manuscript.

Funding: This research was supported by the Ministry of Science and Higher Education of the Russian Federation (Agreement No. 075-15-2022-1238 dated 13 October 2022).

Data Availability Statement: The data presented in this study are available on request from the corresponding author.

Acknowledgments: The authors gratefully acknowledge the help of Dmitrii Pechenitsyn for design and tuning of high-voltage power supply.

Conflicts of Interest: The authors declare no conflict of interest.

References

1. Kornev, J.; Yavorovsky, N.; Preis, S. Generation of active oxidant species by pulsed dielectric barrier discharge in water-air mixtures. *Ozone Sci. Eng.* **2006**, *28*, 207–215. [CrossRef]
2. Malik, M.A. Water purification by plasmas: Which reactors are most energy efficient? *Plasma Chem. Plasma Process.* **2010**, *30*, 21–31. [CrossRef]
3. Bruggeman, P.J.; Kushner, M.J.; Locke, B.R.; Gardeniers, J.G.E.; Graham, W.G.; Graves, D.B.; Hofman-Caris, R.C.H.M.; Maric, D.; Reid, J.P.; Ceriani, E.; et al. Plasma–liquid interactions: A review and roadmap. *Plasma Sources Sci. Technol.* **2016**, *25*, 053002. [CrossRef]
4. Vanraes, P.; Nikiforov, A.Y.; Leys, C. Electrical discharge in water treatment technology for micropollutant decomposition. In *Plasma Science and Technology—Progress in Physical States and Chemical Reactions*; Mieno, T., Ed.; IntechOpen: London, UK, 2016; pp. 429–478. [CrossRef]
5. Mouele, E.S.M.; Tijani, J.O.; Badmus, K.O.; Pereao, O.; Babajide, O.; Zhang, C.; Shao, T.; Sosnin, E.; Tarasenko, V.; Fatoba, O.O.; et al. Removal of pharmaceutical residues from water and wastewater using dielectric barrier discharge methods—A Review. *Int. J. Environ. Res. Public Health* **2021**, *18*, 1683. [CrossRef] [PubMed]
6. Zeghioud, H.; Nguyen-Tri, P.; Khezami, L.; Amrane, A.; Assadi, A.A. Review on discharge plasma for water treatment: Mechanism, reactor geometries, active species and combined processes. *J. Water Process Eng.* **2020**, *38*, 101664. [CrossRef]
7. Zhou, R.; Zhou, R.; Wang, P.; Xian, Y.; Mai-Prochnow, A.; Lu, X.; Cullen, P.J.; Ostrikov, K.; Bazaka, K. Plasma-activated water: Generation, origin of reactive species and biological applications. *J. Phys. D Appl. Phys.* **2020**, *53*, 303001. [CrossRef]
8. Xu, D.; Wang, S.; Li, B.; Qi, M.; Feng, R.; Li, Q.; Zhang, H.; Chen, H.; Kong, M.G. Effects of plasma-activated water on skin wound healing in mice. *Microorganisms* **2020**, *8*, 1091. [CrossRef]
9. Hozák, P.; Scholtz, V.; Khun, J.; Mertová, D.; Vaňková, E.; Julák, J. Further Contribution to the Chemistry of Plasma-Activated Water: Influence on Bacteria in Planktonic and Biofilm Forms. *Plasma Phys. Rep.* **2018**, *44*, 799–804. [CrossRef]
10. Chiappim, W.; Sampaio, A.G.; Miranda, F.; Fraga, M.; Petraconi, G.; da Silva Sobrinho, A.; Kostov, K.; Koga-Ito, C.; Pessoa, R. Antimicrobial effect of plasma-activated tap water on *Staphylococcus aureus*, *Escherichia coli*, and *Candida albicans*. *Water* **2021**, *13*, 1480. [CrossRef]
11. Perez, S.M.; Biondi, E.; Laurita, R.; Proto, M.; Sarti, F.; Gherardi, M.; Bertaccini, A.; Colombo, V. Plasma activated water as resistance inducer against bacterial leaf spot of tomato. *PLoS ONE* **2019**, *14*, e0217788. [CrossRef]
12. Dimitrakellis, P.; Giannoglou, M.; Xanthou, Z.M.; Gogolides, E.; Taoukis, P.; Katsaros, G. Application of plasma-activated water as an antimicrobial washing agent of fresh leafy produce. *Plasma Process. Polym.* **2021**, *18*, e2100030. [CrossRef]
13. Naumova, I.K.; Subbotkina, I.N.; Titov, V.A.; Khlyustova, A.V.; Sirotkin, N.A. Effect of water activated by non-equilibrium gas-discharge plasma on the germination and early growth of cucumbers (*Cucumis sativus*). *Appl. Phys.* **2021**, *4*, 40–46. [CrossRef]
14. Thirumdasa, R.; Kothakot, A.; Annapurec, U.; Siliveru, K.; Blundell, R.; Gatt, R.; Valdramidisgh, V.P. Plasma activated water (PAW): Chemistry, physico-chemical properties, applications in food and agriculture. *Trends Food Sci. Technol.* **2018**, *77*, 21–31. [CrossRef]
15. Herianto, S.; Hou, C.-Y.; Lin, C.-M.; Chen, H.-L. Nonthermal plasma-activated water: A comprehensive review of this new tool for enhanced food safety and quality. *Compr. Rev. Food Sci. Food Saf.* **2021**, *20*, 583–626. [CrossRef] [PubMed]
16. Takahata, J.; Takai, K.; Satta, N.; Takahashi, K.; Fujio, T. Improvement of growth rate of plants by bubble discharge in water. *Jpn. J. Appl. Phys.* **2015**, *54*, 01AG07. [CrossRef]
17. Takaki, K.; Takahata, J.; Watanabe, S.; Satta, N.; Yamada, O.; Fujio, T.; Sasaki, Y. Improvements in plant growth rate using underwater discharge. *J. Phys. Conf. Ser.* **2012**, *418*, 012140. [CrossRef]
18. Cao, Y.; Qu, G.; Li, T.; Jiang, N.; Wang, T. Review on reactive species in water treatment using electrical discharge plasma: Formation, measurement, mechanisms and mass transfer. *Plasma Sci. Technol.* **2018**, *20*, 103001. [CrossRef]
19. Brisset, J.L.; Pawlat, J. Chemical effects of air plasma species on aqueous solutes in direct and delayed exposure modes: Discharge, post-discharge and plasma activated water. *Plasma Chem. Plasma Process.* **2016**, *36*, 355–381. [CrossRef]
20. Tarabová, B.; Lukeš, P.; Hammer, M.U.; Jablonowski, H.; von Woedtke, T.; Reuter, S.; Machala, Z. Fluorescence measurements of peroxynitrite/peroxynitrous acid in cold air plasma treated aqueous solutions. *Phys. Chem. Chem. Phys.* **2019**, *21*, 8883–8896. [CrossRef]
21. Bruggeman, P.J.; Locke, B.R. Assessment of potential applications of plasma with liquid water. In *Low Temperature Plasma Technology—Methods and Applications*; Chu, P.K., Lu, X., Eds.; CRC Press: Boca Raton, FL, USA, 2014; pp. 367–399.
22. Locke, B.R.; Sato, M.; Sunka, P.; Hoffmann, M.R.; Chang, J.S. Electrohydraulic discharge and nonthermal plasma for water treatment. *Ind. Eng. Chem. Res.* **2006**, *45*, 882–905. [CrossRef]
23. Dors, M.; Mizeraczyk, J.; Mok, Y.S. Phenol oxidation in aqueous solution by gas phase corona discharge. *J. Adv. Oxid. Technol.* **2006**, *9*, 139–143. [CrossRef]
24. Sun, B.; Sato, M.; Sid Clements, J. Optical study of active species produced by a pulsed streamer corona discharge in water. *J. Electrost.* **1997**, *39*, 189–202. [CrossRef]
25. Sorokin, D.A.; Skakun, V.S.; Sosnin, E.A.; Sukhankulyev, D.T.; Panarin, V.A.; Surnina, E.N. Effective regime for obtaining activated water plasma and its application in pre-sowing seed treatment. In Proceedings of the XII All-Russian Conference on Physical Electronics PE-2022, Makhachkala, The Republic of Dagestan, 19–22 October 2022.

26. Tang, Q.; Lin, S.; Jiang, W.; Lim, T.M. Gas phase dielectric barrier discharge induced reactive species degradation of 2,4-dinitrophenol. *Chem. Eng. J.* **2009**, *153*, 94–100. [CrossRef]
27. Mouele, E.S.M.; Tijani, J.O.; Badmus, K.O.; Pereao, O.; Babajide, O.; Fatoba, O.O.; Zhang, C.; Shao, T.; Sosnin, E.; Tarasenko, V.; et al. A critical review on ozone and co-species, generation and reaction mechanisms in plasma induced by dielectric barrier discharge technologies for wastewater remediation. *J. Environ. Chem. Eng.* **2021**, *9*, 105758. [CrossRef]
28. Hladik, J. (Ed.) Physics of Electrolytes. In *Transport Processes in Solid Electrolytes and in Electrodes*; Academic Press: London, Uk, 1972; Volume 1, 516p.

Disclaimer/Publisher's Note: The statements, opinions and data contained in all publications are solely those of the individual author(s) and contributor(s) and not of MDPI and/or the editor(s). MDPI and/or the editor(s) disclaim responsibility for any injury to people or property resulting from any ideas, methods, instructions or products referred to in the content.

Communication

Anisotropy Analysis of the Permeation Behavior in Carbon Dioxide-Assisted Polymer Compression Porous Products

Takafumi Aizawa

Research Institute for Chemical Process Technology, National Institute of Advanced Industrial Science and Technology, 4-2-1 Nigatake, Miyagino-ku, Sendai 983-8551, Japan; t.aizawa@aist.go.jp; Tel.: +81-22-237-5211

Abstract: The carbon dioxide-assisted polymer compression method is used to create porous polymer products with laminated fiber sheets that are crimped in the presence of carbon dioxide. In this method, fibers are oriented in the sheet-spread direction, and the intersections of the upper and lower fibers are crimped, leading to several intersections within the porous product. This type of orientation in a porous material is anisotropic. A dye solution was injected via a syringe into a compression product made of poly(ethylene terephthalate) nonwoven fabric with an average fiber diameter of 8 μm. The anisotropy of permeation was evaluated using the aspect ratio of the vertical and horizontal permeation distances of a permeation area. The aspect ratio decreased monotonically with decreasing porosity; it was 2.73 for the 80-ply laminated product with a porosity of 0.63 and 2.33 for the 160-ply laminated product with a porosity of 0.25. A three-dimensional structural analysis using X-ray computed tomography revealed that as the compression ratio increased, the fiber-to-fiber connection increased due to the increase in adhesion points, resulting in decreased anisotropy of permeation. The anisotropy of permeation is essential data for analyzing the sustained release behavior of drug-loaded tablets for future fabrication.

Keywords: carbon dioxide-assisted polymer compression; permeation; anisotropy; porosity; X-ray computed tomography

Citation: Aizawa, T. Anisotropy Analysis of the Permeation Behavior in Carbon Dioxide-Assisted Polymer Compression Porous Products. *Technologies* **2023**, *11*, 52. https://doi.org/10.3390/technologies11020052

Academic Editors: Marina A. Volosova, Sergey N. Grigoriev and Anna A. Okunkova

Received: 9 March 2023
Revised: 29 March 2023
Accepted: 30 March 2023
Published: 3 April 2023

Copyright: © 2023 by the author. Licensee MDPI, Basel, Switzerland. This article is an open access article distributed under the terms and conditions of the Creative Commons Attribution (CC BY) license (https://creativecommons.org/licenses/by/4.0/).

1. Introduction

Polymers have become an indispensable material in our daily life because of their lightness and durability [1,2]. Porous polymers are also an important material; they are lightweight due to the pores, and they can be used to absorb liquid, filter liquid or gas, and absorb shocks and sounds [3–5]. In addition, porous polymers can be used as heat insulators [6]. In general, there are several methods to generate porous polymer materials, such as using physical blowing agents [7], using templates [8], and the freeze-drying method [9]. For methods that include the use of gases, one mixes high-pressure gases to generate porous polymers during injection molding [10,11].

The carbon dioxide-assisted polymer compression (CAPC) method, which uses CO_2 to plasticize and bond fiber polymers at room temperature, is a fast and simple method for producing porous polymers that fit the molds [12]. This method features the elimination of the need for a heater as it can be applied at room temperature. In addition, as the method uses vapor-pressure CO_2, it only requires a ball valve to introduce the gas, which eliminates the need for a pump. Compared to the equipment configuration of a typical supercritical fluid process, it is very simple; it only needs a press machine as the main equipment, with no pumps or heaters.

Fibrous polymers are used as the raw material in this method. The use of fibrous-polymer sheets is advantageous as a drug-loaded porous material can be easily prepared from it, for example, loading the drug in the center of a sheet in advance [13] or by placing the drug in the center of a fiber sheet and putting it in a mold [14].

For the raw material of fibers, it is better to use nonwoven fabrics due to their mass production and low price [15,16]. When the fiber sheets are laminated, the fibers are spread in the direction of the fiber sheet and the overlapping portions of the fibers are crimped together by point bonding. In this case, the resulting porous body will have anisotropic properties in the direction of fiber layering and its perpendicular direction. The anisotropic properties include both peel strength and puncture strength. These anisotropic properties are also desirable for filter applications.

The peel and penetration resistance behaviors and the filtering properties have already been investigated [17,18]. Although scanning electron microscopy has been used to observe porous materials, the observation was limited to the surface. In order to understand the three-dimensional shape, samples were cut and immersed in liquid nitrogen to observe the cross-section, but the effect of cutting on the cross-section can lead to unreliable results.

The relationship between the structure and properties of porous materials made of fibers has been extensively investigated [19,20]. In applications where liquids and gases are required to permeate, the void structure of the porous material plays a crucial role, and this information is necessary to analyze the flow within the porous material. Products containing powder in their center have been successfully fabricated using the CAPC method [14]. By applying this technology, it is possible to create tablets with a certain amount of drug loaded in their center. When considering the sustained release of a drug from a tablet with a drug-loaded center, the anisotropy of permeation affects the rate of sustained release from the top, bottom, and side surfaces, and this is essential for tablet design.

This study aims to quantify the anisotropy of permeation and interpret its relationship to porosity in terms of structure. The anisotropy was quantified by conducting dye solution permeation experiments and calculating the aspect ratio of the vertical and horizontal permeation distances of the permeation area. A three-dimensional structure was obtained using X-ray computed tomography (X-ray CT). Since the anisotropy is a result of fiber orientation, it was hypothesized that an increase in the number of oriented fibers per unit volume would increase the anisotropy. This study tests this hypothesis.

2. Materials and Methods

The TK3 poly(ethylene terephthalate) (CAS RN: 25038-59-9) pellets manufactured by Bell Polyester Products, Inc. (Houfu, Japan) were processed into a nonwoven fabric by the melt-blowing method [21] at Nippon Nozzle Co. Ltd. (Kobe, Japan). The nonwoven fabric was fabricated at a basis weight of 30 g/m^2, with an average fiber diameter of 8 μm. The nonwoven fabric was punched out to 18 mm in diameter using a punch, and the sample sets of 80 (0.637 g), 100 (0.796 g), 120 (0.956 g), 140 (1.115 g), and 160 sheets (1.274 g) were prepared for the permeation tests. In addition, the sample sets of 8 (0.062 g), 10 (0.078 g), 12 (0.093 g), 14 (0.109 g), and 16 (0.125 g) were prepared for X-ray CT.

The CAPC processing apparatus is shown in Figure 1. The piston was mounted on the servo press machine JP-1504 (JANOME Co., Hachioji, Japan). The corresponding procedure of CAPC processing went as follows: a sample with a specified number and weight was placed in a cylindrical high-pressure container directly below the piston under atmospheric conditions. The piston was then lowered to the 7.5 mm position, and CO_2 was injected and exhausted 3 times to replace the air in the container with CO_2. Then, CO_2 was injected again at vapor pressure (6 MPa at 23 °C), and the piston was lowered to the 5 mm position and pressed for 10 s. After pressing, the CO_2 was slowly exhausted for 30 s through the metering valve V_4, and was then released into the atmosphere by opening the valve V_2. The piston was then raised and a 5 mm thick sample could be prepared. The samples for X-ray CT were prepared in the same manner, but with CO_2 introduced when the piston was at the 0.75 mm position and pressed when the piston was at the 0.5 mm position. The CAPC process was performed at 23 °C.

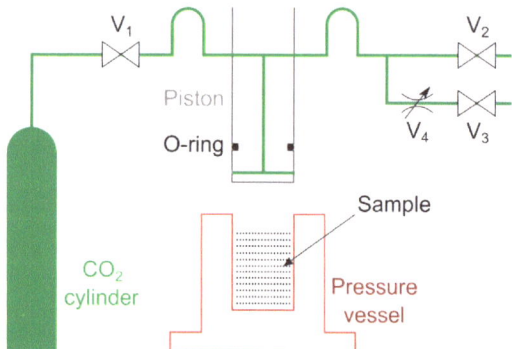

Figure 1. Schematic of the CO_2-assisted polymer compression equipment. V_1 is the introduction valve; V_2 is the exhaustion valve for rapid exhaustion; V_3 is the exhaustion valve for slow exhaustion; and V_4 is the metering valve.

The permeation tests were performed using a Brilliant Blue (CAS RN: 3844-45-9) solution at a concentration of 0.1 wt% in an aqueous solution of 20 wt% ethanol (CAS RN: 64-17-5). Brilliant Blue and ethanol were purchased from FUJIFILM Wako Pure Chemical Co. (Osaka, Japan) and used without further purification. The temperature of the permeation solution was 20 °C. The setup of the permeation test is shown in Figure 2. The dye solution was placed in a syringe and set in the syringe pump Fusion 100 (Chemyx Inc., Stafford, TX, USA), and 0.01 mL of the solution was introduced at a rate of 0.05 mL/min. The sample was set vertically in the sample holder as shown in Figure 2, with the tip of the syringe in horizontal contact with the center of the sample. The liquid extruded from the tip of the needle on the syringe was immediately absorbed by the sample due to the high wettability of the 20 wt% ethanol aqueous solution to the poly(ethylene terephthalate) fiber. The experiments were performed five times under each condition to obtain the averaged results.

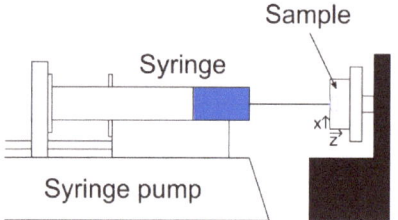

Figure 2. Setup of the apparatus for the dye solution permeation test. The sample is placed vertically, and the z-axis of the sample is defined as the direction of stacking nonwoven fabrics during CO_2-assisted polymer compression preparation.

The sectioning was performed with the ultrasonic cutter ZO-80 (Honda Electronics Co. Ltd., Toyohashi, Japan) installed vertically in the numerical-controlled computer-aided manufacturing KitMill BT100 (Originalmind Inc., Okaya, Japan). The movement speed of the ultrasonic cutter was accurately controlled to 1 mm/s. The cut cross-sections were observed with the optical microscope RH-2000 (Hirox Co. Ltd., Suginami-ku, Japan). Since the permeation area was not able to fit into the images even at the lowest magnification, the stage was moved to take the images, which resulted in partial overlaps. Then, the overlap was automatically recognized and combined using the Microsoft Image Composite Editor (Microsoft Co., Redmond, WA, USA). The aspect ratio of the permeation area was calculated by fitting the edges of the pigment into the combined image with an ellipse. The experiment was performed five times, and the average of the results was used to analyze the aspect ratio.

A three-dimensional structural evaluation using X-ray CT was performed using the nano3DX submicron-resolution X-ray CT scanner (Rigaku Co., Akishima, Japan) at Industrial Technology Institute, Miyagi Prefectural Government (ITIM, Sendai, Japan). The resolution was 0.629 μm with a reconstruction range of 1.288 mm. The image was analyzed by taking the histogram of the brightness, which showed a lower peak for the pore peak and a higher peak for the solid peak. The pore-solid threshold was set to be the center of the two peaks.

The surface observation by scanning electron microscopy was performed using the TM-1000 (Hitachi High-Tech Co., Minato-ku, Japan).

The materials and instruments used in the experiments in this study are summarized in Tables 1 and 2.

Table 1. List of materials.

Name	CAS RN	Model Number	Manufacture Details
Poly(ethylene terephthalate)	25038-59-9	TK3	Bell Polyester Products, Inc., Houfu, Japan
Brilliant Blue	3844-45-9	027-12842	FUJIFILM Wako Pure Chemical Co., Osaka, Japan
Ethanol	64-17-5	054-00461	FUJIFILM Wako Pure Chemical Co., Osaka, Japan

Table 2. List of instruments.

Name	Model Number	Manufacture Details
Press machine	JP-1504	JANOME Co., Hachioji, Japan
Syringe pump	Fusion 100	Chemyx Inc., Stafford, USA
Ultrasonic cutter	ZO-80	Honda Electronics Co. Ltd., Toyohashi, Japan
Numerical-controlled computer-aided manufacturing	KitMill BT100	Originalmind Inc., Okaya, Japan
Optical microscope	RH-2000	Hirox Co. Ltd., Suginami-ku, Japan
X-ray computed tomography scanner	nano3DX	Rigaku Co., Akishima, Japan
Scanning electron microscope	TM-1000	Hitachi High-Tech Co., Minato-ku, Japan

3. Results and Discussion

The cross-sectional results of the dye solution-permeated samples are shown in Figure 3. All images were elliptical and well fitted. In addition, there was no effect of gravity for this experiment, and the dye permeated equally well both downward and upward from the syringe. The results of fitting and averaging all data are shown in Table 3. The standard deviation is also included to show the data variability. Reproducible results were obtained from the dye solution injection test, indicating that the anisotropy of permeation can be evaluated using the quantitative value of the aspect ratio of the cross-section of the permeation area. The experimental results showed that with the increase in the number of layers, the range of permeation increased both vertically and horizontally, simply due to the decrease in porosity. In addition, the aspect ratio became small, indicating that it was easier for the material to permeate deeper as the number of layers increased. The hypothesis that an increase in the number of oriented fibers per unit volume would increase the anisotropy was denied via the experimental results.

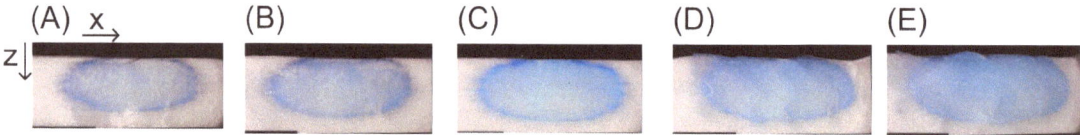

Figure 3. Cross-sectional observation of the samples permeated with dye solution. (**A**) 80-ply laminated product, (**B**) 100-ply laminated product, (**C**) 120-ply laminated product, (**D**) 140-ply laminated product, and (**E**) 160-ply laminated product.

Table 3. Dye solution permeation results for the samples made of nonwoven fabric through CO_2-assisted polymer compression.

Number of Sheets [-]	Weight [g]	Thickness [mm]	Porosity [-]	Permeation Width [mm] *	Permeation Depth [mm] *	Volume of Permeation [mL]	Aspect Ratio [-]
80	0.637	5.00	0.63	6.29 ± 0.21	2.31 ± 0.22	0.0478	2.73
100	0.796	5.00	0.53	6.58 ± 0.54	2.46 ± 0.17	0.0562	2.67
120	0.956	5.00	0.44	6.61 ± 0.21	2.53 ± 0.19	0.0578	2.61
140	1.115	5.00	0.35	6.74 ± 0.15	2.71 ± 0.15	0.0645	2.48
160	1.274	5.00	0.25	7.05 ± 0.03	3.02 ± 0.07	0.0788	2.33

* Error indicates the standard deviation.

From these results, the permeation mechanism is discussed by first focusing on the extent of permeation and the volume of pores. The porosity of the porous sample was derived from the weight of the sample, the volume of the cylinder (1.27 mL) that is obtained from the diameter (1.8 cm) and height (0.5 cm) of the cylinder of the porous sample, and the density of the polymer (1.34 g/mL). For example, for a sample of 80-ply, the weight of a solid cylinder with a diameter of 1.8 cm and a thickness of 0.5 cm should be 1.704 g. However, the actual weight was 0.637 g. The weight discrepancy (1.067 g) is caused by the existence of pores, and it can be used to derive the associated porosity (0.63). Since permeation takes place in the voids, the volume of the voids in this ellipsoid body was obtained by calculating the volume of the ellipsoid body from the fitted width and depth and multiplying it by the porosity of the porous material. By injecting 0.01 mL of the dye solution, the occupancy rate of the dye solution relative to the void was calculated to be 34% for 80-ply samples and 50% for 160-ply samples. This occupancy rate suggests that the dye solution permeates among the fibers while wetting the fiber surface, and simultaneously fills the voids due to the capillary effect.

Next, the cause of lateral permeation and the decrease in aspect ratio is discussed based on the results from the X-ray CT images. The scanning electron microscope images of the surface of a 0.5 mm thick CAPC product are shown in Figure 4. The porosities calculated from the sample size, weight, and solid density were 0.64, 0.54, 0.45, 0.36, and 0.27 for the 8-, 10-, 12-, 14-, and 16-ply samples, respectively. The X-ray CT images of a sample cropped to 503 µm in length and width are shown in Figure 5. The size of the reconstructed X-ray CT image and the cropped area are shown in Figure 6. For X-ray CT, a cut-out was made to avoid the center of rotation because the reconstruction accuracy near the center of rotation is low. In the porous material, the fiber direction was spread in the xy-plane, and both the permeation that wetted the fiber surface and filled the voids (such as the capillary action) tended to spread in the lateral direction. In contrast, for the cross-section, as the number of laminated sheets increased, the area where fibers adhered to each other in the z-direction increased, and wetting across the fibers occurred from the adhered area to the z-direction while the voids were also connected in the z-direction, leading to increased permeation in the z-direction. This phenomenon was also observed from the cross-sectional analysis model in the previous simulation of adhesive strength, indicating that more bonding and higher adhesive strength can be found in the z-direction with higher compression [17].

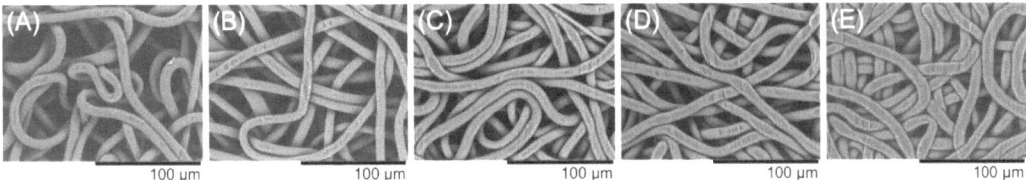

Figure 4. Scanning electron microscope images. (**A**) 8-ply laminated product, (**B**) 10-ply laminated product, (**C**) 12-ply laminated product, (**D**) 14-ply laminated product, and (**E**) 16-ply laminated product.

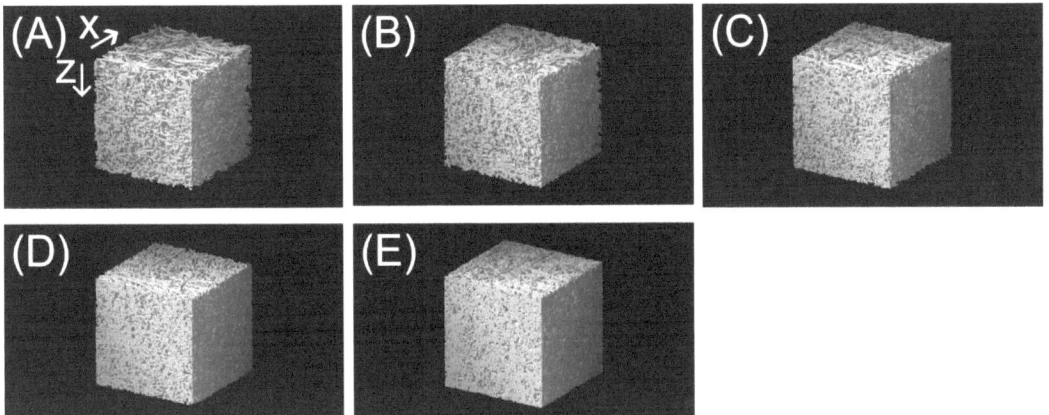

Figure 5. X-ray computed tomography images. The thickness of each image (in the z-direction) is 0.500 mm and the lengths in the x- and y-directions are 0.503 mm. (**A**) 8-ply laminated product, (**B**) 10-ply laminated product, (**C**) 12-ply laminated product, (**D**) 14-ply laminated product, and (**E**) 16-ply laminated product.

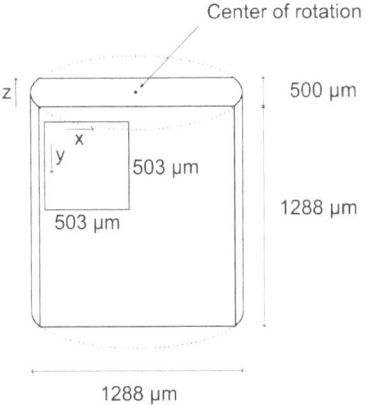

Figure 6. Size of the reconstructed X-ray computed tomography image and the cropped area shown in Figure 5.

For a more detailed examination, the cross-sections perpendicular to the z-axis and y-axis are shown in Figures 7–9 for 8-ply to 16-ply samples. The horizontal axis represents the x-axis in each figure, and the vertical axis indicates the y-axis in the upper figure and the z-axis in the lower figure. The lower figure depicts the cross-section cut along the red line of the upper figure, and the upper figure depicts the cross-section cut along the red line of the lower figure. The left, center, and right figures show the cross-section shifted by 10 μm. From the scanning electron microscope images (Figure 4), it seems that there were more fibers than the calculated results because the information in the depth direction was missing. However, from the X-ray CT slice image in the xy-plane, it is clearly shown that the voids match the porosity. This is a limitation of the scanning electron microscope as it is a two-dimensional measurement method. The slice images in Figures 7–9 clearly show that the fibers are present along the xy-plane, although they appear to move slightly up and down along the z-axis. In addition, the adhesion points (intersections between upper and lower fibers) along the z-axis increased as the number of laminated sheets increased. The slice image in the xz-plane shows that there were many unglued free fibers in the 8-layer product, whereas there were almost no free fibers in the 16-layer product.

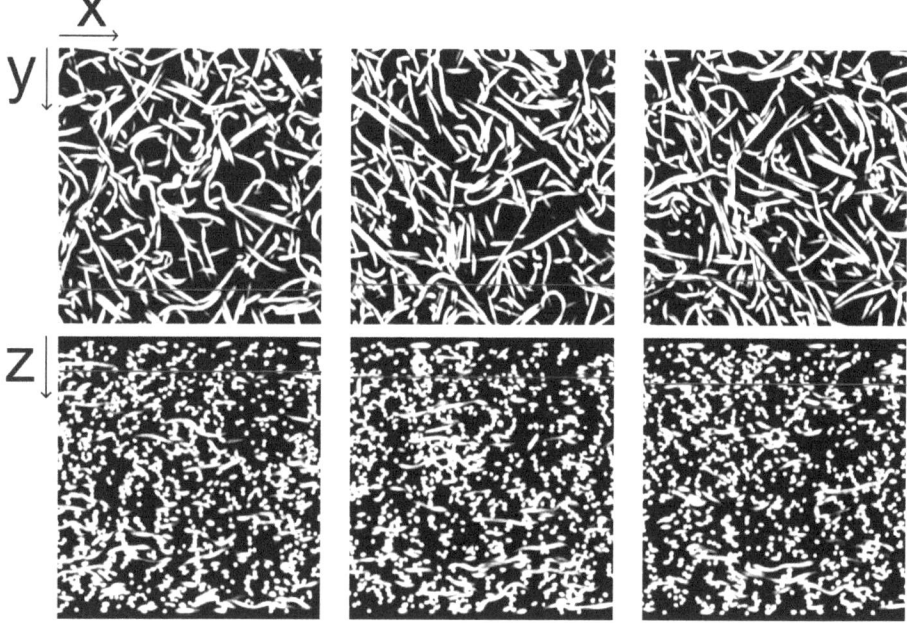

Figure 7. Cross-sectional X-ray computed tomography images of the 8-ply laminated product. The upper images are the xy-plane images cut by the red lines in the lower images. The lower images are the xz-plane images cut by the red lines in the upper images. The left, center, and right cut positions are all shifted by 10 μm.

Figure 8. Cross-sectional X-ray computed tomography images of 12-ply laminated product. The upper images are the xy-plane images cut by the red lines in the lower images. The lower images are the xz-plane images cut by red lines in the upper images. The left, center, and right cut positions are all shifted by 10 μm.

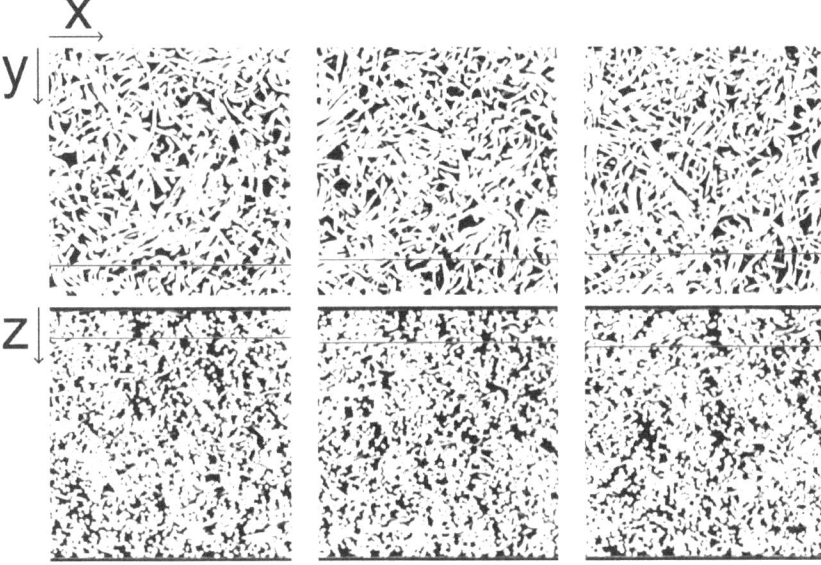

Figure 9. Cross-sectional X-ray computed tomography images of 16-ply laminated product. The upper images are the xy-plane images cut by the red lines in the lower images. The lower images are the xz-plane images cut by red lines in the upper images. The left, center, and right cut positions are all shifted by 10 μm.

The aspect ratio of the permeation in the fiber-spreading direction to the fabric-lamination direction was 2.73 for the 80-ply laminated CAPC porous material with a porosity of 0.63. As shown in the cross-section of a large porosity material (Figure 7), this can be interpreted as little fiber being bonded between the top and bottom when porosity is large, where permeation is more likely to spread in the lateral direction along the fiber. This can be interpreted as reflecting the lateral spread of permeation along the fibers when porosity is large. However, the aspect ratio decreased to 2.33 for the 160-ply laminated CAPC porous material with a porosity of 0.25. In the cross-section of the small porosity state (Figure 9), considerable fiber bonding was observed, which can cause a decrease in the aspect ratio of the permeation or a decrease in the anisotropy because the permeation fluid on one fiber surface is more likely to travel to the surface of another overlapping fiber.

The results of the evaluation of the anisotropy of permeation obtained in this study will provide essential data for designing filters and drug-loaded tablets using CAPC porous materials in future studies.

4. Conclusions

The permeation behavior of dye solutions in carbon dioxide-assisted polymer compression (CAPC) porous materials consisting of poly(ethylene terephthalate) with an average fiber diameter of 8 μm was analyzed. The anisotropy of the permeation behavior in CAPC porous materials was evaluated using the aspect ratio of the dye solution permeation of the fiber-spread direction via the fabric-lamination direction. Cross-sectional images were obtained using X-ray computed tomography. The results and interpretations of the experiments can be summarized as follows:

- The aspect ratio of the permeation of the porous material with a porosity of 0.63 was 2.73, while the aspect ratio of the porous material with a porosity of 0.25 decreased to 2.33.
- The aspect ratio decreased monotonically with decreasing porosity, indicating that the decrease in porosity reduced the anisotropy of permeation.
- The hypothesis that an increase in the number of oriented fibers per unit volume could increase the anisotropy is denied.
- Cross-sectional images of samples with high porosity showed less fiber-to-fiber bonding, and the number of fiber bonding points increased with decreasing porosity.
- Since permeation of the dye solution occurs along the fiber surface, it is interpreted that more bonded fibers promote permeation between the upper and lower fiber surfaces, resulting in less anisotropy of permeation.

Thus, the objective of this paper, to quantitatively evaluate the anisotropy of permeation and to understand the phenomenon by linking it to the structure of the sample, has been achieved.

In this research, experiments were conducted using limited conditions of dye solution permeation with a slow injection rate to emphasize the anisotropy of the structure and examine it. However, for the actual design of the component, the permeation rate and the amount of permeation are also essential factors to consider. Functional components, such as filters and tablets, are important industrial components. Therefore, structural anisotropy is an essential property for the design of filters and drug-loading tablets using CAPC porous materials in future studies.

Funding: This research was supported by JSPS KAKENHI Grant Number 22H01379.

Institutional Review Board Statement: Not applicable.

Informed Consent Statement: Not applicable.

Data Availability Statement: All data generated or analyzed during this study are included in this published article.

Conflicts of Interest: The author declares no conflict of interest.

References

1. Young, R.J.; Lovell, P.A. *Introduction to Polymers*, 3rd ed.; CRC Press: Boca Raton, FL, USA, 2011; ISBN 978-0849339295.
2. Fried, J.R. *Polymer Science and Technology*, 3rd ed.; Prentice Hall: New Jersey, NJ, USA, 2014; ISBN 978-0137039555.
3. Ishizaki, K.; Komarneni, S.; Nanko, M. *Porous Materials—Process Technology and Applications*; Springer Science + Business Media: Dordrecht, NL, USA, 2014; ISBN 978-1461376637.
4. Mills, N. *Polymer Foams Handbook, Engineering and Biomechanics Applications and Design Guide*; Butterworth-Heinemann: Oxford, UK, 2007; ISBN 978-0750680691.
5. Otaru, A.J. Review on the acoustical properties and characterisation methods of sound absorbing porous structures: A focus on microcellular structures made by a replication casting method. *Met. Mater. Int.* **2020**, *26*, 915–932. [CrossRef]
6. Rizvi, A.; Chu, R.K.M.; Park, C.B. Scalable fabrication of thermally insulating mechanically resilient hierarchically porous polymer foams. *ACS Appl. Mater. Interfaces* **2018**, *10*, 38410–38417. [CrossRef] [PubMed]
7. Suethao, S.; Shah, D.U.; Smitthipong, W. Recent progress in processing functionally graded polymer foams. *Materials* **2020**, *13*, 4060. [CrossRef] [PubMed]
8. Silverstein, M.S. Emulsion-templated porous polymers: A retrospective perspective. *Polymer* **2014**, *55*, 304–320. [CrossRef]
9. Riyajan, S.A.; Sukhlaaied, W. Fabrication and properties of a novel porous material from biopolymer and natural rubber for organic compound absorption. *J. Polym. Environ.* **2019**, *27*, 1918–1936. [CrossRef]
10. Jiang, J.; Li, Z.; Yang, H.; Wang, X.; Li, Q.; Turng, L.S. Microcellular injection molding of polymers: A review of process know-how, emerging technologies, and future directions. *Curr. Opin. Chem. Eng.* **2021**, *33*, 100694. [CrossRef]
11. Zhang, L.; Zhao, G.; Wang, G. Formation mechanism of porous structure in plastic parts injected by microcellular injection molding technology with variable mold temperature. *Appl. Therm. Eng.* **2017**, *114*, 484–497. [CrossRef]
12. Aizawa, T. A new method for producing porous polymer materials using carbon dioxide and a piston. *J. Supercrit. Fluids* **2018**, *133*, 38–41. [CrossRef]
13. Wakui, Y.; Aizawa, T. Analysis of sustained release behavior of drug-containing tablet prepared by CO_2-assisted polymer compression. *Polymers* **2018**, *10*, 1405. [CrossRef] [PubMed]
14. Aizawa, T.; Matsuura, S.-I. Fabrication of enzyme-loaded cartridges using CO_2-assisted polymer compression. *Technologies* **2021**, *9*, 85. [CrossRef]
15. Russell, S.J. *Handbook of Nonwovens*, 2nd ed.; Woodhead Publishing: Cambridge, UK, 2022; ISBN 978-0128189122.
16. Elise, R. *Nonwoven Fabric, Manufacturing and Applications*; Nova Science Publishers: New York, NY, USA, 2020; ISBN 978-1536175875.
17. Aizawa, T. Peel and penetration resistance of porous polyethylene terephthalate material produced by CO_2-assisted polymer compression. *Molecules* **2019**, *24*, 1384. [CrossRef] [PubMed]
18. Aizawa, T.; Wakui, Y. Correlation between the porosity and permeability of a polymer filter fabricated via CO_2-assisted polymer compression. *Membranes* **2020**, *10*, 391. [CrossRef] [PubMed]
19. Reddy, V.S.; Tian, Y.; Zhang, C.; Ye, Z.; Roy, K.; Chinnappan, A.; Ramakrishna, S.; Liu, W.; Ghosh, R. A review on electrospun nanofibers based advanced applications: From health care to energy devices. *Polymers* **2021**, *13*, 3746. [CrossRef] [PubMed]
20. Liang, M.; Wu, H.; Liu, J.; Shen, Y.; Wu, G. Improved sound absorption performance of synthetic fiber materials for industrial noise reduction: A review. *J. Porous Mater.* **2022**, *29*, 869–892. [CrossRef]
21. Drabek, J.; Zatloukal, M. Meltblown technology for production of polymeric microfibers/nanofibers: A review. *Phys. Fluids* **2019**, *31*, 091301. [CrossRef]

Disclaimer/Publisher's Note: The statements, opinions and data contained in all publications are solely those of the individual author(s) and contributor(s) and not of MDPI and/or the editor(s). MDPI and/or the editor(s) disclaim responsibility for any injury to people or property resulting from any ideas, methods, instructions or products referred to in the content.

Article

Information-Analytical Software for Developing Digital Models of Porous Structures' Materials Using a Cellular Automata Approach

Igor Lebedev, Anastasia Uvarova and Natalia Menshutina *

Department of Chemical and Pharmaceutical Engineering, Mendeleev University of Chemical Technology of Russia, 125047 Moscow, Russia; lebedev.i.v@muctr.ru (I.L.); anastasia.uvarova2@yandex.ru (A.U.)
* Correspondence: chemcom@muctr.ru; Tel.: +7-495-495-0029

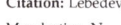

Citation: Lebedev, I.; Uvarova, A.; Menshutina, N. Information-Analytical Software for Developing Digital Models of Porous Structures' Materials Using a Cellular Automata Approach. *Technologies* **2024**, *12*, 1. https://doi.org/10.3390/technologies12010001

Academic Editors: Sergey N. Grigoriev, Alexander Tsouknidas, Marina A. Volosova and Anna A. Okunkova

Received: 31 October 2023
Revised: 11 December 2023
Accepted: 18 December 2023
Published: 20 December 2023

Copyright: © 2023 by the authors. Licensee MDPI, Basel, Switzerland. This article is an open access article distributed under the terms and conditions of the Creative Commons Attribution (CC BY) license (https://creativecommons.org/licenses/by/4.0/).

Abstract: An information-analytical software has been developed for creating digital models of structures of porous materials. The information-analytical software allows you to select a model that accurately reproduces structures of porous materials—aerogels—creating a digital model by which you can predict their properties. In addition, the software contains models for calculating various properties of aerogels based on their structure, such as pore size distribution and mechanical properties. Models have been implemented that allow the description of various processes in porous structures—hydrodynamics of multicomponent systems, heat and mass transfer processes, dissolution, sorption and desorption. With the models implemented in this software, various digital models for different types of aerogels can be developed. As a comparison parameter, pore size distribution is chosen. Deviation of the calculated pore size distribution curves from the experimental ones does not exceed 15%, which indicates that the obtained digital model corresponds to the experimental sample. The software contains both the existing models that are used for porous structures modeling and the original models that were developed for different studied aerogels and processes, such as the dissolution of active pharmaceutical ingredients and mass transportation in porous media.

Keywords: information-analytical software; modeling; cellular automata; digital models; porous materials; aerogels

1. Introduction

The development of new materials with the required properties is an urgent task in various fields of science and industry. However, the final properties of a material depend on many parameters, which require numerous experimental studies; this leads to significant resource costs and slows down rate of development of new material.

In this work, a software complex for creating digital models of porous materials—aerogels—was proposed.

Aerogel and materials based on it are characterized by high thermal insulation (thermal conductivity coefficient 0.014–0.022 W/(m·K)) and sound insulation properties (sound absorption coefficient 0.3–0.5) and low density (0.005–0.1 kg/m^3) [1]. The use of materials with such characteristics makes it possible to reduce the required thickness of thermal insulation several times or significantly reduce heat losses [2]. This makes it possible to use aerogels in the aerospace industry (low-density, highly efficient thermal insulation of fuel tanks, other components of aircraft and launch vehicles), the oil refining industry (highly efficient thermal insulation of oil and gas pipelines), the chemical industry (thermal insulation of cryogenic installations) and civil and industrial construction in the conditions of the Far North and the Arctic.

In addition, the internal structure of aerogels allows them to be used as a carrier matrix for various active substances: active pharmaceutical ingredients, biopolymers, cells and metal compounds [3–5]. Aerogels can be inorganic, organic and hybrid.

Inorganic aerogels are those whose structures consist of inorganic matter. The most common aerogels are based on silicon dioxide, SiO_2. In addition to it, aerogels based on metal oxides, such as Al_2O_3, TiO_2, ZrO_2, etc., are quite common. The particles that form the structure of inorganic aerogels (globules), as a rule, are monodisperse and spherical, with a diameter of 2 to 20 nm, depending on the conditions of the sol-gel process.

Organic aerogels are very diverse and vary greatly depending on the material. The structure of organic aerogels can be formed either by spherical particles of matter—globules, as mentioned above—or they can be fibrous (for example, cellulose-based aerogels). Organic aerogels have the basic properties of aerogels: high specific surface area, open pore system, low density. At the same time, they have several additional important qualities: they have high loading rates of active substances and can be biodegradable and biocompatible. One example is aerogels based on polysaccharides or alginates. Such aerogels can be used as a drug delivery system. Thus, organic aerogels are widely used in the food and pharmaceutical industries, cosmetology and other industries. For example, aerogels based on chitosan and silver nanoparticles have great hemostatic efficiency, which can more quickly stop bleeding and, in addition, disinfect the wound. Aerogels based on polysaccharides, sodium alginate, pectin, starch, chitosan, cellulose, resorcinol–formaldehyde aerogels and others are used.

Hybrid aerogels have a structure that is formed by two components—organic and inorganic. The main advantage of hybrid aerogels is that there are many ways to improve their functional characteristics. Among hybrid aerogels, the most studied are silicon/biopolymer aerogels.

However, the development of new materials with specified properties is always associated with a large number of experimental studies, which leads to an increase in cost and development time. In such cases, the creation of digital models of material structures that can be used to predict their properties will partially replace natural experiments with computational ones, which will reduce the number of necessary experiments.

One of the approaches that is widely used to generate digital structures of materials is cellular automata modeling [6,7]. Due to their flexibility and versatility, cellular automata can be used in many areas [8–11]. Cellular automata use makes it possible to use special software and hardware products that can modify the rules and initial states of cellular automata and visualize their evolution.

Unlike standard approaches using differential equations, cellular automata models make it possible to reflect the essence of the physical and chemical process, directly reflecting the sequence of steps that they contain.

The cellular automata approach provides great advantages in describing systems consisting of a large number of basic units. Such features of cellular automata models as discreteness of space-time and locally defined behavior of each cell make them convenient and practical for implementation within the framework of parallel computing, namely, for creating fine-grained locally parallel algorithms [12]. Table 1 shows different applications of cellular automata for modeling a large number of phenomena.

Table 1. Cellular automata models applications.

Application	References
Diffusion (mass transfer)	[10]
Corrosion	[13,14]
Melting	[15]
Medicine	[12]
Structure modeling	[7,16–18]
Mechanical properties	[8]

One of the features of aerogel structures is their mesoporosity—their properties largely depend on mesopores, defined as pores that have a diameter from 2 to 50 nm [19]. This imposes certain restrictions on the choice of methods for modeling structures since the presence of these pores must be explicitly taken into account. In this case, methods for modeling structures are determined by the choice of basic elements of the model and the corresponding assumptions, from which the scale at which the structure is generated follows. These models reproduce the structure at the nano-, meso- and macro levels.

The main problem in modeling nanomaterials is the gap between the modeling of individual structural elements at the atomic level and macroscopic properties, which are determined by the behavior of groups of structural elements. Modeling at the nanoscale does not allow direct analysis of the properties of nanomaterials, since the size of the resulting digital structures is too small. In addition, it is a difficult task to translate the large amount of data that is generated during modeling at the nanoscale into the physical parameters of the material, which determine its behavior at the macrolevel [20].

Macroscale models consider the structure on a large scale, not allowing mesopores to be identified and taken into account as individual structural elements. This problem can be solved by developing mesoscale models.

Modeling at the mesoscale can be performed for those systems that contain several elements of micro- or nanostructure. This makes it possible to highlight structural features to establish the properties of the material at the macrolevel [21,22]. Thus, when modeling mesoporous structures such as aerogels, it is most advisable to use modeling methods at the mesoscale, when a globule or a group of globules that form the final structure is chosen as the base element of the model. Such methods make it possible to obtain digital structures ranging in size from one hundred to several thousand nanometers with relatively low requirements for computing power.

At the mesoscale, the system is small enough to be considered at the macrolevel, where the structure becomes homogeneous, but large enough to consider the main structural features of the sample. At the mesoscale, the system has dimensions from hundreds to thousands of nanometers.

The aim of the study to create a universal and convenient tool which can be used to study and develop new nanoporous materials by creating digital mesoscale models of the structures. To solve this problem, the following tasks must be solved:

1. Developing and implementing models of digital nanoporous structures.
2. Creating a tool for the calculation of digital nanoporous structures' properties to estimate correspondence between digital models and experimental samples.
3. Developing and implementing models of processes in digital nanoporous structures.

To solve these tasks, the original information-analytical software (IAS), which implements mesoscale models using a cellular automaton (CA) approach to model nanoporous materials and their properties, is suggested.

2. Methods

The main idea of the CA approach is that the system under study is divided into identical cells—elementary volumes (or areas, for the two-dimensional case). At each moment, each cell has one specific state. The state of each cell depends on the states of neighboring cells (the neighborhood). The cell changes its state every discrete time step based on local transition rules. The locality of the rules makes it possible to take into account the heterogeneity of the structure, when the composition and geometry of the material have a significant impact on its properties. The transition rules can be based on theoretical or statistical dependencies [23].

The advantages of the CA approach are the simplicity and locality of the rules for the evolution of the system under study in time and the possibility of organizing high-performance parallel computing, which can significantly speed up the simulation calculations. Cellular automata are increasingly used to solve various practical problems. This is mainly due to the increase in available computing power and the development of high-performance parallel computing technologies. In structure modeling at mesoscale, CA approach allows for considering the heterogeneity of porous materials and, thus, detailed modeling of its geometry without the complication of calculations.

Another advantage of the CA approach is its ease of integration with other models. In [24], the CA approach was used in combination with the lattice Boltzmann method. CA models are convenient for scaling and for creating multiscale models.

To define a cellular automaton, it is enough to determine the structure of the discrete space and the cells of which it consists. The lattice can be different—Cartesian lattice, hexagonal lattice, triangular, etc. The cells of a cellular automaton are determined by a set of possible states, a neighborhood, and transition rules. By changing these parameters, you can obtain different cellular automata for different types of problems.

At each time step, each cell can be in only one state from a strictly defined set. This set is called the alphabet. The simplest set (alphabet) is Boolean, in which a cell can have only one of two states—0 or 1. In early cellular automata, the alphabet could be integer or symbolic, but now there are no restrictions on the alphabet, which increases the flexibility of the created models. When solving practical problems, as a rule, these states are associated with real physical quantities or phenomena, which may include the state of matter (solid/liquid/gaseous), particle concentrations, linear velocities, temperature, etc. In addition, cell states can be characterized not only by well-defined discrete parameters (e.g., crystalline or amorphous) but also by continuously varying values (e.g., linear velocity, concentration, crystal orientation in space, etc.) [25].

2.1. Porous Structures Models

When modeling porous materials at the mesoscale, aggregation methods are widely used (Figure 1) [26].

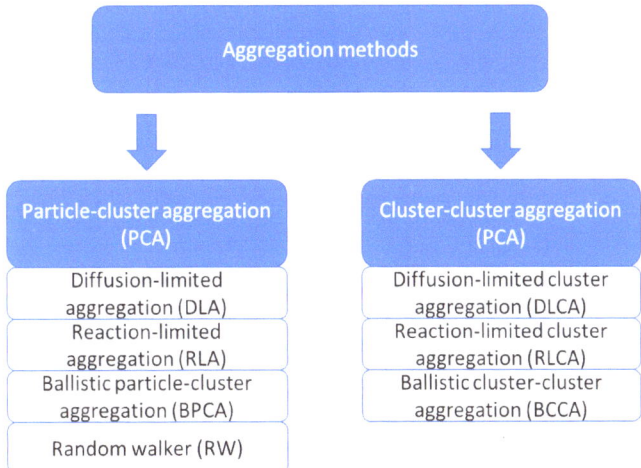

Figure 1. Aggregation methods.

Aggregation methods can simulate the process of formation of porous structures as a result of the thermal motion of molecules and the Brownian motion of particles caused by it. Aggregation models with certain assumptions simulate the process of particles' chaotic motion and aggregation into a single structure.

It is important to note that aerogels are fractal structures. Fractal structures are structures that have a complex shape, consisting of fragmented parts, each of which is a smaller copy of the whole. Therefore, the main property of fractal structures is self-similarity. Fractal systems have a non-continuous, loose and branched structure and are formed in a large number of physical processes accompanied by the association of solid particles of similar sizes.

The aerogel structure is formed as a result of a sol-gel process, during which, for example, for inorganic aerogels, primary globules of small diameter are formed and then aggregate into larger clusters (except in the case of fibrous aerogels, which will be discussed separately).

The self-similar structure, which results from the thermal motion of molecules and the resulting Brownian motion of particles in the system, can be mathematically described using the term multifractal. In the structure of the aerogel, presumably, several levels of hierarchy can be distinguished: primary globule (an ensemble of molecules that make up the basic geometric unit of the aerogel framework), secondary globule (an ensemble of primary globules that make up the second level of hierarchy in the geometry of the aerogel frame), tertiary globule (an ensemble of secondary globules that make up the third level of hierarchy in the geometry of the aerogel frame).

Depending on the reagents and conditions of the sol-gel process, the particle sizes at each level of order can have different values and, in some cases, after the third level of the hierarchy, the globules can exceed the size of 5 microns. Exceeding this threshold means that the particle is no longer subject to thermal motion and no longer participates in the further formation of a Brownian mass multifractal aggregate.

In aggregation methods, the base element of the model is a particle (globule for aerogel) of material. The main idea of aggregation methods is that particles of material are placed in space. They move and aggregate into a single cluster, forming a frame of a porous material. Aggregation methods are divided into "particle–cluster aggregation" (PCA), when moving particles aggregate with a fixed clusterization center, and "cluster–cluster aggregation" (CCA), when all particles in the system are mobile and aggregate with each other, forming local mobile clusters, which then aggregate into a single structure [27].

In addition, aggregation methods are classified according to the discreteness of the model space—the system can be divided into discrete cells (lattice methods) or form a continuous space (off-lattice methods). Lattice aggregation methods can be successfully implemented using a cellular automata approach.

The principle of the PCA models work is as follows: one or more stationary clusterization centers are placed on the field, after which a moving particle is generated in a random place in the field; this particle makes a random movement until it is near the center. After that, aggregation occurs—the moving particle itself becomes part of the cluster and a new moving particle is generated [28]. The process is carried out until the structure reaches the specified porosity (Figure 2).

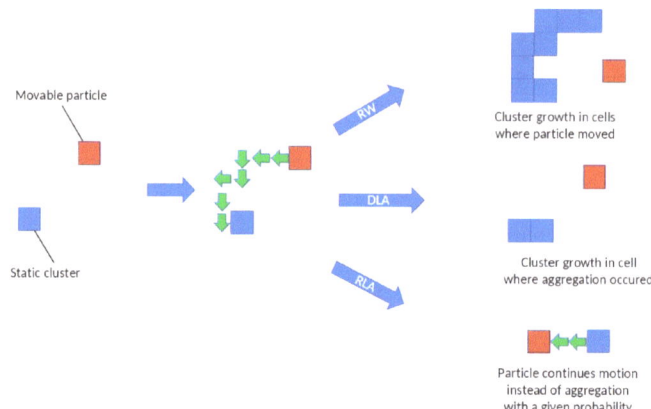

Figure 2. Illustration of the operation of the DLA, RLA and RW methods.

In the Figure 2 the blue and red cells represent static clusters and movable particles respectively. Depending on the aggregation principle, there are different types of PCA models: in DLA, a particle aggregates whenever there is a clusterization center near it. In the RLA model, aggregation occurs with a given probability. RW works as DLA, but after aggregation, both the particle itself and the places where it was before aggregation become clusterization centers. The BPCA model works as DLA, but the particles motion direction does not change.

The main idea of cluster–cluster models is as follows: a given number of moving particles is placed on the field, determined by the porosity of the structure. Particles begin to move chaotically, aggregating into a single cluster after collision. Clusters consisting of several particles continue to move. The process is carried out until all the particles are aggregated into a single cluster (Figure 3).

Movable particles

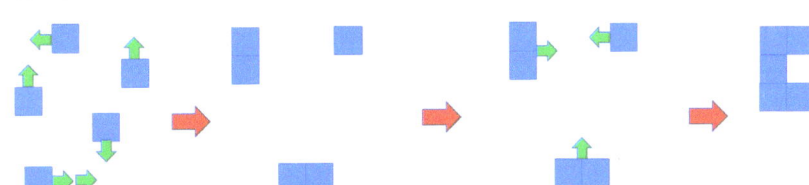

Figure 3. Illustration of the work of the DLCA method.

In the Figure 3 the blue cells represent movable particles. The considered models can be used for the generation of structures that are formed by spherical globules; however, for describing structures formed by fibers, they have insufficient accuracy. The description of fibrous structures is possible with the use of a model based on Bezier curves.

Bezier curves are widely used in designing products of a given shape, for example, car hulls, as well as in several mathematical problems, such as calculating the trajectory of motion [29,30]. Using Bezier curves in the modeling of porous materials makes it possible to obtain digital structures of porous materials, the frameworks of which consist of nanosized fibers. These materials include aerogels based on chitosan and cellulose [31].

The principle of a model based on Bezier curves is as follows: a Bezier curve is plotted on the simulation field, after which a fiber is generated on those sections through which the curve passes. Bezier curves are plotted until the digital structure reaches the specified porosity.

Bezier curves are built using two or more control points. The number of control points specified determines the order of the curve. Two control points define a Bezier curve as a linear curve (first-order curve—a straight line), three control points define a quadratic curve (second-order curve), four control points define a cubic curve (third-order curve). In the developed software complex, third-order Bezier curves are used to determine the shape of the fiber [32]:

$$B(t) = (1-t)^3 P_0 + 3t(1-t)^2 P_1 + 3t^2(1-t)P_2 + t^3 P_3, \; t \in [0,1] \quad (1)$$

where P0, P1, P2, P3 are control points, which contain the set of three-dimensional coordinates, and t is the curve plotting step.

P0 and P3 indicate the beginning and the end of the curve; these points are chosen on the different field edges. P1 and P2 are points with random coordinates between P0 and P3. Bezier curve passes through P0 and P3 but does not pass through P1 and P2 (Figure 4).

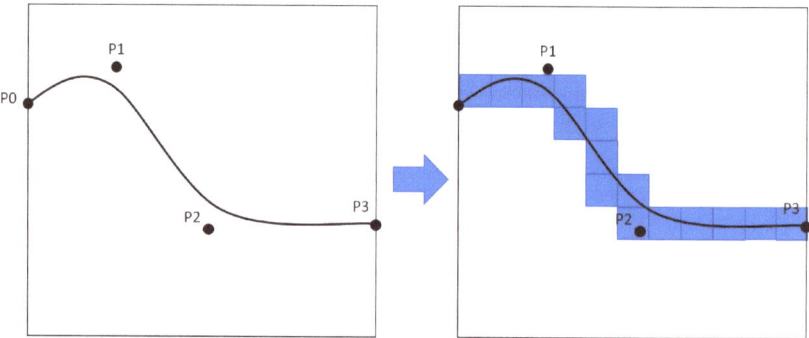

Figure 4. Fiber built using a Bezier curve.

Since there are a large number of different models and methods of porous materials, it seems a promising task to develop a software complex that combines various aggregation methods, a cellular automaton approach for the development of new porous materials, namely, aerogels, an analysis of their properties, and various processes using them.

2.2. The IAS Structure

The IAS was developed as a convenient and efficient tool for modeling the porous materials' structures and properties, which can be used by a wide range of users who do not have programming skills. Figure 5 shows the diagram of the developed software complex.

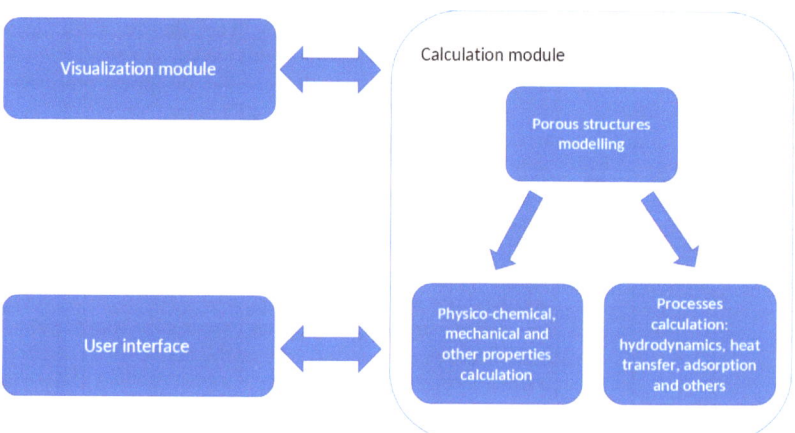

Figure 5. Structure of the IAS.

The calculation block contains a module for modeling porous structures, which contains the main aggregation methods using the cellular automaton approach: diffusion-limited aggregation (DLA), reaction-limited aggregation (RLA), aggregation with a set clusterization centers (multiDLA), diffusion-limited cluster aggregation (DLCA), random walker (RW). In addition, it implements an original cellular automaton fibrous porous structures model based on the Bezier curves [33].

The module for calculating physicochemical, mechanical and other properties contains models that allow for calculating the properties of aerogels from their digital structure, for example, pore size distribution and strength properties.

Software complex was developed with C# language. The user interface and visualization module were developed with Windows Forms framework, but the calculation module can be used independently.

3. Results

The process calculation module allows the user to generate digital structures to calculate various processes in them—hydrodynamics, mass transfer and others.

Some examples of modeling with the developed software complex are considered below.

3.1. Porous Structures Modeling

The module for modeling porous structures consists of the following submodules: a particle–cluster model, a cluster–cluster model and a model based on Bezier curves. Each of them has a separate interface window and is intended for modeling porous structures with different characteristics (Figure 6).

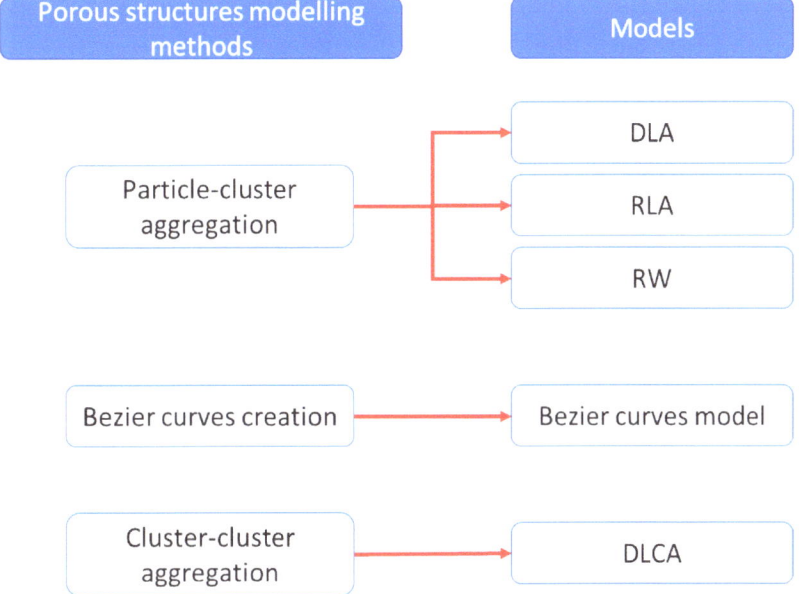

Figure 6. Structure of the module for modeling porous bodies.

The particle–cluster module simulates porous structures formed as a result of a sol-gel process and supercritical drying. These structures are a network of spherical globules or fibers. Figure 7 shows the interface window of the particle–cluster module.

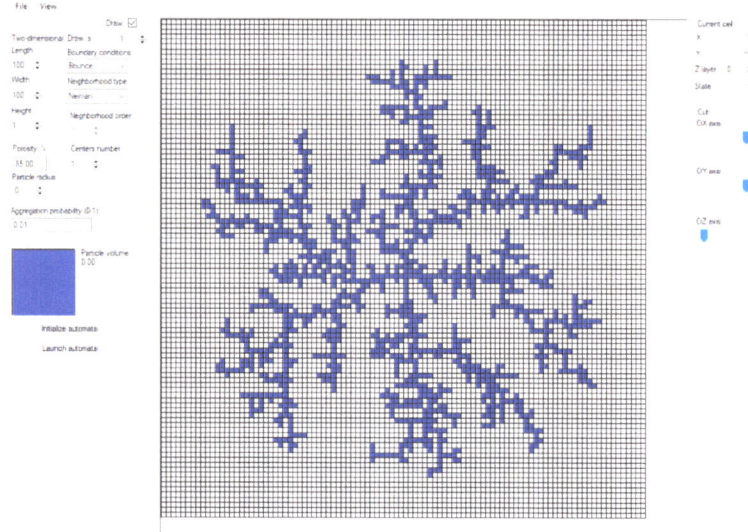

Figure 7. Interface window of the particle–cluster module.

As input parameters, the module requires the geometric size of the simulation field, the particle diameter, the number of clusterization centers (in this case, aggregation with multiple clustering centers is implemented—multiDLA), the probability of aggregation upon collision (if it is less than 100%, then the RLA method is implemented) and the porosity of the structure. To generate a structure using the random walker (RW) model, a separate parameter in the generation settings must be set.

Figure 8 shows the porous structure obtained by DLA at various stages of generation.

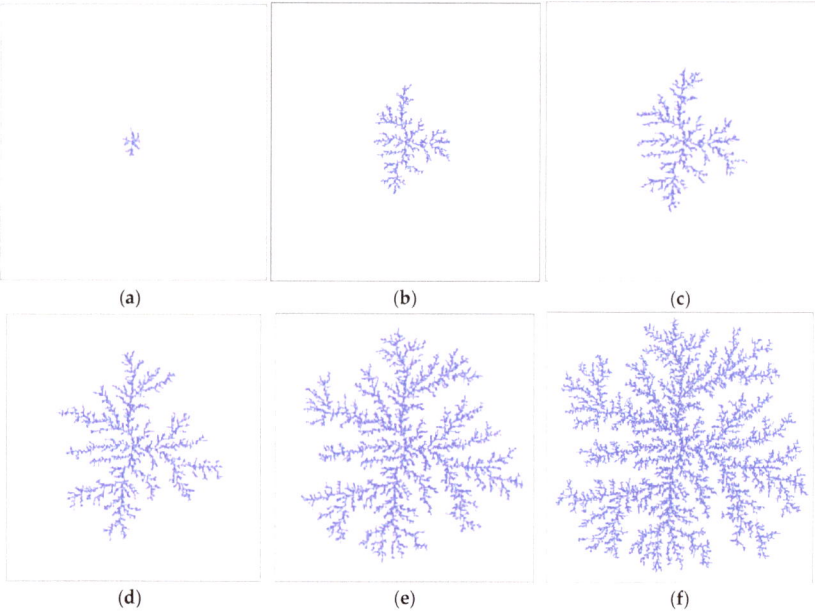

Figure 8. Different steps (**a**–**f**) of generating a porous structure by the DLA method.

The structure in Figure 8 has a 500 × 500 nm size and 85% porosity.

In addition, the type of cell neighborhood can be chosen—von Neumann or Moore. If the Moore neighborhood is chosen, the cell can move and aggregate along the diagonal; if the von Neumann neighborhood is chosen, it cannot [34].

The software complex allows for obtaining both two-dimensional and three-dimensional structures and real-time visualization.

Figure 9 shows 2D and 3D digital structures of porous structures obtained using various models of the particle–cluster module.

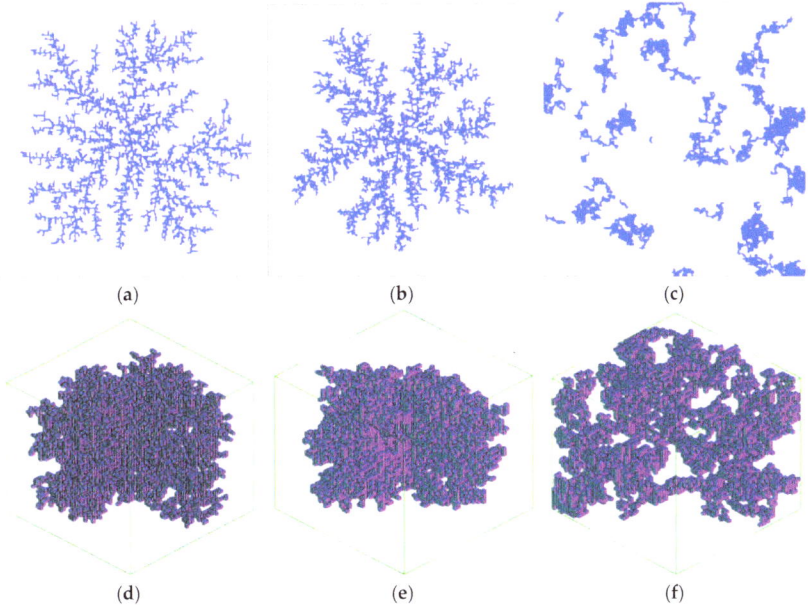

Figure 9. Two-dimensional and three-dimensional digital porous structures obtained are by: (**a**–**c**) DLA, RLA and RW methods for 2D structures; (**d**–**f**) DLA, RLA and RW methods for 3D structures.

The size of the structures in Figure 9 is 200 × 200 nm for 2D cases and 50 × 50 × 50 nm for 3D cases. The porosity of the structures is 85% for two-dimensional and 90% for three-dimensional.

The models are implemented with the ability to vary the size of the particles, which makes it possible to consider their size and shape (Figure 10).

The cluster–cluster module allows for modeling of porous structures formed by spherical globules. Inorganic aerogels based on silicon dioxide, hybrid aerogels based on silicon–resorcinol–formaldehyde and organic aerogels based on egg white have this kind of structure. Figure 11 shows the cluster–cluster module interface.

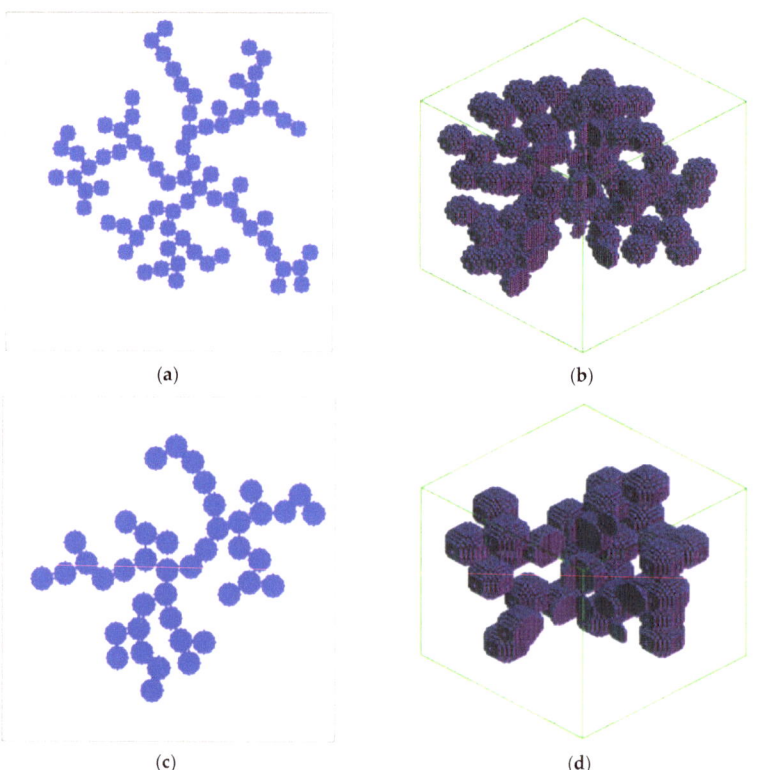

Figure 10. Porous structures obtained by the DLA method at various particle diameters and 85% porosity: (**a**,**b**) 9 nm particle diameter; (**c**,**d**) 15 nm particle diameter.

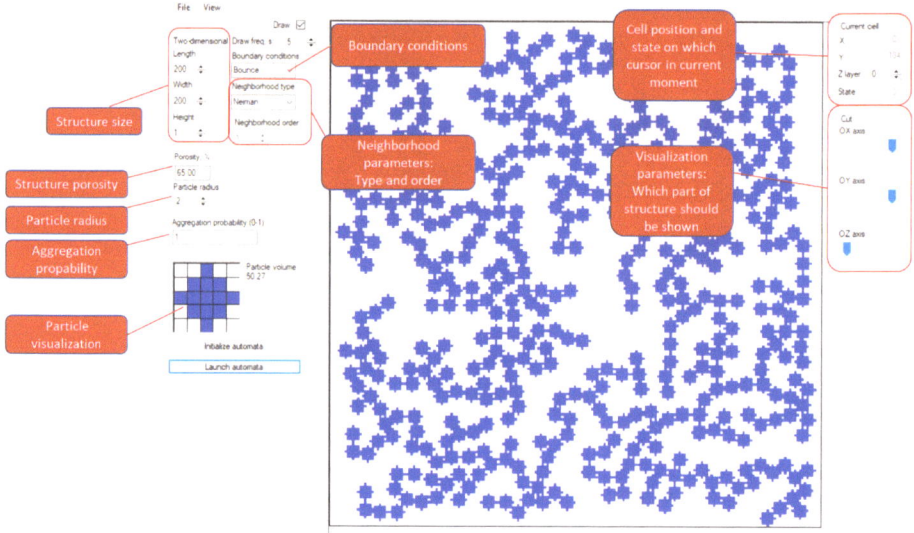

Figure 11. Cluster–cluster module interface window.

As input parameters, the module requires the geometric size of the simulation field, the particle diameter and the porosity of the final structure.

As in the case of the particle–cluster module, different neighborhoods of the cell (von Neumann or Moore) and the diameter of the particles can be chosen. Figure 12 shows two-dimensional and three-dimensional structures obtained by the DLCA method for various model parameters.

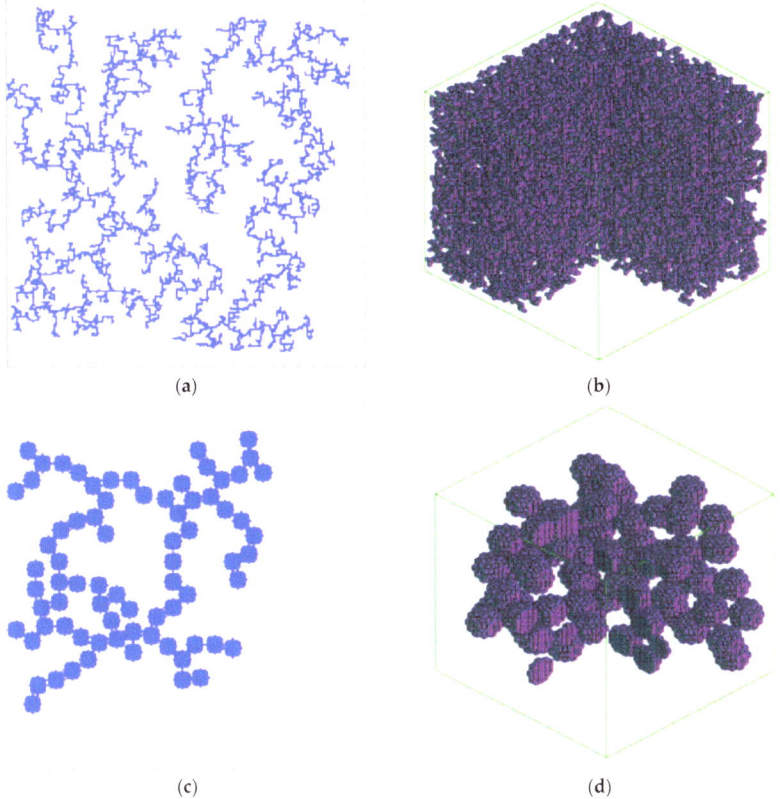

Figure 12. Porous structures obtained by the DLCA method at different particle diameters and 85% porosity: (**a**,**b**) 1 nm particle diameter; (**c**,**d**) 9 nm particle diameter.

The Bezier curves module implements a cellular automaton model based on Bezier curves. As input parameters, the module requires the geometric size of the simulation field, the porosity of the material, and the diameter of the fiber. Figure 13 shows the digital porous structures obtained using a model based on Bezier curves.

The final result of all the considered models' calculations is a digital discrete porous structure, which can be saved to a file and loaded later. All the software complex's modules are integrated so that structure files generated in one module can be loaded in any other.

The software complex allows the user to model structures using various models and vary their parameters, making it possible to select the most appropriate model and parameters for a particular type of material.

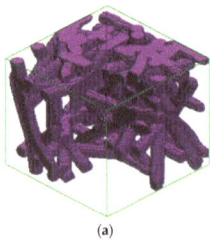 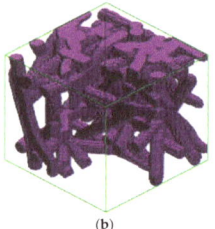

Figure 13. Fibrous porous structures obtained using a model based on Bezier curves for various fiber diameters and a porosity of 70%: (**a**) 7 nm fiber diameter; (**b**) 11 nm fiber diameter.

These models presented do not predict the structural characteristics of the samples. The generation of structures using cellular automata models is carried out to obtain a digital twin, the structural characteristics of which correspond to the existing experimental sample. The purpose of obtaining such structures is to predict the properties of the materials under study. Since structural characteristics determine the physicochemical and mechanical properties of nanostructures, it can be argued that a digital twin, the structural properties of which correspond to the experimental sample, has the same properties. Consequently, the calculation of physicochemical and mechanical properties for a digital copy makes it possible to predict them for real samples.

3.2. Aerogel Properties Calculation

Pore size distribution is one of the most important aerogel properties because it directly influences material parameters such as sorption capacity and thermal conductivity. Thus, for comparing digital structures and experimental samples, the module for calculating a digital structure's pore size distribution was implemented.

This module is based on the algorithm described in [35] and the main idea that pores of all possible diameters, from larger to smaller, are sequentially entered into each point of the digital structure. If the pore is successfully inscribed inside the structure, then its diameter and volume are recorded. The calculation goes until the diameter of the inscribed pores reaches a diameter equal to the size of one cell. As a digital structure has discrete pores, the inside of it consists of a set of cells (Figure 14a). After a pore is placed, its volume can be calculated, so, as a result, it is possible to calculate the occupied volume for all pore diameters (Figure 14b).

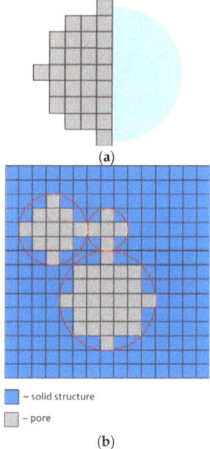

Figure 14. Pore size distribution algorithm illustration: pore presentation as a set of cells (**a**) and discrete pores found inside digital structure (**b**).

The result of the algorithm is the pore size distribution curve (Figure 15).

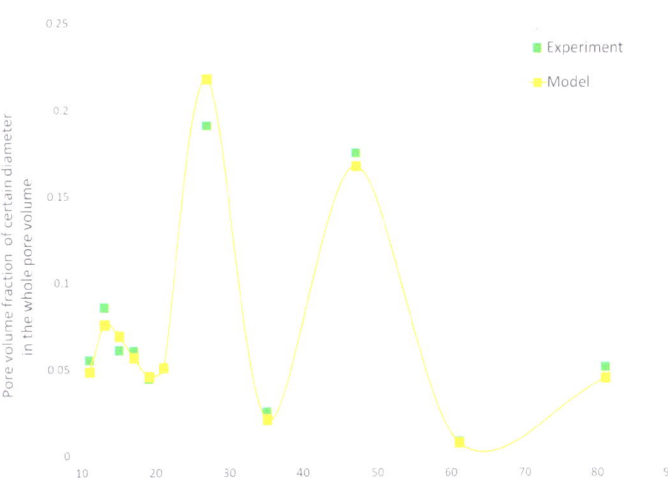

Figure 15. Experimental and calculated curves of the pore size distribution of the porous fibrous structure of an aerogel based on chitosan.

If the experimental and calculated pore size distribution curves correspond, it can be concluded that the generated structure and the experimental sample correspond too; thus, a digital model of the material is obtained. This digital structure can be further used to calculate sample properties such as thermal conductivity, mechanical and sorption properties [36]. With models based on presented aggregation models and the Bezier curve model, digital porous aerogel structures were generated for silicon dioxide aerogels [35] and organic aerogels with structures formed with spherical globules [36] and organic aerogels formed with nanofibers [33]. Deviation on the calculated and experimental pore size distribution curves did not exceed 15%, which indicates that the digital structures correspond with the experimental ones and can be used further for properties calculations.

3.3. Processes Calculation

In the developed software complex, the possibility of modeling hydrodynamics using the lattice Boltzmann method (LBM) is implemented.

A feature of the LBM is that it does not use the Navier–Stokes equations; rather, it models the flow of a Newtonian fluid using a discrete form of the Boltzmann equation [37–39].

The main idea of the LBM is that the system is divided into identical cells. Each cell represents the volume of the simulated fluid containing fluid particles. Fluid particles can only move between cells, and in one discrete time step, a particle can only move to a neighboring cell. Each time step is divided into a phase of streaming step, when fluid particles move to neighbor cells, and a collision step, when the collision of particles inside the cell is calculated by the discrete form of the Boltzmann equation [40]:

$$f_i\left(\vec{r} + \vec{e_i}t^*, t + t^*\right) = f_i\left(\vec{r}, t\right) - \Omega_i \quad (2)$$

where t^*—discrete time step, i—motion direction index, f_i—number of particles moving in direction i, Ω_i—collision operator in direction i, $\vec{r}$—cell radius vector.

The motion direction index presents one of the possible discrete directions in which fluid particles can move. The collision operator is an operator that simplifies the calculation of collisions between particles inside a cell. There can be different collision operators depending on the specific study. In addition, it is possible to calculate multiphase and multicomponent systems.

The LBM approach is suitable for hydrodynamics in nanoporous structures calculations because discrete cells allow for the consideration of complex geometry consisting of pores and a solid skeleton. Also, in the LBM system is divided by cells, allows for combining it with CA models.

Figure 16 shows the interface window for hydrodynamics modeling with the LBM method.

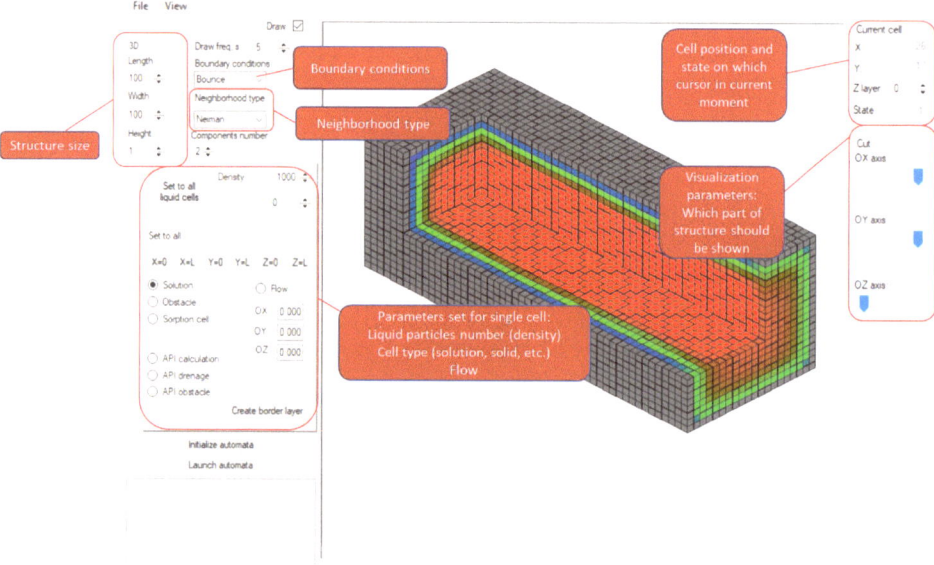

Figure 16. Interface window of the module for calculating hydrodynamics.

The developed software module allows for multicomponent systems modeling. Also, the digital structures generated earlier can be used as input data (Figure 17). This feature allows for modeling mass transport inside porous samples.

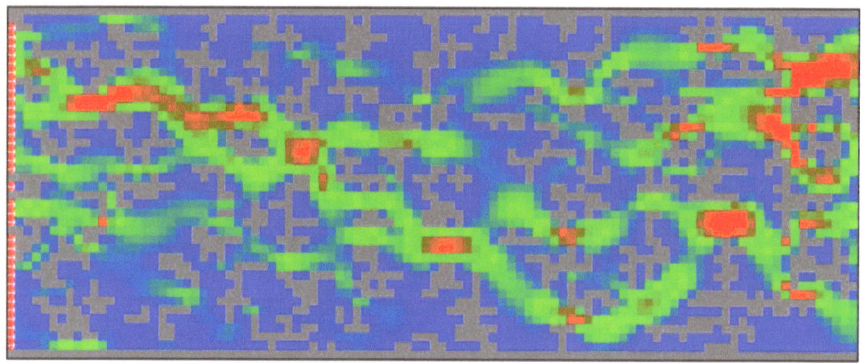

Figure 17. Calculation of hydrodynamics in a porous structure obtained by DLA methods.

In the Figure 17, the gradient from blue to red shows the relative strength of the flow from less to stronger, respectively. The grey cells represent solid area.

The module allows for interaction with the simulation field, creating arbitrary structures or changing existing ones, including directly during simulation (Figure 18).

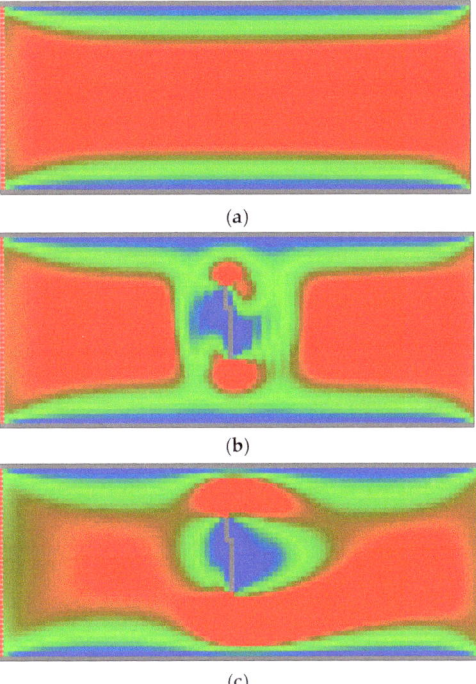

Figure 18. Modeling of the Poiseuille flow in a pipe (**a**), creating an obstacle (**b**), changing the flow by the created obstacle (**c**).

In the Figure 18, the gradient from blue to red shows the relative strength of the flow from less to stronger, respectively. The grey cells represent solid area.

In addition, the LBM can be combined with CA models, forming a single complex model for different processes, which significantly expands its application areas. The developed software uses a hybrid model that combines the lattice Boltzmann method and the cellular automata approach, which allows for predicting the processes of active pharmaceutical ingredients (API) release from the aerogel particles and its flow in various media. Flow is simulated with the LBM, and the release process is simulated with a cellular automata model.

The main idea of the API release model is as follows. Each cell can have one of two states: aerogel particle and solution. Each aerogel particle cell has a given amount of API in it. Every time step, the aerogel cells near the solution cells release API in the solution cells. After that, API transportation in liquid media (solution cells) is calculated according to the LBM method (Figure 19).

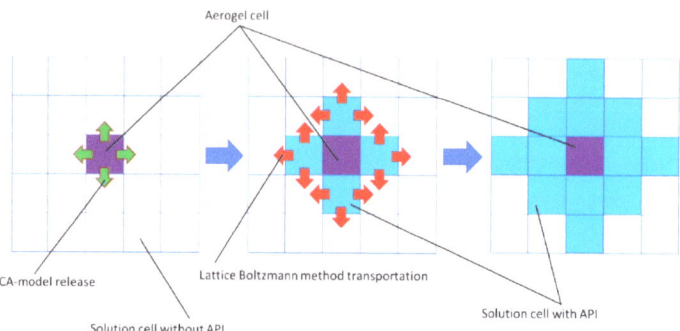

Figure 19. Stages of calculating the API release from aerogel.

In the Figure 19 the purple cell represents aerogel with API inside, the white cells represent solvent without API and the blue cells represent solution with released API.

This model was used in [6] to model melatonin release from chitosan-based aerogel particles and showed high accuracy.

The advantage of this model is that it allows the user to obtain both the release curve and API concentration distribution at each time moment.

Another process that is calculated with the software complex is the API particles deposited on human tissues dissolution process. The dissolution process includes API dissolution and its diffusion in liquid media. For this process, two different CA models operate on cells field—the dissolution CA model and the diffusion CA model. This CA has three possible cell states: active pharmaceutical ingredient (API), solution and air (empty). Empty cells are added for cases when not all the particle area is in contact with a solution. The basic idea of the model is as follows: a deposited particle is placed on the field, which is a collection of cells with an active pharmaceutical ingredient (API) state.

After this, the system simulates the dissolution and mass transfer (diffusion) of the dissolved API. API state cells represent the elementary volume of the API in solid form, and solution cells represent cells containing solvent and dissolved API. Within the model, the dissolution process represents the transfer of the API mass from the API cell to the cell with the solution. The diffusion process in the model means that the API moves from a solution cell with more API to a solution cell with less API. After the API cell dissolves some amount of its API, it transfers to the solution cell. The calculation is carried out until there is not a single cell left with the API state—this means that all the API particles have passed into the dissolved state. For API particles that are not completely immersed in the solution, further calculations are made at each time step, when the immersed portion of the particle has dissolved and the particle is immersed further (Figure 20).

In the Figure 20 the yellow cells represent API particle, the grey cells represent air cells (where solution can't flow) and the gradient from blue to red shows the relative amount of the API in cell from less to stronger, respectively.

Figure 20 shows that there is an air space where dissolution is not observed, so the particle interacts with solvent that is not all over its area. After the bottom areas, a completely dissolved particle moves and other areas can interact with the media. Both the lattice Boltzmann method and the dissolution model can work in two- or three-dimensional space.

The developed software is implemented in C# on the .NET framework.

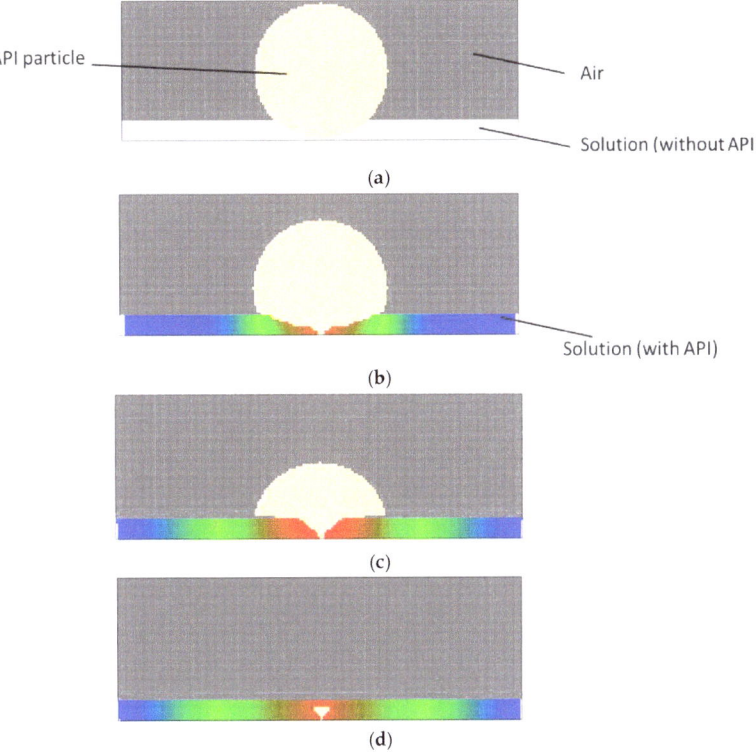

Figure 20. Stages (**a**–**d**) of calculating the API particle dissolution.

4. Results

In this work, a software complex to create digital models of porous materials was developed. The proposed software makes it possible to simulate the porous structures of aerogels using various models. The models are implemented with the possibility of a wide variation of input parameters, which allows for choosing a model for each type of aerogel, considering the features of the current sample, and for studying the dependence of the material's different properties on its structure and calculating the properties of a material based on its digital structure, creating a digital model.

In addition, the developed software allows for modeling processes, for example, hydrodynamics inside digital porous structures using the lattice Boltzmann method and the CA particle dissolution model. The LBM can be combined with cellular automata models, which allows for the calculation of various processes inside porous structures, such as sorption, mass transfer processes and dissolution. Additionally, software modules can be expanded with new cellular automata and other discrete models.

With the suggested IAS, there were developed various aerogels of different types: silicon dioxide, silica–resorcinol–formaldehyde, polyamide, carbon, chitosan, cellulose and protein, and the following properties were predicted: thermal conductivity, electrical conductivity, mechanical properties, sorption and solubility. These models allow for establishing a connection between structure geometry and its properties, which further allows for the development of materials with required properties.

The IAS is a universal decision for studying and developing materials with required properties. It allows the use of cellular automata models (both original developments and independent implementation of existing models) with wide possibilities for varying their parameters and adding new modules, so it is a perspective tool for studying and developing new nanoporous materials. The developed software will reduce the required number of full-scale experiments by partially replacing them with computational ones, which will reduce time and costs for creating new materials with the required properties.

Author Contributions: Conceptualization, N.M.; data curation, A.U.; formal analysis, A.U.; investigation, A.U. and I.L.; methodology, I.L.; project administration, N.M.; software, I.L.; validation, A.U.; visualization, I.L.; writing—original draft, I.L.; writing—review and editing, N.M. and A.U. All authors have read and agreed to the published version of the manuscript.

Funding: The research was carried out with the financial support of the Ministry of Science and Higher Education of the Russian Federation within the framework of scientific topics FSSM-2022-0004.

Institutional Review Board Statement: Not applicable.

Informed Consent Statement: Not applicable.

Data Availability Statement: Data is contained within the article.

Conflicts of Interest: The authors declare no conflicts of interest.

References

1. Smirnova, I.; Gurikov, P. Aerogel Production: Current Status, Research Directions, and Future Opportunities. *J. Supercrit. Fluids* **2018**, *134*, 228–233. [CrossRef]
2. Berthon-Fabry, S.; Hildenbrand, C.; Ilbizian, P.; Jones, E.; Tavera, S. Evaluation of Lightweight and Flexible Insulating Aerogel Blankets Based on Resorcinol-Formaldehyde-Silica for Space Applications. *Eur. Polym. J.* **2017**, *93*, 403–416. [CrossRef]
3. Wang, L.; Sánchez-Soto, M.; Abt, T. Properties of Bio-Based Gum Arabic/Clay Aerogels. *Ind. Crops Prod.* **2016**, *91*, 15–21. [CrossRef]
4. Wan, C.; Li, J. Graphene Oxide/Cellulose Aerogels Nanocomposite: Preparation, Pyrolysis, and Application for Electromagnetic Interference Shielding. *Carbohydr. Polym.* **2016**, *150*, 172–179. [CrossRef] [PubMed]
5. Wang, X.; Zhang, Y.; Jiang, H.; Song, Y.; Zhou, Z.; Zhao, H. Fabrication and Characterization of Nano-Cellulose Aerogels via Supercritical CO_2 Drying Technology. *Mater. Lett.* **2016**, *183*, 179–182. [CrossRef]
6. Lian, Y.; Gan, Z.; Yu, C.; Kats, D.; Liu, W.K.; Wagner, G.J. A Cellular Automaton Finite Volume Method for Microstructure Evolution during Additive Manufacturing. *Mater. Des.* **2019**, *169*, 107672. [CrossRef]
7. Teferra, K.; Rowenhorst, D.J. Optimizing the Cellular Automata Finite Element Model for Additive Manufacturing to Simulate Large Microstructures. *Acta Mater.* **2021**, *213*, 116930. [CrossRef]
8. Aniszewska, D.; Rybaczuk, M. Mechanical Properties of Silica Aerogels Modelled by Movable Cellular Automata Simulations. *Mater Today Commun.* **2021**, *27*, 102432. [CrossRef]
9. Svyetlichnyy, D.; Krzyzanowski, M.; Straka, R.; Lach, L.; Rainforth, W.M. Application of Cellular Automata and Lattice Boltzmann Methods for Modelling of Additive Layer Manufacturing. *Int. J. Numer. Methods Heat Fluid Flow* **2018**, *28*, 31–46. [CrossRef]
10. Khatami, D.; Hajilar, S.; Shafei, B. Investigation of Oxygen Diffusion and Corrosion Potential in Steel-Reinforced Concrete through a Cellular Automaton Framework. *Corros. Sci.* **2021**, *187*, 109496. [CrossRef]
11. Bozkurt, H.; Karwowski, W.; Çakıt, E.; Ahram, T. A Cellular Automata Model of the Relationship between Adverse Events and Regional Infrastructure Development in an Active War Theater. *Technologies* **2019**, *7*, 54. [CrossRef]
12. Salguero, A.G.; Capel, M.I.; Tomeu, A.J. Parallel Cellular Automaton Tumor Growth Model. In Proceedings of the PACBB 2019: 13th International Conference on Practical Applications of Computational Biology & Bioinformatics, Ávila, Spain, 26–28 June 2019; pp. 175–182.
13. Reinoso-Burrows, J.C.; Toro, N.; Cortés-Carmona, M.; Pineda, F.; Henriquez, M.; Galleguillos Madrid, F.M. Cellular Automata Modeling as a Tool in Corrosion Management. *Materials* **2023**, *16*, 6051. [CrossRef] [PubMed]
14. Zenkri, M.; di Caprio, D.; Raouafi, F.; Féron, D. Cathodic Control Using Cellular Automata Approach. *Mater. Corros.* **2022**, *73*, 1631–1643. [CrossRef]
15. Krzyzanowski, M.; Svyetlichnyy, D. A Multiphysics Simulation Approach to Selective Laser Melting Modelling Based on Cellular Automata and Lattice Boltzmann Methods. *Comput. Part. Mech.* **2022**, *9*, 117–133. [CrossRef]
16. Gu, C.; Lu, Y.; Cinkilic, E.; Miao, J.; Klarner, A.; Yan, X.; Luo, A.A. Predicting Grain Structure in High Pressure Die Casting of Aluminum Alloys: A Coupled Cellular Automaton and Process Model. *Comput. Mater. Sci.* **2019**, *161*, 64–75. [CrossRef]
17. Ruan, X.; Li, Y.; Zhou, X.; Jin, Z.; Yin, Z. Simulation Method of Concrete Chloride Ingress with Mesoscopic Cellular Automata. *Constr. Build. Mater.* **2020**, *249*, 118778. [CrossRef]
18. Rolchigo, M.; Stump, B.; Belak, J.; Plotkowski, A. Sparse Thermal Data for Cellular Automata Modeling of Grain Structure in Additive Manufacturing. *Model. Simul. Mater. Sci. Eng.* **2020**, *28*, 065003. [CrossRef]
19. Everett, D.H. Manual of Symbols and Terminology for Physicochemical Quantities and Units, Appendix II: Definitions, Terminology and Symbols in Colloid and Surface Chemistry. *Pure Appl. Chem.* **1972**, *31*, 577–638. [CrossRef]

20. Stoneham, A.M.; Harding, J.H. Not Too Big, Not Too Small: The Appropriate Scale. *Nat. Mater.* **2003**, *2*, 77–83. [CrossRef]
21. Faller, R. Automatic Coarse Graining of Polymers. *Polymer* **2004**, *45*, 3869–3876. [CrossRef]
22. Müller-Plathe, F. Scale-Hopping in Computer Simulations of Polymers. *Soft Mater.* **2002**, *1*, 120016739. [CrossRef]
23. Menshutina, N.V.; Kolnoochenko, A.V.; Lebedev, E.A. Cellular Automata in Chemistry and Chemical Engineering. *Annu. Rev. Chem. Biomol. Eng.* **2020**, *11*, 87–108. [CrossRef] [PubMed]
24. Lebedev, I.; Uvarova, A.; Mochalova, M.; Menshutina, N. Active Pharmaceutical Ingredients Transportation and Release from Aerogel Particles Processes Modeling. *Computation* **2022**, *10*, 139. [CrossRef]
25. Raabe, D. Cellular Automata in Materials Science with Particular Reference to Recrystallization Simulation. *Annu. Rev. Mater. Res.* **2002**, *32*, 53–76. [CrossRef]
26. Abdusalamov, R.; Scherdel, C.; Itskov, M.; Milow, B.; Reichenauer, G.; Rege, A. Modeling and Simulation of the Aggregation and the Structural and Mechanical Properties of Silica Aerogels. *J. Phys. Chem. B* **2021**, *125*, 1944–1950. [CrossRef] [PubMed]
27. Lin, M.Y.; Lindsay, H.M.; Weitz, D.A.; Klein, R.; Ball, R.C.; Meakin, P. Universal Diffusion-Limited Colloid Aggregation. *J. Phys. Condens. Matter* **1990**, *2*, 3093–3113. [CrossRef]
28. Meakin, P. Formation of Fractal Clusters and Networks by Irreversible Diffusion-Limited Aggregation. *Phys. Rev. Lett.* **1983**, *51*, 1119–1122. [CrossRef]
29. Louzazni, M.; Al-Dahidi, S. Approximation of Photovoltaic Characteristics Curves Using Bézier Curve. *Renew Energy* **2021**, *174*, 715–732. [CrossRef]
30. Chen, C.; He, Y.; Bu, C.; Han, J.; Zhang, X. Quartic Bézier Curve Based Trajectory Generation for Autonomous Vehicles with Curvature and Velocity Constraints. In *2014 IEEE International Conference on Robotics and Automation (ICRA)*; IEEE: Piscataway, NJ, USA, 2014; pp. 6108–6113.
31. Takeshita, S.; Sadeghpour, A.; Malfait, W.J.; Konishi, A.; Otake, K.; Yoda, S. Formation of Nanofibrous Structure in Biopolymer Aerogel during Supercritical CO2 Processing: The Case of Chitosan Aerogel. *Biomacromolecules* **2019**, *20*, 2051–2057. [CrossRef]
32. AL Satai, H.; Zahra, M.M.A.; Rasool, Z.I.; Abd-Ali, R.S.; Pruncu, C.I. Bézier Curves-Based Optimal Trajectory Design for Multirotor UAVs with Any-Angle Pathfinding Algorithms. *Sensors* **2021**, *21*, 2460. [CrossRef]
33. Lebedev, I.; Lovskaya, D.; Mochalova, M.; Mitrofanov, I.; Menshutina, N. Cellular Automata Modeling of Three-Dimensional Chitosan-Based Aerogels Fiberous Structures with Bezier Curves. *Polymers* **2021**, *13*, 2511. [CrossRef] [PubMed]
34. Wolfram, S. Computation Theory of Cellular Automata. *Commun. Math. Phys.* **1984**, *96*, 15–57. [CrossRef]
35. Menshutina, N.V.; Kolnoochenko, A.V.; Katalevich, A.M. Structure Analysis and Modeling of Inorganic Aerogels. *Theor. Found. Chem. Eng.* **2014**, *48*, 320–324. [CrossRef]
36. Menshutina, N.; Lebedev, I.; Lebedev, E.; Paraskevopoulou, P.; Chriti, D.; Mitrofanov, I. A Cellular Automata Approach for the Modeling of a Polyamide and Carbon Aerogel Structure and Its Properties. *Gels* **2020**, *6*, 35. [CrossRef]
37. Shan, X.; Yuan, X.-F.; Chen, H. Kinetic Theory Representation of Hydrodynamics: A Way beyond the Navier–Stokes Equation. *J. Fluid Mech.* **2006**, *550*, 413. [CrossRef]
38. McDonough, J.M. Lectures in Elementary Fluid Dynamics: Physics, Mathematics and Applications. In *Mechanical Engineering Textbook Gallery*; University of Kentucky: Lexington, KY, USA, 2009; p. 164.
39. Wang, L.; Su, T. A Comprehensive Study on the Aerodynamic Characteristics of Electrically Controlled Rotor Using Lattice Boltzmann Method. *Aerospace* **2023**, *10*, 996. [CrossRef]
40. Qian, Y.H.; D'Humières, D.; Lallemand, P. Lattice BGK Models for Navier-Stokes Equation. *Europhys. Lett. (EPL)* **1992**, *17*, 479–484. [CrossRef]

Disclaimer/Publisher's Note: The statements, opinions and data contained in all publications are solely those of the individual author(s) and contributor(s) and not of MDPI and/or the editor(s). MDPI and/or the editor(s) disclaim responsibility for any injury to people or property resulting from any ideas, methods, instructions or products referred to in the content.

Article

A Machine-Learning-Based Approach to Critical Geometrical Feature Identification and Segmentation in Additive Manufacturing

Alexandre Staub [1,2,*], Lucas Brunner [3], Adriaan B. Spierings [2] and Konrad Wegener [1]

[1] Institute of Machine Tools and Manufacturing, ETH Zurich, CH-8092 Zurich, Switzerland
[2] Inspire, Innovation Centre for Additive Manufacturing Switzerland (ICAMS), CH-9014 St. Gallen, Switzerland
[3] Department of Computer Science, ETH Zurich, CH-8092 Zurich, Switzerland
* Correspondence: staub@inspire.ethz.ch

Citation: Staub, A.; Brunner, L.; Spierings, A.B.; Wegener, K. A Machine-Learning-Based Approach to Critical Geometrical Feature Identification and Segmentation in Additive Manufacturing. *Technologies* **2022**, *10*, 102. https://doi.org/10.3390/technologies10050102

Academic Editor: Eugene Wong

Received: 23 May 2022
Accepted: 13 September 2022
Published: 16 September 2022

Publisher's Note: MDPI stays neutral with regard to jurisdictional claims in published maps and institutional affiliations.

Copyright: © 2022 by the authors. Licensee MDPI, Basel, Switzerland. This article is an open access article distributed under the terms and conditions of the Creative Commons Attribution (CC BY) license (https://creativecommons.org/licenses/by/4.0/).

Abstract: Additive manufacturing (AM) processes offer a good opportunity to manufacture three-dimensional objects using various materials. However, many of the processes, notably laser Powder bed fusion, face limitations in manufacturing specific geometrical features due to their physical constraints, such as the thermal conductivity of the surrounding medium, the internal stresses, and the warpage or weight of the part being manufactured. This work investigates the opportunity to use machine learning algorithms in order to identify hard-to-manufacture geometrical features. The segmentation of these features from the main body of the part permits the application of different manufacturing strategies to improve the overall manufacturability. After selecting features that are particularly problematic during laser powder bed fusion using stainless steel, an algorithm is trained using simple geometries, which permits the identification of hard-to-manufacture features on new parts with a success rate of 88%, showing the potential of this approach.

Keywords: SLM; LPBF; scanning strategies; machine learning; segmentation

1. Introduction

Laser powder bed fusion has shown great potential to change the way metallic parts are produced, allowing a freedom in part design not achievable with conventional manufacturing. However, the process requires supporting structures in overhanging areas, and the user must follow certain strict design guidelines to successfully manufacture the final complex component. Such guidelines were presented by Klahn et al. [1] on function-driven modeling and by Leutenecker-Twelsiek et al. [2] on the requirements to be considered when designing for AM. In this respect, Calignano [3] presented guidelines for support optimization for both aluminum and titanium alloys, representing an additional manual step in the build preparation process, requiring know-how in the whole AM process chain.

The current state-of-the-art commercial machines support a wide variety of processing parameters and scanning strategies. However, the selected set of manufacturing parameters will be applied with no distinction to all layers during the manufacturing process, and to all geometrical features. Hence, a productive parameter set (i.e., with a high energy input) could be used for its obvious advantages. This kind of parameter set will nevertheless display difficulties in resolving fine details such as lattice structures, and will cause severe degradation in overhanging areas, even with the use of contour scans, as shown by Staub et al. [4].

This is mainly due to the low thermal conductivity of the surrounding powder within the powder bed, as supported by the experiments involving process monitoring and simulation presented by Dursun et al. [5]. However, different parameter sets, even if suboptimal for the manufacturing of the whole part, could be beneficial to reduce the use of supporting structures, improve the surface quality, and allow lower overhanging angles with respect to the build plate. Notably, Chen et al. [6] showed through simulations using the finite volume method and experimental results the opportunity to improve the overhanging surface quality by tuning the scanning parameters. Ultimately, such process improvements could improve the productivity of the process, and more generally of the process chain, as the removal of the supporting structure is a costly task in several AM processes, including LPBF. Hence, there is a necessity to allow for parameter modification within certain features of a 3D model.

So far, the different commercial solutions for the preparation of build jobs do not offer the option to freely allocate specific manufacturing strategies to individual 3D part features. Even though some basic geometries have been identified, such as overhang, this does not allow the segmentation of the part or the tailoring of manufacturing parameters at these specific locations. The efforts made by Siemens towards a path optimizer [7] led to improved surface quality on a single demonstrator, yet this is far from being a standard tool. The application of core–shell strategies, as presented by Niendorf et al. [8], is not considered sufficient, as some features require specific scanning strategies considering not only the type of a feature, but also its orientation regarding the build direction, its thickness, or its volume. On the other hand, other commercial solutions such as Aconity3D GmbH (Germany) rely on a completely open architecture and allow the user to modify every single vector of the scanning path for each layer. This is also not satisfying for the treatment of large 3D files, as the tailoring of the scanning path should be done by taking the 3D geometry into account rather than single layers, one after the other. The recent work by Chen et al. [9] highlighted the efficiency of a part-specific scanning strategy by showing reductions in residual stresses due to the optimized scanning path.

Druzgalski et al. [10] highlighted the necessity of software that pre-corrects build files by adapting the process parameters. The authors proposed the extraction of features based on the sliced data, by analyzing scanning vectors, extracting risky vectors, and adapting parameters to these vectors and the 3 subsequent layers. This tailoring is supported by previous process monitoring data and a database of pre-simulated scanning strategies.

Shi et al. [11] proposed an approach for a manufacturability analysis and feature identification based on the heat kernel signature. This approach considers the loss of heat over time of a known source, e.g., the geometry to be manufactured, completely independently of the AM process. This method can describe the shapes and topological characteristics of a feature through its heat diffusion potential. The work showed promising results yet did not elaborate on the necessary time for computing the heat kernel signatures on complex geometries.

Finally, the literature presents an extensive overview of possible methods of process parameter tuning for the improvement of specific critical features but lacks scientific inputs on the recognition and segmentation of these feature from a complete part. The aim of this work is to identify specific features, segment them from the original geometry, and allow for an automated treatment for further parametrization in the build processor. The developed method can take into account complex 3D geometries based on the STL file standard. Throughout this work, the part orientation is considered as fixed from the beginning. The output is a set of different features, which altogether represent the part to be manufactured. This heuristic approach focuses on LPBF-critical features but can be used for the improvement of any AM process.

2. Materials and Methods

2.1. Features Definition

Specific scanning strategies to improve the part quality for specific features have been successfully tested, notably in early the work by Clijsters [12], where the first scanning adaptation strategies were successfully implemented to manufacture free-overhanging structure with limited down-face surface degradation. In the most recent work by Illies [13], a thermal process simulation showed potential process parameter modifications to improve the quality of critical features. The lack of opportunity to recognize and segment the features from the core geometries has led these studies to become dead ends. To demonstrate the use of a novel approach for feature recognition and segmentation, different critical features must be defined, such as those proposed by Wegener et al. [14] and described in Table 1. The last column of Table 1 presents the parameter ranges in which the geometry is allowed to be modified for the machine learning approach presented in the following section. The criterion for criticality is defined for a standard grade of steel used in LPBF (e.g., SS316L, 17-4PH, maraging steel 300). Outside of the parameter range defined here, the features are either not critical or require a supporting structure, as an adaptation of the scanning strategy will not be enough to overcome the manufacturing difficulty. Even though this work is centered on limitations encountered while manufacturing steel in LPBF, it could be transferred to other metals and materials easily, as well as other processes, such as direct metal deposition. In such cases, the range of each feature, as well as its criterion for criticality, will need to be adapted.

Table 1. The features considered in this study and their parameters.

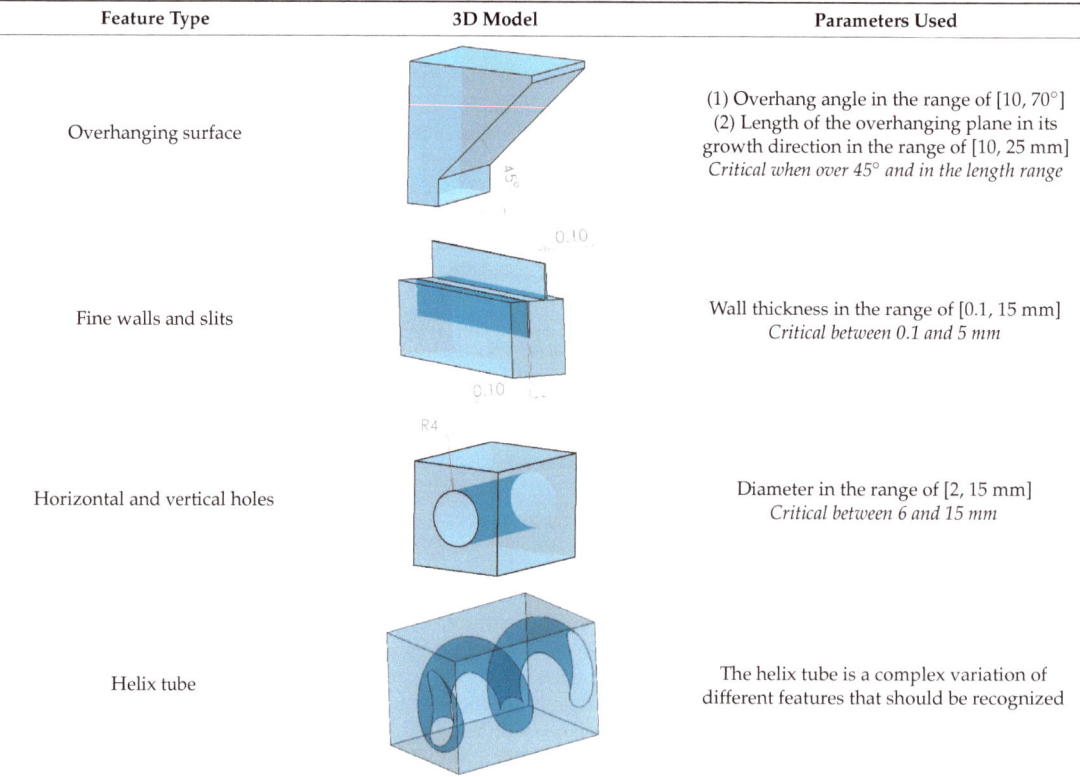

Feature Type	3D Model	Parameters Used
Overhanging surface		(1) Overhang angle in the range of [10, 70°] (2) Length of the overhanging plane in its growth direction in the range of [10, 25 mm] *Critical when over 45° and in the length range*
Fine walls and slits		Wall thickness in the range of [0.1, 15 mm] *Critical between 0.1 and 5 mm*
Horizontal and vertical holes		Diameter in the range of [2, 15 mm] *Critical between 6 and 15 mm*
Helix tube		The helix tube is a complex variation of different features that should be recognized

2.2. Suitable Approaches for 3D Feature Identification

For the detection of critical features, multiple methods can be applied. However, the focus of this work is on identifying the 3D features in the 3D geometry and reducing the computational resources needed to a minimum. Hence, layer-by-layer approaches, double slicing, and the like are eliminated due to their high complexity and excessive computing time.

As most of the geometries used in AM are mostly available in the STL file format (i.e., a surface description of the 3D part, discretized with triangles), classifying triangles as critical in the triangle mesh by only taking the triangle-normal into account could be an option. However, complex parts are composed of thousands to millions of triangles and cannot be handled properly. Such approaches will result in successful results for overhanging structures but will not allow the identification of more complex features, such as fine walls and helix tubes, as they deliver no information on the overall geometry and the feature.

A second approach assumes the existence of a database with known critical features. Iterative cloud point (ICP) registration, as proposed by Besl and McKay [15], is a technique used to match clouds of points by sequentially changing the orientations of both clouds until the points match together. The main applications for this method are the mapping of 3D scans on ideal objects and the recognition of the forms of known dimensions. This method could be used to match objects in a database onto a complex object. However, this method has limitations in the scalability of each feature. Indeed, depending on the overall size of the 3D model as well as the size of each feature, the scaling down or up of each feature of the database will be suboptimal in terms of the computing time, which will become exponential. In addition, pre-study tests showed that even at a constant scale, the identification of certain features would remain unsuccessful using ICP when presented with numerous features to be recognized.

Finally, it is proposed to study the opportunity to train a machine learning algorithm based on the simplest geometrical description of each individual feature (Table 1). This would allow for a simple definition of the features and an easily upgradable algorithm for new features depending on the material and process. Hence, materials that can be difficult to manufacture using LPBF such as Al- or Cu-based alloys will represent good candidates for the further implementation of this framework. A workflow combining the automatic generation of a database of geometries as per the definition in Table 1, the training of the machine learning algorithm based on the criticality of each feature, and the recognition and segmentation of an unknown geometry is presented in Figure 1. It is worth underlining that the goal of the recognition process is not to identify a 3D feature but rather a critical surface, which is further inflated in the direction of the part to apply segmentation.

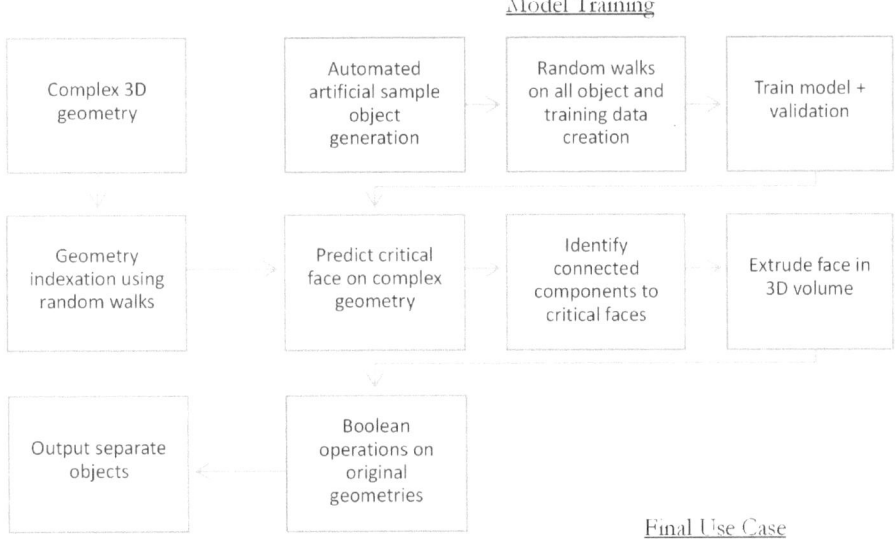

Figure 1. Workflow representation for machine learning training and final use case.

It is not possible to work with the 3D representation of a part as a geometry input for the machine learning approach, because there is no systematic representation of a 3D object (e.g., triangle mesh quality, orientation). Hence, it is necessary to be able to register each geometry by applying mathematical methods. These methods presented below will turn the 3D object into a repeatable and comparable description of the geometry. For this purpose, a random walk algorithm is implemented on the triangle mesh. A random walk is a mathematical process that consists of exploring a mathematical space by going from neighbor to neighbor in a succession of random steps in that direction. In that case, the algorithm will consider the triangle mesh as the mathematical space and will jump to a neighboring triangle randomly (out of the 3 adjacent triangles). To collect the local features of a triangle mesh, the dual graph of the geometry is then created, i.e., a graph composed of nodes, each representing a face of mathematical body, and the link (edge) between the nodes representing adjacent faces. This dual graph is such that every triangle face of the STL file becomes a node, and the neighboring faces are connected with an edge. Note that every node of the dual graph has three degrees (i.e., having three edges) in this construction method, as only watertight objects are considered. To predict if a face is critical, a random walk is started at its node in the dual graph. The random walk outputs the concatenation of the face features. Each face is represented through the 3D coordinates of its corners and the face-normal vector. The face-normal is used to capture angles between neighboring faces.

2.3. Database Generation

A machine learning algorithm needs a training data set, consisting of thousands of individual points, some of them being critical and some not, to ensure a good training and validation process. The training dataset for the machine learning algorithm consists of simple geometries, which can be easily labeled as critical or non-critical. To generate a diverse set of samples, the features are sampled uniformly within the range of valid parameters defined in Table 1. An exception is made for the helix tube, which should be recognized by the intersection of other features. Additional geometries are generated as intersections of the features defined in Table 1, such as intersections of 3-mm-high random walls for walls with thicknesses ranging from 0.2 to 1.5 mm and with lengths ranging between 3 and 10 mm. Such wall intersections contain up to 12 intersecting elements. The generation of this dataset is completely automatized and randomized, and as such the data are representative of a complex geometry, as they consist of thousands of single features. However, the 3D representations of these geometries are not generated, only their mathematical description, as presented in the section above. This dataset, thus, represents a compilation of simple features and a total of 0.5 million random walk starting positions. This dataset is further randomly subdivided into a learning set (90%) and independent validation set (10%). A single training sample consists of a label indicating whether a triangle is critical and 20 random walks with lengths of 20 nodes in the dual graph starting from this triangle.

3. Results and Discussion

3.1. Machine Learning Efficiency

A convolutional neural network (CNN) architecture is trained with the goal of reaching a critical feature detection rate higher than 85%. To train deeper networks, residual layers (Resnet), as proposed by Szegedy et al. [16], are implemented. As a loss function, the binary cross entropy function is used. To model the problem as a regression predicting how critical a face is, the mean squared error is used as a predictor. The modeling as a regression problem is especially suitable for an additional manual step, where the final user can adjust the critical region by manually tuning an additional parameter indicating the cutoff for the critical region. An illustration of a geometry with surfaces of different critical levels is shown in Figure 2, where red represents highly critical and blue represents non-critical, as per Table 1. To accomplish better training, the batch normalization approach proposed by Ioffe and Szegedy [17] is implemented.

Figure 2. Representation of different critical levels (not trained) for manual cut-off by the final user, where red is most critical and blue is non-critical.

To ensure repeatability between geometries, the 3D geometry is not used in the CNN, and only the registered transcription of it, namely the dual graph, is used. The CNN architecture takes the dual graph parameters of the indexed geometry as an input layer. It is composed of the different parameters of the triangle mesh (3D coordinates of each triangle corners, the triangle normal vector, and the angles between neighboring triangles), as well as the random walk parameters (number of walks, length of walks). The input layer in this case has dimensions of (20,20,12), according to the description of the random walk in Section 2.2.

Multiple convolutions are applied in successive steps to create a map of the activation, so as to identify where the feature is located. The multiple convolutions are successively applied on the following parameters of the dual graph to create an activation map (also called the feature map in machine learning language, not to be confused with the geometrical features of this work):

1. On the parameters of a single triangle (3D coordinates of each triangle corner, the triangle normal vector and the angles between neighboring triangles);
2. On the random walk length;
3. On the random walk number of walks;
4. On the feature space (created by the first convolutions).

The last dimension is reduced to 8 in step 2 and to 16 in step 3. The tensors from the walk length, the number of walks, and the feature dimension are reshaped between each layer. Finally, an operator with multiple dense layers is applied for the prediction. The continuous reduction of the dimensionality of the hidden space leads to a network achieving a final detection rate of 88% on the validation set presented above, without overfitting, calculated on the validation data set (known critical features to be detected). In addition, a geometry composed of all trained features and an untrained feature (the helix hole) is successfully recognized and segmented (Figure 3).

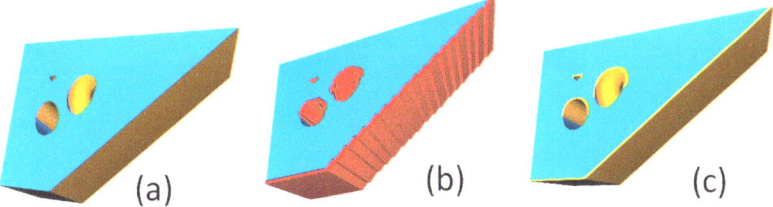

Figure 3. (**a**) The identified critical areas are shown in yellow. (**b**) The marching cubes before Boolean operation are shown in red. (**c**) The developed solution used for segmentation.

The number of random walks and their lengths are set as fixed parameters to avoid any lack of data for the training. However, this produces a relatively large dataset, which needs to be reduced to optimize the training time. Regarding the indexation of the geometry before the machine learning algorithm, an alternative to the random walk approach would be to eliminate the randomness by building a feature tree, as presented by Classen et al. [18]. Such a feature tree would consist of every face rooted at its corresponding triangle in the STL file. The root degree of the feature tree equals three, and all other degrees equal two (some faces might be included multiple times), as the third neighboring triangle can be found on the path to the root. This approach is worth evaluating for further research, as it might capture the local topology without producing a dataset that needs to be reduced afterwards for training.

3.2. Segmentation of the Goemetry and Final Output

In addition to the validation from the generated dataset, the trained neural network is used for the recognition of features on parts containing multiple critical features. A successful geometry with recognition and segmentation, including a free overhang, horizontal hole, helix tube, and long overhanging surface, is presented in Figure 3a,c. This allows us to draw conclusions on the efficiency of the strategy in recognizing critical features on parts containing multiple critical features.

The validation of several other parts shows good recognition for the different features, yet some artifacts (e.g., neighboring triangles in hole openings) or instances of misrecognition are experienced, as per Figure 4. In such cases, the triangles that are not part of the hole (the critical feature to be recognized) are marked as critical. As the machine learning is only effective at 88%, these false positives are typically the margin of error.

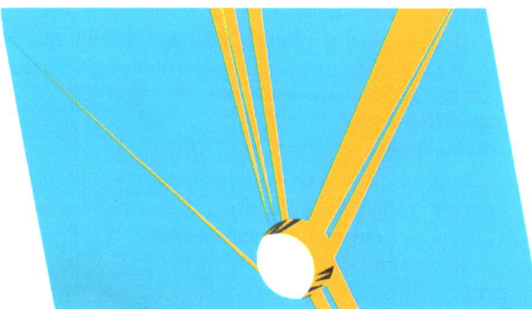

Figure 4. The neighboring triangles of a critical hole are recognized as critical (yellow triangles, outside the hole) when using single-pass recognition.

Because the random walk process is in fact random, the results for different evaluations of the same triangles and their probabilities of being critical can change, but the critical areas always have triangles being recognized as critical. The answer is to conduct multiple evaluations and use the maximum from all evaluations or to filter the results by redistributing the critical behavior among the neighboring triangles to solve such issues. In addition, further training on more geometries and the validation of the results by the user will help the algorithm to avoid such false positives. These false positives are not present anymore in Figure 3, showing the efficiency of the multiple evaluation strategy. The recognition time increases with the increasing numbers of triangles on the model's surface from a few seconds for the part presented in Figure 4 to several minutes for larger models, allowing for multiple evaluations at a reasonable cost.

The application of different process parameter settings to different features while knowing which faces are critical, as per Figure 3a, is insufficient, and the 3D objects are needed. This is accomplished by first generating a 3D volume from the critical surface and by applying Boolean operations between the original geometry and generated features. One object is the non-critical volume, i.e., the blue volume in Figure 3c, while the others are critical, which can be further treated in an AM build processor.

To generate the critical object from the critical surface, a ball cloud pivoting surface reconstruction algorithm, as presented by Bernardini et al. [19], is implemented. This allows the reconstruction of a triangle mesh from the mathematical description of the geometry by rotating a fictive "ball" around the first point selected until it reaches 3 points, then the triangle is formed and the ball is switched to the next point in the cloud of points. It is interesting to note that the marching cubes approach, as per the definition proposed by Lorensen and Cline [20], could represent a suitable and drastically faster solution, but smaller cubes result in gaps on the critical surface or require larger critical regions. As per Figure 3b, this method is not regular and not implementable for the clean resolution of each feature. A stair case effect is present on the overhanging surface, and the holes are poorly defined.

Finally, the implemented method eliminates the above problems by extruding each triangle individually if the number of triangles connected to each other is less than 100 triangles, while otherwise it uses a convex hull (smallest convex shape containing the number of points considered) algorithm-based method to fit the larger amount of critically connected triangles. This technical solution is fast and efficient. A complex structure such as the helix hole is easily modeled with a regular volume from the surface identified as critical, as per Figure 5.

Figure 5. Extracted volume of the critical surface for a helix tube.

4. Conclusions

The final results of this work can be summarized as follow:
- A database of basic and hard-to-manufacture geometries was randomly generated from a known description with limits;
- A CNN algorithm was implemented and fed with the database;
- The final tuned algorithm permitted a success rate of 88% of recognition of typical hard-to-manufacture features in LPBF;
- Untrained complex features (helix tube) can be successfully recognized;
- The segmentation of any feature can be successfully achieved.

Hence, the machine learning approach presented here is suitable for the complex problem of geometry segmentation. However, some process optimization approaches can still be used, notably for the topology indexation and the definition of the algorithm input space. The output of this classification of features is a collection of three-dimensional geometries representing the uncritical volume part, as well as the critical manufacturing features. For processes other than LPBF or other materials than standard steels in LPBF, a new definition of the features and their parameters (i.e., Table 1) is necessary. A new database could then be generated and the model could be retrained. This work is in favor of a richer file format for handling complex 3D models in the AM world, such as the 3mf format. This open-source, XML-based file format would be able to take into account the different features in a single file and let them be further treated by an AM build processor in an automated way. The further development of build processors is necessary to adapt to this new way of handling complex 3D parts. Further research on this topic should include processing capabilities to assign specific scanning strategies to each feature and more complex features recognition tools, such as lattice structures, on which tailored scanning strategies could also be proposed.

Author Contributions: Conceptualization, A.S.; methodology, A.S. and L.B.; software and investigations, L.B.; writing—original draft preparation and editing, A.S.; writing—review, A.B.S. and K.W. All authors have read and agreed to the published version of the manuscript.

Funding: This research received no external funding.

Data Availability Statement: Not applicable.

Conflicts of Interest: The authors declare no conflict of interest.

References

1. Klahn, C.; Leutenecker, B.; Meboldt, M. Design Strategies for the Process of Additive Manufacturing. *Procedia CIRP* **2015**, *36*, 230–235. [CrossRef]
2. Leutenecker-Twelsiek, B.; Klahn, C.; Meboldt, M. Considering Part Orientation in Design for Additive Manufacturing. *Procedia CIRP* **2016**, *50*, 408–413. [CrossRef]
3. Calignano, F. Design optimization of supports for overhanging structures in aluminum and titanium alloys by selective laser melting. *Mater. Des.* **2014**, *64*, 203–213. [CrossRef]
4. Staub, A.; Spierings, A.B.; Wegener, K. Selective Laser Melting at High Laser Intensity: Overhang Surface Characterization and Optimization. In Proceedings of the Direct Digital Manufacturing Conference, Berlin, Germany, 14–15 March 2018.
5. Dursun, G.; Pehlivanogullari, B.; Sen, C.; Orhangul, A. An investigation upon overhang zones by using finite element modelling and in-situ monitoring systems. *Procedia CIRP* **2020**, *93*, 1253–1258. [CrossRef]
6. Chen, H.; Gu, D.; Xiong, J.; Xia, M. Improving additive manufacturing processability of hard-to-process overhanging structure by selective laser melting. *J. Mater. Process. Technol.* **2017**, *250*, 99–108. [CrossRef]
7. Finberg, N. Path Optimization for Laser Powder Bed Fusion Printing. Available online: https://blogs.sw.siemens.com/thought-leadership/2020/12/17/path-optimization-for-laser-powder-bed-fusion-printing/ (accessed on 11 February 2021).
8. Niendorf, T.; Leuders, S.; Riemer, A.; Richard, H.A.; Tröster, T.; Schwarze, D. Highly Anisotropic Steel Processed by Selective Laser Melting. *Metall. Mater. Trans. B* **2013**, *44*, 794–796. [CrossRef]
9. Chen, Q.; Liu, J.; Liang, X.; To, A.C. A level-set based continuous scanning path optimization method for reducing residual stress and deformation in metal additive manufacturing. *Comput. Methods Appl. Mech. Eng.* **2020**, *360*, 112719. [CrossRef]
10. Druzgalski, C.L.; Ashby, A.; Guss, G.; King, W.E.; Roehling, T.T.; Matthews, M.J. Process optimization of complex geometries using feed forward control for laser powder bed fusion additive manufacturing. *Addit. Manuf.* **2020**, *34*, 101169. [CrossRef]
11. Shi, Y.; Zhang, Y.; Baek, S.; De Backer, W.; Harik, R. Manufacturability analysis for additive manufacturing using a novel feature recognition technique. *Comput. Aided Des. Appl.* **2018**, *15*, 941–952. [CrossRef]
12. Clijsters, S. *Development of a Smart Selective Laser Melting Process*; KU Leuven: Leuven, Belgium, 2017.
13. Illies, O. *Simulationsbasierte Thermische Analyse zur Anpassung der Hatching-Strategie beim Selektiven Laserstrahlschmelzen*; Universität Bremen: Bremen, Germany, 2020.
14. Wegener, K.; Spierings, A.; Staub, A. Bioinspired intelligent SLM cell. *Procedia CIRP* **2020**, *88*, 624–629. [CrossRef]
15. Besl, P.J.; McKay, N.D. A method for registration of 3-D shapes. *IEEE Trans. Pattern Anal. Mach. Intell.* **1992**, *14*, 239–256. [CrossRef]
16. Szegedy, C.; Ioffe, S.; Vanhoucke, V.; Alemi, A.A. Inception-v4, Inception-ResNet and the Impact of Residual Connections on Learning. In Proceedings of the Thirty-First AAAI Conference on Artificial Intelligence (AAAI-17), San Francisco, CA, USA, 4–9 February 2017.
17. Ioffe, S.; Szegedy, C. Batch Normalization: Accelerating Deep Network Training by Reducing Internal Covariate Shift. *Proc. Mach. Learn. Res.* **2015**, *37*, 448–456.
18. Classen, A.; Heymans, P.; Laney, R.; Nuseibeh, B.; Tun, T.T. On the structure of problem variability: From feature diagrams to problem frames. In Proceedings of the International workshop on Variability Modeling of Software-intensive Systems, Limerick, Ireland, 16–18 January 2007; pp. 109–118.
19. Bernardini, F.; Mittleman, J.; Rushmeier, H.; Silva, C.; Taubin, G. The ball-pivoting algorithm for surface reconstruction. *IEEE Trans. Vis. Comput. Graph.* **1999**, *5*, 349–359. [CrossRef]
20. Lorensen, W.E.; Cline, H.E. Marching cubes: A high resolution 3D surface construction algorithm. In Proceedings of the Proceedings of the 14th Annual Conference on Computer Graphics and Interactive Techniques, Anaheim, CA, USA, 27–31 July 1987; pp. 163–169.

Review

Surface Quality of Metal Parts Produced by Laser Powder Bed Fusion: Ion Polishing in Gas-Discharge Plasma Proposal

Alexander S. Metel, Sergey N. Grigoriev, Tatiana V. Tarasova, Yury A. Melnik, Marina A. Volosova, Anna A. Okunkova *, Pavel A. Podrabinnik and Enver S. Mustafaev

Department of High-Efficiency Processing Technologies, Moscow State University of Technology "STANKIN", Vadkovsky per. 1, 127055 Moscow, Russia; a.metel@stankin.ru (A.S.M.); s.grigoriev@stankin.ru (S.N.G.); tarasova952@mail.ru (T.V.T.); yu.melnik@stankin.ru (Y.A.M.); m.volosova@stankin.ru (M.A.V.); p.podrabinnik@stankin.ru (P.A.P.); e.mustafaev@stankin.ru (E.S.M.)
* Correspondence: a.okunkova@stankin.ru; Tel.: +7-909-913-1207

Abstract: Additive manufacturing has evolved over the past decades into a technology that provides freedom of design through the ability to produce complex-shaped solid structures, reducing the operational time and material volumes in manufacturing significantly. However, the surface of parts manufactured by the additive method remains now extremely rough. The current trend of expanding the industrial application of additive manufacturing is researching surface roughness and finishing. Moreover, the limited choice of materials suitable for additive manufacturing does not satisfy the diverse design requirements, necessitating additional coatings deposition. Requirements for surface treatment and coating deposition technology depend on the intended use of the parts, their material, and technology. In most cases, they cannot be determined based on existing knowledge and experience. It determines the scientific relevance of the analytical research and development of scientific and technological principles of finishing parts obtained by laser additive manufacturing and functional coating deposition. There is a scientific novelty of analytical research that proposes gas-discharge plasma processing for finishing laser additive manufactured parts and technological principles development including three processing stages—explosive ablation, polishing with a concentrated beam of fast neutral argon atoms, and coating deposition—for the first time.

Keywords: accelerated ions; explosive ablation; fast atoms; glow discharge; surface sputtering

1. Introduction

Additive manufacturing allows producing complex solid structures by direct material deposition that fuses the deposed layer's material with the substrate or previous layer [1]. Most modern research aims to unveil and improve the exploitation and mechanical properties of the steels [2], alloys [3,4], and even oxide ceramics [5,6]. One of the actual directions is related to the work with nanoscaled powders [7]. The approach of the direct growing of solids allows reducing timing on operational steps and the volume of used material during manufacturing [8].

For the continuous development of the additive approach and its application in industry, it is necessary to eliminate its main disadvantage related to the low quality of the produced surface that remains extremely rough with the presence of the unmelted powder granules. It requires finishing operation without losing the exploitation properties of the operational surfaces [9] that can negatively influence the service life of a product [10], especially regarding its responsible applications as a part of aircraft or gas-turbine engine [11].

The choice of materials suitable for additive manufacturing is limited at the technological level, and, in some cases, it does not satisfy the design and technology requirements—wear resistance [12], microstructure [13], and mechanical properties [14]. One of the additional solutions can be in coating deposition [15], which is especially actual in the case of using modern nanoscale multilayered coatings [16] and significantly improves the service life of the product [17]. The coating microhardness and wear resistance can then allow the part's operation in extreme conditions [18].

The surface polishing and coating deposition requirements depend on the intended operational conditions, material, and manufacturing approach in the part production. In most cases, it can be determined by developing the scientific and technological principles of post-processing based the analytical and experimental research.

The authors' previous experience showed that most of the research aims to overview residual stress [19,20], analyze the mechanical and physicochemical properties [21,22], and related microstructure evolution [23], mostly for high-entropy [24], intermetallic, titanium alloys, and steels [25,26]. The first two types of alloys were under particular interest in the context of materials development when titanium alloy, as other metal alloys and steels using in additive manufacturing, showed and proved their engineering prospects in particular applications [27]. Simultaneously, alloys based on titanium with a high strength-to-weight ratio and excellent corrosion resistance show their superior technological and exploitation properties for some critical applications, including aviation and biomechanical industries [28]. It should be noted that titanium has its particular properties among heat-resistance and anti-corrosion behavior (due to formed TiO_2 thin film) that make it difficult to process by mechanical milling and lathing [29]. However, it is easily welded and melted with a laser beam in the inert atmosphere, and its final properties depend mainly on the pureness of precursors. The field of its application will only grow in the following decades.

At the same time, the questions of finishing and developing the principles of disruptive technology to improve the surface quality of parts produced by laser additive manufacturing stay unveiled.

The study presents an overview of the research domain related to the main trends in the application of the additively manufactured parts by a laser, which are used and proposed by several research groups finishing operation methods with their advantages and disadvantages in application to the complex geometry parts and simplicity of used equipment, development of the innovative approach, and their technological principals in the finishing of additively produced parts using ion polishing in gas-discharge plasma.

This analytical research's scientific novelty is determined by an overview of surface finishing methods and their influence on the functionality and operation ability of the product responsible surfaces, which are still not fully and completely overviews, by the proposal of using ion polishing in gas-discharge plasma for the finishing and theoretical development of three stages of detailed post-processing for the first time.

The study's scientific tasks are as follows:

- Determining particularities in the surface quality problem (surface properties and roughness parameters) of metal parts produced by additive manufacturing methods from various metallic alloys—steels, cobalt, nickel, aluminum, and titanium alloys in the context of airspace industry application,
- Classification existed methods to improve exploitation properties and surface quality of the parts produced by laser additive manufacturing,
- Analyses of the last achievement in implementing finishing technologies depending on its nature—thermal, electrochemical, mechanical, and combined methods,
- Determining finishing methods that were not covered by the experimental research for additively manufactured parts but have a potentially valuable impact on surface quality,
- Developing the technological principles of ion polishing in gas-discharge plasma for finishing laser additively manufactured parts to improve their surface quality in the context of resistance to abrasive wear.

It should be noted that this paper aimed to review the state of science on the existing post-processing methods and last achievements in the field. The proposed idea of the ion-polishing method for additively manufactured parts is known and was never proposed before. That can be proved by the conducted overview of the research domain.

2. Problem Statement
2.1. Prospects and Surface Quality Problem

The use of laser additive manufacturing technologies has great potential for companies manufacturing products in the aerospace industry [30], especially in the production of parts with complex geometries [31]. If we look at the parts for aviation purposes, manufactured over the past decade, then complex-profile parts stand out against others' background [32], and their use is growing every year [33]. Besides, many problems arise when these parts are manufactured using traditional processing methods [34]. One of the critical issues is the cost reduction by simplifying the production and manufacturing process. The solution of both functional and aesthetic problems can be carried out using additive manufacturing technologies (AM). Structural parts consisting of several components can be grown layer by layer as a single solid part using AM technology. It is an important issue in terms of cost and time. Fewer parts can significantly reduce assembly and other costs. It is possible to design parts that can be easily produced with AM technologies when they are extremely difficult to shape by traditional machining methods—for example, thin-walled complex-shaped parts with complex internal surfaces.

Thanks to the technological capabilities of additive manufacturing, designers can optimize the strength-to-weight ratio of a part [35] and minimize the use of consumables and provide a reduction in processing costs [36], waste disposal, material transportation costs, reduction of storage costs for raw materials, and for the direct production of final products, which reduces overall production costs [37]. There is no necessity for additive manufactured parts to buy various types of workpieces but various granulometry powder [38].

It should be noted that using a laser beam source or another concentrated energy flow for melting or sintering powders gives the advantages that are not available for other printing technologies:

- The ability to work with metal alloys, polymers, ceramics, and metal alloys or polymers reinforced with ceramics [6,7,39–41],
- The ability to produce a ready part with high operational properties and service life [4,19,20,31],
- The ability to produce high-precision parts for the needs of medicine, jewelry, and even watch production [42].

The application of a laser beam expander or profiler allows even significantly improving the efficiency of production (up to 30%) [43–45] and extending the potential additive manufacturing market.

Many companies that are leaders in the aviation industry have begun production testing of various aircraft parts, taking advantage of the additive manufacturing technology. Boeing manufactured various thermoplastic parts using commercially available laser sintering technologies for the 737, 747, 777, and 787 commercial aircrafts [38,46]. Boeing was estimated to have earned an estimated \$3 million in revenue for every produced 787 Dreamliner aircraft [47]. Although some of the produced parts are complex, manufacturing processes are accomplished by eliminating production constraints in a shorter timeframe, at a lower cost [48], and with the required performance [49].

GE Aviation is currently successfully manufacturing fuel injector parts for LEAP engines ("Leading Edge Aviation Propulsion" engines produced by CFM International, Cincinnati, OH, USA) using selective laser melting. Each LEAP engine is equipped with 19 fuel injectors. All engine fuel supply components have been land-tested and approved for use in civil aircraft [50].

However, with all the attractiveness of additive technologies, already available examples of their successful application and prospects, they should not be idealized. All world manufacturers of equipment for AM are currently working on many shortcomings. In particular, such disadvantages include the porosity of the formed workpiece and its increased roughness, which must be brought to the required technical level by machining operations, mainly by milling.

An illustration of it is presented in Figure 1, which presents the average values of microroughness height with a thickness of the sintered layer of 50 μm (in terms of R_a roughness parameter) for the surface of products made from various powder alloys on machines for laser powder bed fusion (LPBF) or selective laser melting (SLM) [51–56], laser cladding [57–59] or even cold spray [60]. It can be seen that the formed surface roughness is far from the quality parameters of the surface layer, which must be satisfied by critical engineering products. Even optimizing the melting or sintering parameters will not exclude the need for subsequent machining.

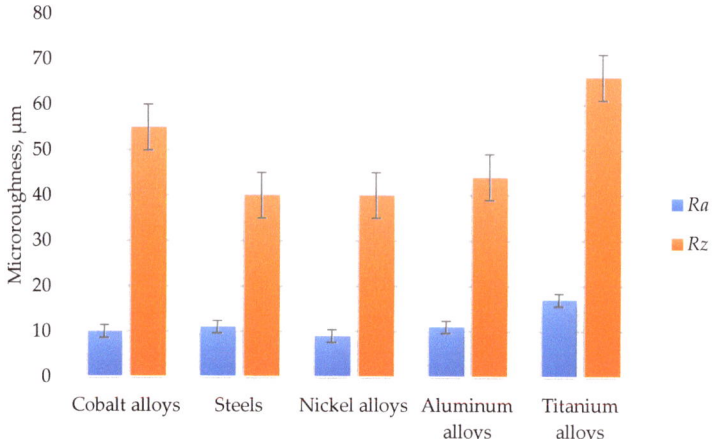

Figure 1. Arithmetic mean deviation (R_a) and ten-point height (R_z) of the product surfaces produced by laser powder bed fusion (selective laser melting) for various types of metals and alloys.

The products produced by laser additive manufacturing has particular geometry [61,62] that can be characterized as having an "edge effect" related to the used hatching strategy [63]. It mostly depends on the character of heat distribution through the layers and strongly depends on the main controlled process factors of laser power, scanning speed, powder layer thickness, and hatching parameters.

The edge defect can be identified by the roughness diagram [64] in addition to the used optical monitoring methods [65]. At the same time, the experiments showed that the used scan strategy influences its pronouncedness [66].

Another problem of the parts produced by laser additive manufacturing is an effect of aliasing [67] that can be especially pronounced for tiny objects [68] and depends on the layer thickness and granulometry of used granules [69], and thermal deformation [70]. A larger thickness of the layer increases the "aliasing effect" (build angle) when a larger diameter of the used powder granules increases the layer's thickness.

Several post-processing methods based on the varied nature of the material destruction that can be applied for the parts obtained by the laser additive manufacturing methods are presented in Figure 2.

The methods can be divided by influencing the exploitation properties and surface quality, including roughness parameters and resistance to abrasive wear. Among the methods influencing the exploitation properties of the parts produced by laser additive manufacturing, quenching and tempering allow to reduce the grain of the metal alloy and redistribute the residual stresses in the layers of the product, making the structure more monolithic than layered. However, such techniques do not always lead to an increase in the service life of their product in friction pairs and strongly depend on the material and factors of product growth. Hot isotactic pressing stays aside with its proven improvement of material ductility (by ≈20–25% for titanium alloys) but requires developing unique forms for a complex geometry product.

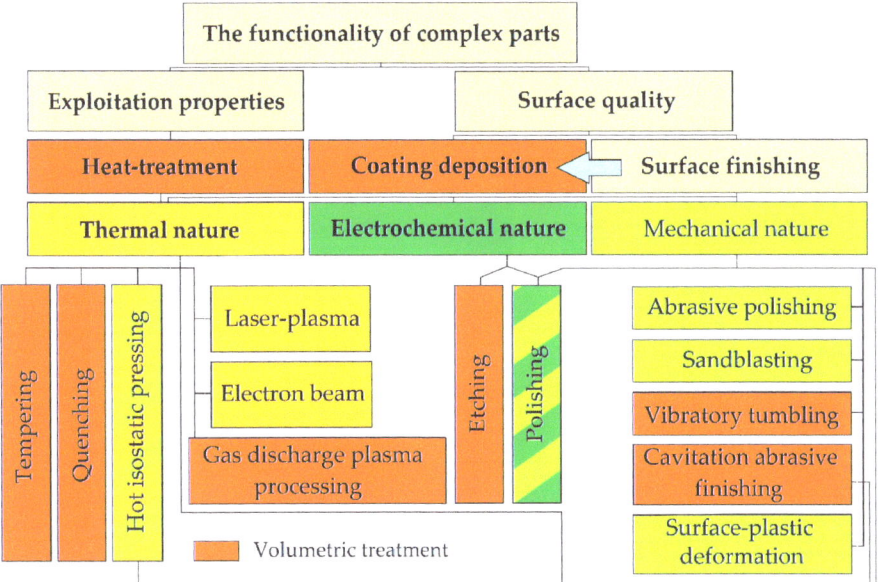

Figure 2. Technologies' classification improving the exploitation properties and surface quality of the complex parts produced by additive manufacturing based on its nature.

The use of a coating to improve the surface layer's performance properties is a known technique and can be carried out both chemically and by plasma vapor deposition. Nevertheless, a coating that allows increasing the product's service life several times requires preliminary cleaning of the product from unmelted granules and reducing the layer's waviness due to the use of various techniques.

Methods for improving the product's roughness parameters can be divided into thermal, electrochemical, and mechanical, depending on the nature of the destruction of surface irregularities. The latter group of methods is most widespread due to their relative cheapness and availability in any production.

Furthermore, the latest quantitative achievements in the field of application of these methods concerning products produced by additive manufacturing are considered, the possibility of their application for parts with complex geometry are assessed, those methods that researchers have not experimentally approbated are identified, and alternative technology principles for improving the surface properties and roughness parameters are developed.

2.2. Research Methodology

The analytical research was conducted, taking into account the basic principles of electrical and thermal physics, physics of plasma and concentrated energy fluxes, and available theoretical and practical data related to the research subject of post-processing methods for the metal parts produced by additive manufacturing methods based on a laser.

The research object is a complex geometry part produced by laser additive manufacturing from the metallic alloy powder with a powder diameter of 20–80 μm for the airspace industry. That has requirements of arithmetic mean deviation (R_a) of less than 3.2 μm for the parts with an overall size in plane less than 20 mm and less than 6.3 μm for the parts with an overall size in plane less than 200 mm for working in the conditions of abrasive wear in friction pair.

3. Analyses of Surface Finishing Methods

3.1. Mechanical Methods

Mechanical polishing and sandblasting is widely used to reduce the porosity and surface roughness of metals that allow one to obtain both the developed surface morphology of the metal and significantly reduce surface roughness depending on the used abrasive, reducing residual stress and fatigue properties. Mechanical abrasive polishing for complex geometry parts requires handling to control geometry over all the part but lapping in places. Sandblasting is a universal technique that removes all unmelted granules and unifies the geometry's waviness in a brief period (5–10 min for each part with an overall size of less than 20 mm) without requiring a unique tool. Using any abrasives during polishing or sandblasting has several disadvantages: abrasive particles remaining on the part surface and specific geometry after processing, such as scratches in the direction of movement of the abrasive and particle impact craters (cavities).

Unevenly distributed residual stresses are one of the actual problems for laser additive manufactured parts with the maximum σ in the perpendicular direction to the solid growing direction of 205 ± 15 MPa for AlSi10Mg alloy [71]. For Inconel 718, improved roughness increased the fatigue properties at 650 °C by ≈50% [72]. A combined application of hot isostatic pressing, sandblasting, abrasive polishing, and chemical etching for Ti-6Al-4V improves the yield strength and ultimate tensile strength (UTS) by ≈2–3 times [73].

Vibration tumbling is more suitable for mass production and allows complex geometry treatment by tumbling bodies such as ceramics prisms or cork-like clean chips in water medium (for dust binding) that improve the surface roughness parameter R_a by ≈3–4 times, abrasive wear resistance by ≈18–20% for chrome–nickel anti-corrosion steels [74] (Figures 3–5). For high carbon steels, the improved wear resistance is explained by the increased surface hardness caused by the stress-induced transformation of residual austenite into untempered martensite during wear, while maintaining acceptable toughness in the subsurface layers prevents brittle cracking [75]. The hardness of tumbling bodies determines the character of formed cavities and surface properties—the use of metal tumbling bodies results in the surface's strengthening by pressure, when the most significant reduction of roughness parameters and mass loss occurs in the first minutes of processing [76]. It should be noted that dry tumbling is dangerous for the human respiratory system and is strongly not recommended for any production.

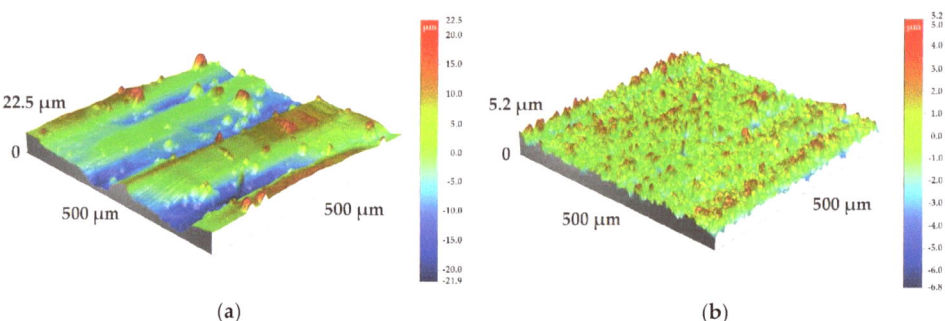

Figure 3. Topology of the laser additively produced part surface made of 12Cr18Ni9Ti austenite chrome–nickel stainless steel (analog of AISI 321) powder with granules of ≈20 µm (by a Dektak XT stylus profilometer (Bruker Nano, Inc., Billerica, MA, USA) with a vertical accuracy of 5 Å (0.5 nm) and a tip radius of 12.5 µm): (**a**) after production; (**b**) after vibratory tumbling.

Ultrasonic cavitation abrasive finishing allows complex treatment of the part and based on the combined thermochemical and mechanical nature of mechanolysis (thermodynamic cavitation mechanism in the homogeneous liquids), sound and radiation pressure, acoustic streams, and sound capillary effect and abrasives' action. The cavitation intensity is determined based on L. D. Landau and E. M. Lifshitz's theory of fluctuations [77,78]. The cavitation is realized by both fluctuations of bulk and a vapor bubble with a similar probability proven by the relationship between the tensile strength of liquids and the bulk fluctuations [79]. The bubble system's energy is emitted in the form of an acoustic wave and is dissipated by viscosity. The local energy of a transient bubble follows a step function in time, being nearly conserved for most of each cycle of oscillation but decreasing rapidly and significantly at bubble inception and the end of collapse due to the emission of steep pressure waves or shock waves [80].

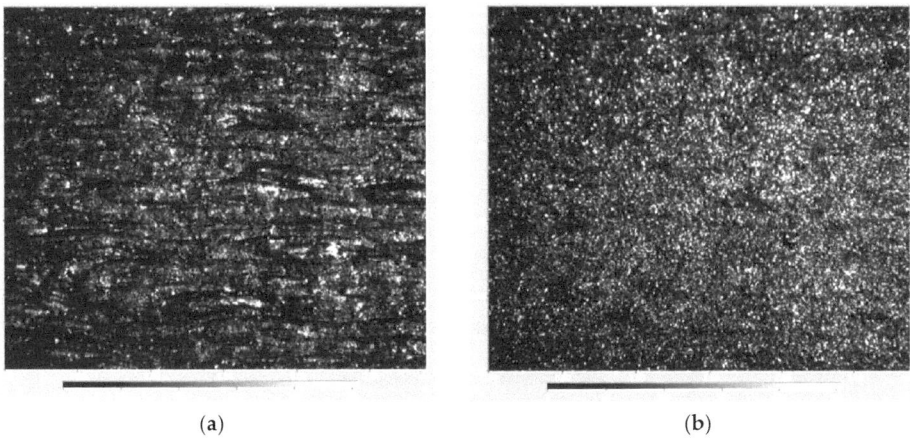

(a) (b)

Figure 4. Topology of the laser additively produced part surface made of 12Cr18Ni9Ti austenite chrome–nickel stainless steel (analog of AISI 321) powder with granules of ≈20 μm (by a MikroCAD-lite 3D measuring system, GFMesstechnik, Notzingen, Germany): (**a**) after production; (**b**) after vibratory tumbling.

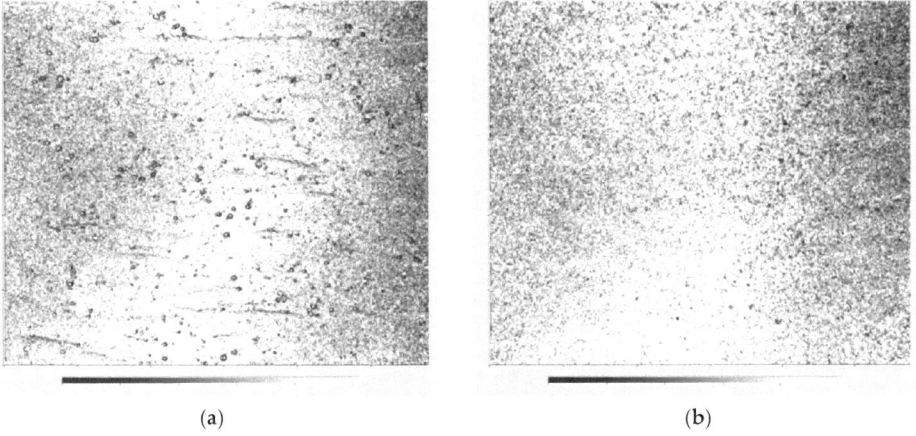

(a) (b)

Figure 5. Three-dimensional (3D) presentation of the laser additively produced part surface made of 12Cr18Ni9Ti austenite chrome–nickel stainless steel (analog of AISI 321) powder with granules of ≈20 μm (by a MikroCAD-lite 3D measuring system, GFMesstechnik, Notzingen, Germany): (**a**) after production; (**b**) after vibratory tumbling.

The mechanism of destruction of the structure material as a whole can be represented according to the Kornfeld–Suvorov hypothesis (the hypothesis was proposed by soviet scientists M.O. Kornfeld and L.I. Suvorov in the 1940s) as follows (Figure 6): when the cumulative microjet upon the collapse of a cavitation bubble upon collision with a solid body exhibits the behavior of a solid body and describes the laws for a solid body. When a collapsing bubble is exposed to a streamlined surface, produced by cumulative streams (the process is impulsive and non-stationary), surface deformation occurs, and because of a decrease in the fatigue strength of the material, chipping and knocking out of individual particles occurs as well. From this moment and further, the intensity of erosion increases sharply, and hydrodynamic factors that determine destruction within the framework of the shock wave theory begin to play a significant role: vibration, acoustic radiation, and in certain cases, chemical and corrosion factors, etc. At the initial stage (plastic deformation), the process is possible to describe by solving the hydrodynamic penetration problem taking into account the strength of the material.

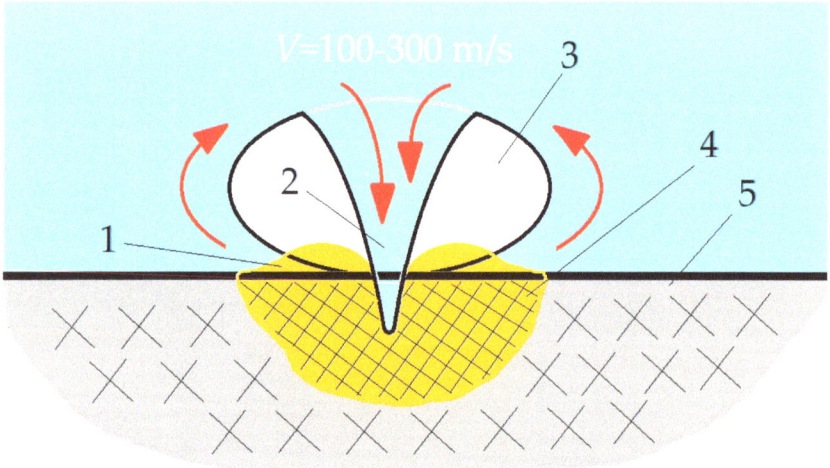

Figure 6. Scheme of cavitation (erosion) wear, where (**1**) is extruded by cumulative microjet material at the stage of plastic deformation; (**2**) is a cumulative microjet; (**3**) is a collapsing bubble; (**4**) is plastic deformation zone; and (**5**) is surface to be machined.

Experiments have shown that all materials, even the strongest and hardest, are subject to cavitation destruction [81]. In this regard, the question of what characteristics of a material determine its erosion resistance has been repeatedly investigated. E.P. Georgievskaya [82] proposed an empirical formula for defining the concept of deformation energy introduced by Thiruvengadam [83,84] as a parameter characterizing the resistance of a material to erosion [85]:

$$S_e = (T + Y)^{\frac{\varepsilon}{2}} \quad (1)$$

where S_e is a calculated value of deformation energy; T is temporary resistance; Y is the yield point; and ε is relative extension.

The energy of deformation is the power of absorption of energy per unit volume of the metal before the moment of fracture formation. The development of cavitation erosion is influenced by the conditions of flow around surfaces to a large extent: speed, pressure, pressure gradient, temperature. It was found that at a fixed value of the cavitation number, the erosion rate strongly depends on the flow rate. Activation of solid and liquid systems, leading to a change in their physical and chemical properties, reactivity, defective (impurity) structure, etc., can be carried out by various external influences: weak and strong. Such effects include, in particular, mechanical, magnetic, ultrasonic treatment, radiation exposure (for example, irradiation with gamma quanta and ion beams), as well as heat treatment. Activation methods can be subdivided into methods that destroy samples as a whole (dispersion) and do not destroy but change only the defective structure. Studies of hydromechanical treatment of water (as a sufficiently strong effect) have shown that water's subsequent activity is manifested both at the macroscale and at the microlevels (at the molecular and submolecular level).

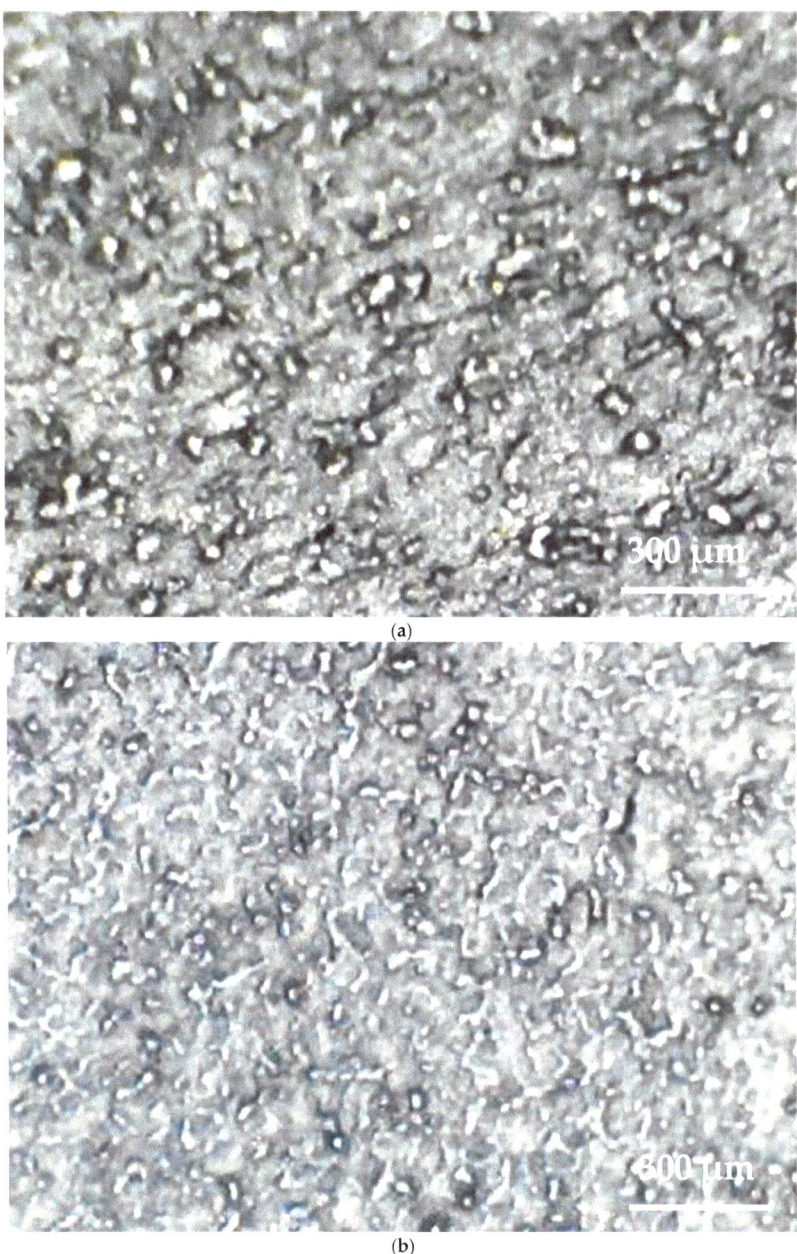

Figure 7. Microphotograph of laser additively produced part surface made of 20Cr13 corrosion-resistant steel of the martensitic class (analog of AISI 420) powder with granules of ≈20 μm (optical microscopy, ×300): (**a**) after production; (**b**) after ultrasonic cavitation abrasive finishing.

The essence of the hydrodynamic impact can be summarized by the action of two mechanisms: the propagation of shock waves near the collapsing cavitation microbubble and the shock action of cumulative microstructures in the case of the asymmetric collapse of cavitation microbubbles. Moreover, in this context, the method of obtaining cavitation microbubbles does not matter. These main mechanisms are accompanied by an increase in temperature and pressure near the bubble, making the local area around it a unique reactor for carrying out various reactions and processes.

The phenomenon under consideration underlies the specific properties and related phenomena occurring in water subjected to mechanical (hydrodynamic) action. The modified (or, as it is called, activated) water resulting from hydromechanical treatment can intensify many technological processes by about 30%. Here, by the term "activated" water, we mean the generally accepted concept of an active medium, that is, a substance in which the distribution of particles (atoms, molecules, ions) over energy states is not equilibrium, and at least one pair of energy levels undergoes population inversion. The activation of solid and liquid systems, leading to a change in their physical and chemical properties, reactivity, defective (impurity) structure, etc., can be carried out by various external influences: weak and strong. Such influences include, in particular, mechanical, magnetic, ultrasonic treatment, radiation exposure (for example, irradiation with gamma quanta and ion beams), as well as heat treatment. Activation methods can be subdivided into methods that destroy samples as a whole (dispersion) and do not destroy but change only the defective structure.

Studies of hydromechanical treatment of water (as a rather strong effect) have shown that water's subsequent activity is manifested both at the macroscale and at the microlevels (at the molecular and submolecular level) (Figure 7) [86]. It indicates that the bubble's symmetric collapse creates high-intensity fields of pressure of 5–10,000 atm. and temperatures up to 2000 °C [87], which are related to the ultrasonic waves in the working area and were even reported for finishing by a lapping tape [88]. In the case of asymmetric bubble collapse, near the surface, the collapse pattern changes significantly—the collapse occurs with the formation of a high-speed cumulative microjet, and the formation mechanism is described in sufficient detail in the literature. In the asymmetric bubble collapse far from the surface, microjets can be formed far from the surface and hit one another. It should be noted that at the moment of collapse of the bubbles' system near the surface, the formation of streams between adjacent bubbles is equally probable.

Hydrogen bonds' presence and development largely explain the unique properties and paradoxes of liquid water. The hydrogen bond has a cooperative nature in H_2O molecules and largely determines the water structure under various external conditions. Hydrogen bonds are about ten times stronger than intermolecular interactions typical for most other liquids. In the general case, it can be said that the interactions of a large number of molecules, ensembles of molecules, the organization of a particular structure that determines the properties of water and, accordingly, its reactivity, are determined by the collective forces of Van der Waals [89]. These forces are known as dispersive, long-range forces. They cover regions above 1000 Å and determine the stability of a particular structure, physical sorption, etc. The relaxation time for a number of processes in water at 20 °C, t of 10^{-11}–10^{-13} s. The processes of energy transfer and recharge with the participation of water molecules, noble and active gases, and even the dissociation of water molecules become possible because the duration of the final stage of bubble collapse is about 10^{-9}–10^{-8} s. Thus, under the action of hydrodynamic cavitation as a strong effect, the decomposition (mechanolysis) of water occurs. An excited water molecule can dissociate along with radiation and dissipation of excess energy into heat:

$$H_2O^* \rightarrow H\uparrow + OH^*, \qquad (2)$$

$$OH^* \rightarrow OH + h\nu. \qquad (3)$$

As a result of cavitation action, the concentration of O_2 increases during mechanochemical reactions of the following type due to the mechanolysis of water on H and OH:

$$\dot{O}H + \dot{O}H \rightarrow H_2O_2, \tag{4}$$

$$\dot{O}H + H_2O_2 \rightarrow HO_2 + H_2O, \tag{5}$$

$$\dot{O}H + HO_2 \rightarrow H_2O + O_2 \uparrow. \tag{6}$$

At the same time, there is a change in the structure of water with the formation of free hydrogen bonds, which determines its increased activity and reagent ability. In the case of aqueous systems, activation and the mechanolysis of water consist of changing the degree of uniformity of distribution of impurities over the volume of the system, aggregation, and disaggregation (dispersion) of impurities, as well as in a change in their active state. The most important feature of water systems, in particular, is the heterogeneity of impurities, which can change significantly during cavitation. Under the influence of cavitation in an aqueous solution containing inert and active gases, various chemical reactions are possible.

Cavitation treatment (as opposed to, for example, magnetization, exposure to various fields of electromagnetic origin, etc.) gives stable repeated results in obtaining water modified in the process of mechanolysis, which is reproducible regardless of place and time. Along with those indicated in the cavitation cavity, transformation reactions occur with radicals with the participation of chemically active gases and radicals' recombination in the time of 10^{-6}–10^{-7} s. As a result of these processes, after the collapse of the cavitation bubble, the products of radical decomposition of H_2O molecules are detected using the method of spin traps, and recombination of radicals passes into solution, which leads to the accumulation of molecular O_2, H_2O_2, and other compounds in water. The high rate of reactions is evidence that they occur directly in the bubble collapse zone. As a result, along with microturbulent mixing and activation of the surface of aqueous semi-finished products, the process of mechanolysis of water during its hydromechanical treatment makes it possible to create and use cavitation technology to intensify various technological processes and serve as a basis for the development of new applications.

The mechanical factor is the main factor in the destruction of materials, and the known hypotheses and studies of this mechanism reflect one or another side of this complex process, complementing each other. However, the factors considered secondary (thermal effects, thermo- and hydrodynamic, chemical and electrochemical processes, etc.) and accompanying the collapse of bubbles have not been sufficiently studied. Water, simple in its chemical composition, has various anomalous properties due to its natural structural features. The main difficulty in studying the high water structure is the comparability of the potential energy of intermolecular interaction (forces of Coulomb interaction of charges, hydrogen bonds) with the kinetic energy of thermal motion.

According to the experimentally validated two-structure model [90,91], water is a mixture of ice-like and close-packed (disordered) structures. The impact of external factors on water structure is expressed in a change in the parameter characterizing the structural equilibrium shift. One of the strong physical factors affecting water is hydrodynamic cavitation, especially its bubble form, or bubble wake supercavitation. The kinetics of cavitation action is as follows. Fields of high pressures (up to 1000 MPa) and temperatures (up to 1000–2000 °C) are formed during the collapse of a cavitation microbubble in a local volume near it and inside. At the same time, rarefaction–compression waves are generated in the liquid and cumulative microjets with speeds of 100–500 m·s^{-1} are formed near the solid boundaries of the flow. Hydrodynamic cavitation is accompanied by the intense turbulent mixing processes, dispersion of liquid and solid components of the flow, various chemical reactions initiated by the collapse of cavitation microbubbles. Thus, the liquid region in a small neighborhood of the collapsing microbubble and the bubble itself is a kind of unique microreactor in which various chemical and technological processes are possible.

The method deserves attention since several works are devoted to researching the cavitation effect on laser additively manufactured parts from nickel alloys such as Inconel 625 [92–94] and corrosion-resistant chrome-nickel steels [86]. One of the most visible disadvantages of the technology is the features and traces of cavitation erosion on the part surfaces that can reduce the part's operational life in a complex unit.

At the same time, ultrasonic plastic deformation can positively influence the properties of the surface and subsurface layer of the part produced by laser additive manufacturing. The tool passes' geometry is determined by manually or by used equipment—turning or milling machine [15], but it can be produced as well manually for individual production [75]. The subject deserves additional research, since the effect of plastic deformation on the surfaces of laser additively manufactured parts is not covered enough but certainly has many advantages in comparison with mechanical cutting methods related to the formation of a hardened subsurface layer with increased microhardness and creating favorable residual compressive stresses (Figure 8).

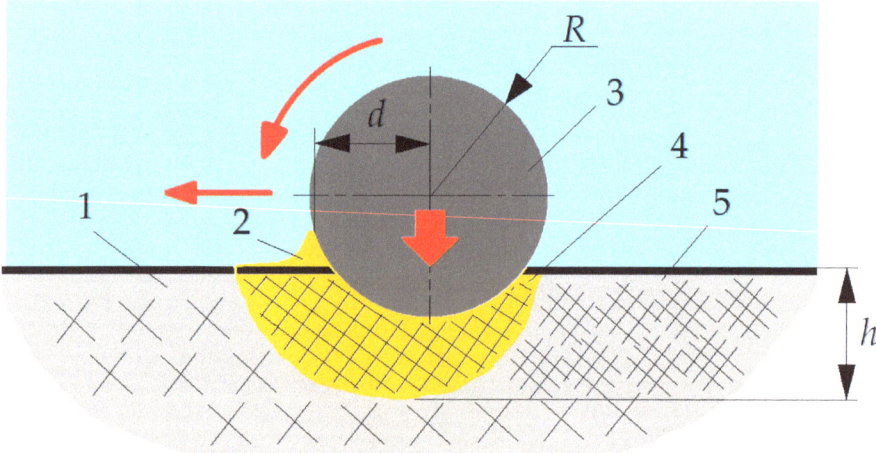

Figure 8. Scheme of surface plastic deformation, where (**1**) is the surface to be machined; (**2**) is extruded by a roller/tip material at the stage of plastic deformation; (**3**) is a roller/tip; (**4**) is plastic deformation zone; (**5**) is a hardened subsurface layer; R is the radius of a roller; d is feed; h is hardened sublayer depth.

When hardening titanium alloys with σ_B of 650–900 MPa with brushes, the microhardness of the subsurface layer increases by 15–30% with a hardened layer thickness of 0.1–0.3 mm. Studies have shown that brushing leads to a 1.2–1.4-fold decrease in the arithmetic mean deviation of the micro-profile of the ground surfaces up to mirror-like quality [95]. The surface microhardness of the investigated hardened steels increases by 10–30%. The wear of these parts is reduced by 30–40% in comparison with the ground surfaces, while the running-in time decreases by 1.5–2.2 times, which has a beneficial effect on the increase in wear resistance.

3.2. Electrochemical Methods

The chemical and electrochemical etching of metals are also used, which consist of using a specific electrolyte for each of them and direct or pulsed current allowing control of the etching speed [96] and the etched layer's thickness to reduce the surface roughness [97]. Following the combined Faraday's law, the volume V of the dissolved metal during electrolysis is directly proportional to the volume electrochemical equivalent K_v of this metal, the current I and the time t:

$$V = K_v \cdot I \cdot t. \tag{7}$$

The volumetric electrochemical equivalent K_v of a metal depends on its valence and atomic mass and constant. In practice, the dissolved metal volume does not always correspond to the volume calculated. With a specific combination of process parameters—current density at the anode, determined by the ratio of the current to the anode area, the type of metal being processed, the composition and rate of electrolyte renewal in the interelectrode gap—the volume of the dissolved metal relative to its calculated value might decrease. In some cases, the process anodic dissolution completely stops due to the formation of poorly soluble oxide films on the anode surfaces.

If there is a sufficient amount of activating anions in the electrolyte, such as chlorine anions Cl^-, oxygen is displaced from the oxide film and destroyed without additional electrical energy consumption. In such processes called active, electrical energy is spent directly on the anode metal's electrochemical dissolution. If the electrolyte lacks activating anions, additional electrical energy is spent on these films' electrochemical anodic dissolution. In this case, the efficiency of the processes is significantly reduced and called passive.

Active anodic dissolution differs from passive dissolution in the features of the reactions taking place at the anode. The anode metal's good solubility characterizes active dissolution since side reactions do not occur, except for the main one—anodic dissolution. For example, the active dissolution of the metal occurs during electrochemical etching. In passive dissolution, part of the electrical energy is spent on side reactions that remove hardly soluble oxide films from the anode surfaces. For example, passive dissolution of metal occurs during electrochemical abrasive polishing. Under certain conditions, an increase in the current density relative to its optimal value can lead to the formation of oxide films of a complex composition, which do not dissolve during classical electrolysis [98]. The complete passivation occurs and the transition of the surface layer of the metal from the active state to the passive state, at which point the process of anodic dissolution stops [99]. The pH of the used electrolyte in electrochemical abrasive polishing of cobalt influences the thickness of the complex oxide film: the thickness increases when the pH is varied in the range of 5.0–8.0 and decreases when the pH is 8.0–9.0, which is combined with increased wear [100].

3.3. Beam Polishing Methods

Prof. A.M. Chirkov and his colleagues proposed to solve the problematic roughness of additively manufactured parts using the laser-plasma polishing of a metal surface [101]. They are ignited in metal vapor, and a surface laser plasma is supported in a continuous optical discharge during laser-plasma polishing of a metal surface above a polished surface using a laser beam [102]. Changing the polishing mode is carried out by moving the plasma's center relative to the polished surface. The method provides for rough polishing of the surface in the mode of deep penetration, volume vaporization, and «finish» polishing of the surface. It provides a significant simplification of process control with high productivity. The disadvantages of this method are the locality of the laser beam, the relatively small spot size (20–100 µm, in the case of using laser beam expander—up to 400 µm), the need to create a protective atmosphere that prevents the oxidation of the material during polishing, and the evaporation of the surface material.

The authors under the supervision of Prof. N.N. Koval proposed polishing the surface of metal parts obtained by additive manufacturing methods with high-current pulsed electron beams [103,104]. They irradiated the surface of metallic samples shaped as $15 \times 30 \times 5$ mm plates produced by selective sintering in a vacuum of titanium powder VT6 (Ti-6Al-4V) with particle sizes 40–80 µm using an electron beam at the unit of the company Arcam (Sweden) [105]. The optimal for the titanium alloy VT6 (Ti-6Al-4V) mode, in which the maximum decrease in surface roughness was observed, has the pulse energy density of 45 J·cm^{-2}, the pulse duration of 200 µs, the number of pulses on the same surface area of 10, and the pulse repetition rate of 0.3 Hz. Investigation of the sample surfaces showed that the roughness parameter R_a (Arithmetic Mean Deviation) was decreased from 11 to 1.1 µm, and R_z (Ten-Point Height)—from 74 to 6 µm. The porosity of the surface layer of the sample has disappeared. Scanning electron microscopy showed that the surface profile of the samples changed significantly. A homogeneous granular structure was formed in the surface layer of titanium alloy VT6 (Ti-6Al-4V), in the composition of which individual powder particles no longer existed. However, the results obtained on small flat samples seem to be difficult to repeat on the part of a complex shape manufactured by the additive method, where some of its parts block the access of the electron beam to other parts. In addition, the electron beam reduces the roughness only to $R_a > 1$ µm, and true polishing implies the achievement of $R_a \approx 0.04$ µm, i.e., the highest 14 surface finish class.

Therefore, after processing the surface of a complex-shaped part made by the additive manufacturing method with an electron beam or a laser beam, it is necessary to continue polishing until the indicated parameters are achieved. One or another coating must be deposited on its surface, depending on its purpose. Thus, the surface treatment of the parts under consideration should include several stages and operational steps.

A group under the leadership of Prof. J. Eckert studied the surface layer's roughness depending on the different methods of surface treatment of stainless powder steels produced by selective laser melting [106]. The scientific group performed a comparative analysis of traditional mechanical methods of treatment (grinding and sandblasting) and energy methods of treatment (electrolytic and plasma polishing) and determined their influence on the surface layer's quality.

The international group of scientists, headed by the leading scientist Prof. J.A. Porro, also deserves attention paid to their excellent work [107]. Scientists under his supervision studied in detail two types of post-treatment of products produced by selective laser melting of powder aluminum alloy—sandblasting with ceramic particles and laser impact treatment. It is shown that not only the surface roughness should be optimized in the process of post-treatment but also the stress state of the surface. These properties significantly influence the durability of products produced by additive technologies.

A well-known group of scientists under the supervision of Prof. J.M. Flynn is engaged in developing a unified methodological approach to the selection of finishing post-treatment to improve the roughness of products produced by selective laser melting [108].

3.4. Ion Polishing Methods

Ion polishing is used to achieve a high surface finishing class [109]. It has been shown that ion polishing the surface of optical glass improves its quality [110]. Ion bombardment of polished glass surfaces with microroughness heights of 5–10 nm does not lead to a deterioration in the quality of optical surfaces when the surface layer of 20 µm thickness is removed [111].

The mechanism of sizing during material removal (spraying) is based on the removal of surface atoms of the workpiece because of exposure to them preformed and accelerated to the required energies of ion beams (Figure 9).

Ions with high kinetic energy are incorporated into the material. During their motion, they experience elastic and inelastic collisions with atomic nuclei and electrons of matter. There is a displacement and excitation of atoms, which is a change in the collision zone's material structure. Bombarding ions are partially reflected from the surface, and they can change their charge state in the process of backscattering. There is a removal from the surface (sputtering) of material atoms, which can also be in the different charge states. The interaction is accompanied by secondary electron emission and electromagnetic radiation, the spectrum of which ranges from infrared to X-ray.

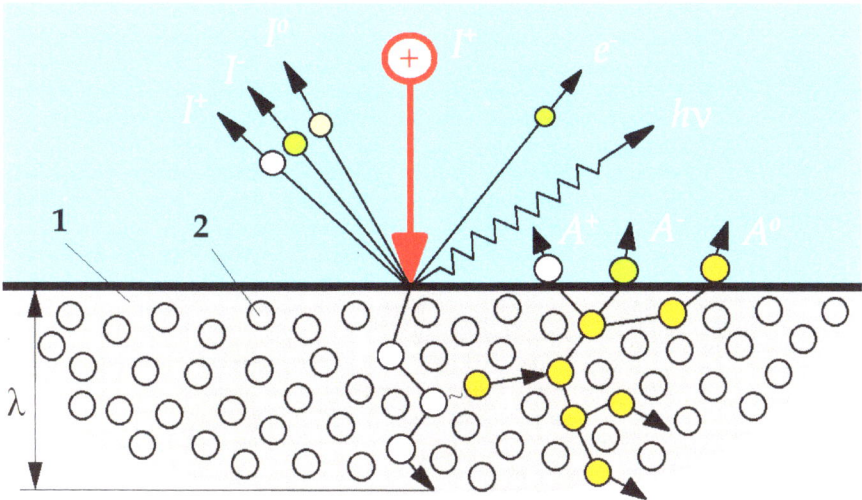

Figure 9. Scheme of a cascade of elastic collisions of atoms in a material under ion bombardment, where I^+, I^-, and I^0 are the bombarding and backscattered ions in different charge states; (**1**) is the surface to be machined; (**2**) is an unexcited atom of the material; A^+, A^-, and A^0 are the sputtered atoms in different charge states; e^- is secondary electrons; $h\nu$ is photons, λ is an ion free path that is less than the defect zone.

Considering that the ion current density of 100 µA·cm^{-2} corresponds to the fall of one ion on the treated surface in 10 s, and the time of elastic interaction of an atom with an ion is comparable to this time, then because of each action, one atom will be removed. An entire monolayer of atoms from an area of 1 cm is sprayed in 1 s, and the layer is 1 µm thick in 1 h. The approximate removal rate is about 3 A·s^{-1}, which depends on the spraying ratio of the processed material S_p:

$$S_p = k \frac{E}{\lambda} \frac{M_1 \cdot M_2}{(M_1 + M_2)^2} \tag{8}$$

where k is the coefficient taking into account the heat of sublimation of the material, M_1 and M_2 are the mass of an accelerated ion and an atom of a solid, respectively, λ is the free path of an ion in the workpiece during deceleration; and E is the energy of the ion.

With the development of ion polishing, a new direction in technology has been formed—ion processing of surfaces and coatings, which allows creating fully controlled processes to form surfaces and layers with specified characteristics [112]. For this treatment, sources of ion beams of the large cross-section are used [113].

The main disadvantage of using ion beams for processing glass and other non-conductive materials is the effect of charge accumulation on the surface and volume. It leads to a change in the electrophysical properties of materials that can be a disadvantage for some product purposes and applications. Another consequence of the effect is the appearance of strong electric fields, which leads to deviation of the ion trajectories and, consequently, to a distortion of the created profile's geometric dimensions. The same effect also occurs on topological non-uniformities of the treated surface, which leads to an increase in its roughness [114]. In this regard, recently, sources of fast neutral atoms and molecules have been used instead of ion sources when processing both conductive and non-conductive materials [109,115]. Sixteen-µm-deep grooves were obtained in 5 h using one of these sources of fast argon atoms with an energy of 3 keV and a mask on the surface of a flat corundum substrate [116]. The etching rate of hard-to-sputter corundum was v = 3.2 µm·h^{-1} and was four times lower than the etching rate by the same beam of stainless steel substrate of v = 13 µm·h^{-1} [117]. It is possible to concentrate a broad beam on a small surface area of the part produced by the additive manufacturing method and, as a result, it increases the processing rate using a source with a concave emission grid. It is necessary to use a positioning device to move it so that all parts of the surface pass sequentially in the area of the emission grid's geometric focus when a part is polishing. It is necessary to use sources of metal atom fluxes that do not contain metal droplets to deposit a coating on the polished surface of a part manufactured by additive manufacturing methods. First, these are planar magnetrons [118]. Powerful pulsed magnetrons have been proposed to increase the deposition rate [119]. They used expensive pulsed power supplies, but no significant increase in the deposition rate was achieved. A serious problem is a decrease in the deposition rate during the synthesis of metal compounds, such as titanium, with a reactive gas such as nitrogen. When the latter is added to argon, a nitride film is formed on the surface of the magnetron target, and the sputtering rate and the deposition rate of the coating are reduced several times.

The deposition rate can be increased by order of magnitude if an uncooled target holder is used that is made of refractory material that does not interact with the target material in the molten state [120]. At an argon pressure of 0.1–0.3 Pa, an ordinary magnetron discharge is ignited, a target made, for example, of copper melts in the holder, and the ionization of its vapors noticeably reduces the discharge voltage [121]. At this moment, the supply of argon to the chamber can be ceased, and the discharge continues in copper vapor [122,123]. In addition to high speed, deposition using a magnetron with an evaporated target is independent of chemically active gases and can be used to synthesize on the surface of a part manufactured by the methods of additive manufacturing a wear-resistant coating, for example, titanium nitride. This coating has already been synthesized by evaporation of titanium in a crucible-anode of a glow discharge [16,124].

A group of scientists led by E.V. Berlin investigated the ultra-high-speed sputtering of a magnetron working in the vapor of a liquid metal target [119], which can be considered as one of the scientific competitors of the proposed idea. Other known scientists proposed processing materials with beams of fast neutral particles [109,125]. Other competitors are a group headed by Prof. A. Anders, who proposed to solve the problem using powerful pulsed magnetrons [118] and a group of scientists led by Prof. I. Musil—they made a significant contribution to the development of coating deposition technology using magnetrons [117].

4. Development of Ion Polishing Principles

It is necessary to develop a method for filling a working vacuum chamber with a uniform plasma at a gas pressure of 0.01–1 Pa to develop explosive ablation in plasma of surface protrusions, polishing with fast atoms, and coating deposition. Thus, it is possible to use a glow discharge between the chamber, which plays the role of a hollow cathode, and the anode located inside it. With a chamber volume of about 0.1 m^3 and anode surface area of 0.001 m^2, the electrons emitted by its surface are accelerated to hundreds of electronvolts in the cathode sheath between the chamber wall and the plasma filling it. They fly through the plasma and are reflected in the cathode sheath at the opposite chamber wall. The chamber is an electrostatic trap for electrons, and they can get to the anode only after hundreds of flights through the plasma. They spend all their energy on the gas's excitation and ionization by their way to the anode at a pressure of 0.01–1 Pa. It allows maintaining a constant glow discharge current of 1–5 A in the indicated gas pressure range [126]. It is also necessary to provide the possibility of pulsed power supply with a short-term increase in discharge current from units to hundreds of amperes [127] for determining the optimal parameters of explosive ablation of surface protrusions when high-voltage pulses with a duration of 0.001–1 µs are applied to a part immersed in the plasma. A study of the surface of titanium, nickel, niobium, aluminum, copper, and lead cathodes showed that after a large number of pulses, the cathode microrelief is formed by the superposition of the same number of microcraters. As the pulse duration decreases, the microcrater size diminishes. The character size of a copper cathode's surface inhomogeneities decreases to

0.1–0.2 µm with a pulse duration of fewer than 1.5 ns. This phenomenon was called the polishing effect [128].

It is necessary to determine the dependence of the amplitude of high-voltage pulses at which breakdowns occur between the surface protrusions of the part and the plasma on its density (current amplitude of glow discharge in the chamber) when studying the removal of powder particles protruding on a part surface that is responsible for the initial roughness parameter R_a up to 30 µm and surface porosity. It is also important to establish the part surface roughness's dependence on the pulse energy that destroys large protrusions. Exceeding an optimal value of the pulse energy can increase the surface roughness instead of a decrease. The optimal parameters' determination can be carried out on small flat samples made by sintering in a vacuum with an electron beam of VT6 (Ti-6Al-4V) grade titanium alloy powder with a particle size of 40–80 µm, which is most common for additive manufacturing. After the explosive ablation of the surface protrusions, the same samples will be used to study polishing their surface with a concentrated beam of fast argon atoms and/or ions at an angle exceeding 60° of incidence to the surface of the sample moved in a vacuum chamber using a positioning device. It is necessary to establish the sample surface roughness's dependence on its initial value, the flux density on accelerated particles' surface, their energy, angle of incidence, and processing time. It is necessary to modernize the previously developed source of fast neutral atoms and replace the flat emissive grid with a concave surface [114]. It will make it possible to concentrate a fast argon atoms beam in a small focal region of the grid, provide access to the narrow fast atom beam to the part cavities' internal surfaces, and significantly increase the etching rate. After determining the optimal conditions for flat samples' processing, a study will be made of the protrusions removal on the outer and inner surfaces of the parts shaped as a hollow cylinder obtained by the methods of additive manufacturing and subsequent polishing of all its surfaces.

It is proposed to use a set of planar magnetrons mounted at the top of the chamber and on its sidewalls to create a uniform flow of metal atoms from all sides to the sample installed in the chamber when studying the coating deposition on obtained by additive manufacturing methods flat and hollow cylindrical samples, after reducing the initial surface roughness by explosive ablation of the protrusions on their surfaces and polishing with a beam of fast argon atoms. The flux of metal atoms vaporized from the surface of the molten target of magnitude is higher than the flux of atoms sputtered by ions from the surface of a solid magnetron target by an order. Therefore, it is proposed to install at the bottom of the chamber a magnetron with a target holder made of refractory material and use it as an uncooled crucible. The magnetron discharge will be used not only to sputter a target but also mainly to heat it in the crucible to melt and vaporize its material. It will make it possible to fill the complex geometry part's cavities with metal vapor more uniformly and in less than an hour to deposit a coating with a thickness of ≈ 10 µm on its surface.

5. Discussion

Additively manufactured parts by a laser have a vast potential in the aviation industry that will grow with the development of solids-growing and finishing technologies. Additive manufacturing took its place in the production industry and market. However, additive manufacturing methods further development has been exhausted until now; it does not have a principal character, and it concentrates on improving details. The development is hampered by the existing obstacles—the post-processing methods' inability to improve the wear resistance of the complex-shaped functional surfaces.

Most of the post-operations based on mechanical abrasive principles have a few disadvantages related to the mechanical wear of the operational surface that can be critical for some applications [105,106,128,129] when plastic deformation methods stay mostly unveiled [15,75]. Simultaneously, methods based on the use of the concentrated energy flows cannot be suitable for polishing complex geometry parts [60,111]. Chemical etching requires special electrolytes and their disposal [95–97,130–132]. The microstructure of surface and subsurface layers of the metal parts after the various post-processing is presented in Figure 10, where mechanical machining and ultrasonic deformation are not available complex-shaped parts.

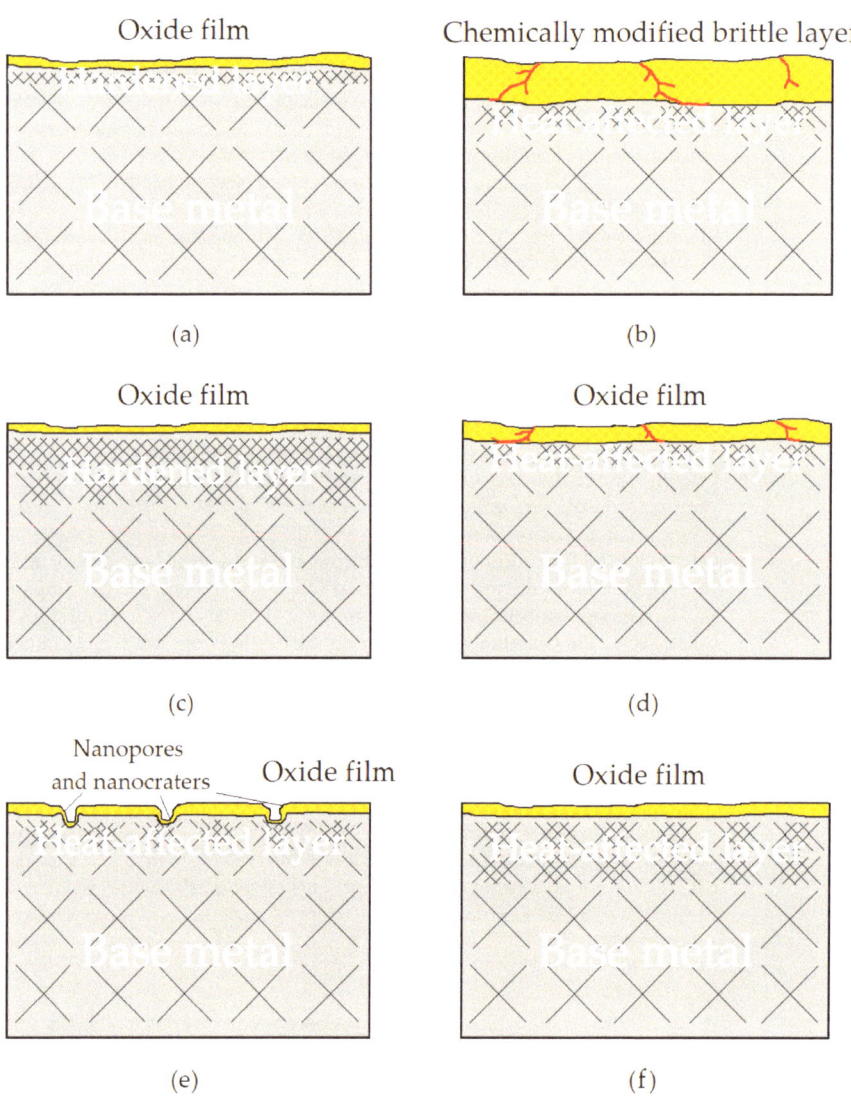

Figure 10. Scheme of surface layers after various types of post-processing: (**a**) mechanical machining; (**b**) chemical etching; (**c**) surface plastic deformation; (**d**) cavitation abrasive finishing; (**e**) laser ablation [133]; (**f**) ion polishing in gas-discharge plasma, where heat-affected layer includes dislocation area, area of increased dislocation density with Cottrell zones (negatively charged acceptors).

The conducted analytical research proposes developing the application of one of the promising approaches in finishing the laser additively manufactured parts—ion polishing in a gas discharge plasma [108–112]. The observation showed that this technology was never proposed before for processing parts after laser-based growing solids or for any other 3D-printing technology. It has strong advantages that improve the operational ability and service life of the responsible surfaces. In addition, it allows the processing of complex-shaped parts and requires sophisticated equipment that is usually settled for tool production.

The developed approach, including the finishing operation in three successive stages, was developed based on previously conducted research [114,125–127,134–136] that can be summarized as follows:

- Explosive ablation of surface protrusions when voltage pulses with an amplitude up to 30 kV and a width of 0.001–1 μs are applied to a detail immersed in the plasma;
- Polishing with a concentrated beam of fast neutral argon atoms at a large angle of incidence to the surface of the part moved in the chamber using a positioning device;
- Coating deposition on the part surface upon sputtering with argon ions of solid magnetron targets and/or the evaporation of a liquid metal magnetron target heated by ions.

The previously observed "polishing effect" of the electron gun's cathode, which forms high-current beams of nanosecond pulse width, consisted in reducing the size of its surface roughness to 0.1–0.2 μm with a decrease in pulse width to 1.5 ns. It cannot be considered an example of protrusion removal on the surface of a part immersed in a plasma due to explosive ablation when high-voltage pulses are applied to it, but it may testify to the feasibility of solving the problem.

Broad beams of fast atoms have already found numerous applications. However, studies of their compression to small transverse dimensions, providing access to complex-shaped parts' internal surfaces and polishing the surfaces, have not yet been carried out.

The coating deposition features using magnetrons are well known, including a several-fold reduction in the target-sputtering rate when using a chemically active gas. The evaporation rate of liquid metal magnetron targets heated by ions is much higher than the sputtering rate of solid targets, and it does not decrease in the presence of a chemically active gas. When using them, one can expect an increase by order of magnitude of the coating deposition rate. However, studies of this method have not yet been carried out but have immense potential and prospects.

The proposed approach will allow:

1. Removal of powder particles 40–100 μm in size used in the manufacture of the part and protruding on its surface, which is responsible for the initial roughness parameter R_a (Arithmetic Mean Deviation) of 30 μm and surface porosity, by explosive ablation of surface protrusions when microsecond pulses of negative voltage up to 30 kV are applied to the part immersed in the plasma.
2. Polishing with a concentrated beam of ions and/or fast argon atoms at an angle of incidence greater than 60° of the surface of the part moved in a vacuum chamber using a positioning device.
3. Coating deposition on the surface of a part immersed in a dense metal plasma obtained by the evaporation of liquid metal magnetron targets.

The attainability of the problem solution for the first processing stage is determined by the known results of studying the surface of the explosive emission cathodes for electron guns forming high-current nanosecond beams. Based on these results, one can hope for a decrease in the roughness parameter R_a to ≈2 μm and a decrease in the surface layer's porosity.

The attainability of the solution of the second processing stage problem is determined by the known results of polishing with ion beam products made of various materials up to 14 surface finishing class. The main problems are the forming a focused beam of fast argon atoms and developing a part positioning device that ensures successive bombardment of all its surface at an angle of incidence of more than 60°.

The attainability of the task solution for the third processing stage is based on already available experimental data on magnetron targets' evaporation. In this case, the magnetron discharge is used not only for sputtering a target but also mainly for its heating in a crucible for melting and evaporation of its material. The flux of metal atoms vaporized from a liquid target's surface is by an order of magnitude higher than the flux of atoms sputtered by ions from a solid target's surface. It makes it possible to more uniformly fill with metal vapor the part's cavities and deposit its surface coatings with a thickness of ≈10 μm in less than an hour with the complex geometry part.

6. Conclusions

All observed technologies have their disadvantages, mostly related to the nature of surface destruction that determines the increased wear of working surfaces of additively manufactured parts even at the stage of post-processing technologies, which has a complex character. Along with the positive effect of plastic deformation and recrystallization of near-surface layers, erosion processes are observed, leading to stress states. That hampers applying the complex metallic parts obtained by laser additive manufacturing for responsible mechanisms and units.

The conducted analytical research provides an innovative approach in finishing the parts produced by laser additive manufacturing based on treatment in gas-discharge plasma. The development approach includes a technology proposal for three principal stages that will allow:

- removal granules from the surfaces of the parts with the size that primarily used in additive manufacturing of 40–100 μm and achieving roughness parameter R_a (Arithmetic Mean Deviation) of 30 μm by microsecond pulses of negative voltage up to 30 kV are applied to the part immersed in the plasma;
- polishing the surface with concentrated ions or fast argon atoms under angle exceeds 60°;
- coating deposition by the evaporation of liquid metal magnetron targets.

The proposed approach has no analogs in the modern industry that allow deducing laser additive manufacturing at a new principal level to reboot the industry's current state.

Author Contributions: Conceptualization, S.N.G.; methodology, A.S.M., T.V.T.; software, E.S.M. and P.A.P.; investigation, Y.A.M.; resources, A.A.O. and P.A.P.; writing—original draft preparation, Y.A.M. and T.V.T.; writing—review and editing, M.A.V.; visualization, A.A.O. and E.S.M.; supervision, S.N.G.; project administration, M.A.V.; funding acquisition, A.S.M. All authors have read and agreed to the published version of the manuscript.

Funding: This research was funded by the Russian Science Foundation, grant number No. 20-19-00620.

Institutional Review Board Statement: Not applicable.

Informed Consent Statement: Not applicable.

Data Availability Statement: The data presented in this study are openly available in [Figures 6 and 10] at [https://doi.org/10.3390/met10111540], reference number [86].

Acknowledgments: The research was done at the Department of High-Efficiency Processing Technologies of MSTU Stankin.

Conflicts of Interest: The authors declare no conflict of interest.

References

1. Sova, A.; Doubenskaia, M.; Grigoriev, S.; Okunkova, A.; Smurov, I. Parameters of the Gas-Powder Supersonic Jet in Cold Spraying Using a Mask. *J. Therm. Spray Technol.* **2013**, *22*, 551–556. [CrossRef]
2. Yadroitsev, I.; Bertrand, P.; Antonenkova, G.; Grigoriev, S.; Smurov, I. Use of track/layer morphology to develop functional parts by selectivelaser melting. *J. Laser Appl.* **2013**, *25*, 052003. [CrossRef]
3. Klein, T.; Schnall, M. Control of macro-/microstructure and mechanical properties of a wire-arc additive manufactured aluminum alloy. *Int. J. Adv. Manuf. Technol.* **2020**, *108*, 235–244. [CrossRef]
4. Shen, C.; Liss, K.D.; Reid, M.; Pan, Z.X.; Hua, X.M.; Li, F.; Mou, G.; Huang, Y.; Dong, B.S.; Luo, D.Z. Effect of the post-production heat treatment on phase evolution in the Fe3Ni-FeNi functionally graded material: An in-situ neutron diffraction study. *Intermetallics* **2021**, *129*, 107032. [CrossRef]
5. Khmyrov, R.S.; Protasov, C.E.; Grigoriev, S.N.; Gusarov, A.V. Crack-free selective laser melting of silica glass: Single beads and monolayers on the substrate of the same material. *Int. J. Adv. Manuf. Technol.* **2016**, *85*, 1461–1469. [CrossRef]

6. Khmyrov, R.S.; Grigoriev, S.N.; Okunkova, A.A.; Gusarov, A.V. On the possibility of selective laser melting of quartz glass. *Phys. Procedia* **2014**, *56*, 345–356. [CrossRef]
7. Grigoriev, S.; Tarasova, T.; Gusarov, A.; Khmyrov, R.; Egorov, S. Possibilities of Manufacturing Products from Cermet Compositions Using Nanoscale Powders by Additive Manufacturing Methods. *Materials* **2019**, *12*, 3425. [CrossRef] [PubMed]
8. Bunnell, D.E.; Bourell, D.L.; Beaman, J.B.; Marcus, H.L. Fundamentals of liquid phase sintering during selective laser sintering. In *Processing and Fabrication of Advanced Materials IV, Proceedings of the Symposium on Processing and Fabrication of Advanced Materials IV, Cleveland, OH, USA, 29 October–2 November 1995*; Srivatsan, T.S., Moore, J.J., Eds.; Minerals, Metals & Materials Society: Warrendale, PA, USA, 1996; pp. 17–26.
9. Alsulami, M.; Mortazavi, M.; Niknam, S.A.; Li, D.S. Design complexity and performance analysis in additively manufactured heat exchangers. *Int. J. Adv. Manuf. Technol.* **2020**, *110*, 865–873. [CrossRef]
10. Wanjara, P.; Gholipour, J.; Watanabe, E.; Watanabe, K.; Sugino, T.; Patnaik, P.; Sikan, F.; Brochu, M. High Frequency Vibration Fatigue Behavior of Ti6Al4V Fabricated by Wire-Fed Electron Beam Additive Manufacturing Technology. *Adv. Mater. Sci. Eng.* **2020**, *2020*, 1902567. [CrossRef]
11. Sotov, A.V.; Agapovichev, A.V.; Smelov, V.G.; Kokareva, V.V.; Dmitrieva, M.O.; Melnikov, A.A.; Golanov, S.P.; Anurov, Y.M. Investigation of the IN-738 superalloy microstructure and mechanical properties for the manufacturing of gas turbine engine nozzle guide vane by selective laser melting. *Int. J. Adv. Manuf. Technol.* **2020**, *107*, 2525–2535. [CrossRef]
12. Tarasova, T.V.; Nazarov, A.P.; Shalapko, Y.I. Abrasive and fretting wear resistance of refractory cobalt alloy specimens manufactured by the method of selective laser melting. *J. Frict. Wear* **2014**, *35*, 365–373. [CrossRef]
13. Tarasova, T.V.; Gusarov, A.V.; Protasov, K.E.; Filatova, A.A. Effect of Thermal Fields on the Structure of Corrosion-Resistant Steels Under Different Modes of Laser Treatment. *Met. Sci. Heat Treat.* **2017**, *59*, 433–440. [CrossRef]
14. Nowotny, S.; Tarasova, T.V.; Filatova, A.A.; Dolzhikova, E.Y. Methods for Characterizing Properties of Corrosion-Resistant Steel Powders Used for Powder Bed Fusion Processes. *Mater. Sci. Forum* **2016**, *876*, 1–7. [CrossRef]
15. Gavrin, V.N.; Kozlova, Y.P.; Veretenkin, E.P.; Logachev, A.V.; Logacheva, A.I.; Lednev, I.S.; Okunkova, A.A. Reactor target from metal chromium for "pure" high-intensive artificial neutrino source. *Phys. Part. Nucl. Lett.* **2016**, *13*, 267–273. [CrossRef]
16. Volosova, M.A.; Gurin, V.D. Influence of vacuum-plasma nitride coatings on contact processes and a mechanism of wear of working surfaces of high-speed steel cutting tool at interrupted cutting. *J. Frict. Wear* **2013**, *34*, 183–189. [CrossRef]
17. Grigoriev, S.N.; Gurin, V.D.; Volosova, M.A.; Cherkasova, N.Y. Development of residual cutting tool life prediction algorithm by processing on CNC machine tool. *Materialwiss. Werkstofftech.* **2013**, *44*, 790–796. [CrossRef]
18. Metel, A.; Grigoriev, S.; Melnik, Y.; Panin, V.; Prudnikov, V. Cutting Tools Nitriding in Plasma Produced by a Fast Neutral Molecule Beam. *Jpn. J. Appl. Phys.* **2011**, *50*, 08JG04. [CrossRef]
19. Fang, Z.C.; Wu, Z.L.; Huang, C.G.; Wu, C.W. Review on residual stress in selective laser melting additive manufacturing of alloy parts. *Opt. Laser Technol.* **2020**, *129*, 106283. [CrossRef]
20. Acevedo, R.B.O.; Kantarowska, K.; Santos, E.C.; Fredel, M.C. Residual stress measurement techniques for Ti6Al4V parts fabricated using selective laser melting: State of the art review. *Rapid Prototyp. J.* **2020**. [CrossRef]
21. Vermilion, M.L.D.; de Oliveira, T.T.; Kreve, S.; Batalha, R.L.; de Oliveira, D.P.; Pauly, S.; Bolfarini, C.; Bachmann, L.; dos Reis, A.C. Analysis of the mechanical and physicochemical properties of Ti-6Al-4 V discs obtained by selective laser melting and subtractive manufacturing method. *J. Biomed. Mater. Res. Part B* **2020**. [CrossRef]
22. Cardoso, R.M.; Kalinke, C.; Rocha, R.G.; dos Santos, P.L.; Rocha, D.P.; Oliveira, P.R.; Janegitz, B.C.; Bonacin, J.A.; Richter, E.M.; Munoz, R.A.A. Additive-manufactured (3D-printed) electrochemical sensors: A critical review. *Anal. Chim. Acta* **2020**, *1118*, 73–91. [CrossRef] [PubMed]
23. Gokuldoss Prashanth, K.; Scudino, S.; Eckert, J. Tensile Properties of Al-12Si Fabricated via Selective Laser Melting (SLM) at Different Temperatures. *Technologies* **2016**, *4*, 38. [CrossRef]
24. Kim, J.; Wakai, A.; Moridi, A. Materials and manufacturing renaissance: Additive manufacturing of high-entropy alloys. *J. Mater. Res.* **2020**, *35*, 19963–19983. [CrossRef]
25. Wang, Y.; Liu, S.; Fan, Y.; He, Z.R. A short review on selective laser melting of H13 steel. *Int. J. Adv. Manuf. Technol.* **2020**, *108*, 2453–2466. [CrossRef]
26. Sing, S.L.; Yeong, W.Y. Laser powder bed fusion for metal additive manufacturing: Perspectives on recent developments. *Virtual Phys. Prototyp.* **2020**, *15*, 359–370. [CrossRef]
27. Volosova, M.A.; Fyodorov, S.V.; Opleshin, S.; Mosyanov, M. Wear Resistance and Titanium Adhesion of Cathodic Arc Deposited Multi-Component Coatings for Carbide End Mills at the Trochoidal Milling of Titanium Alloy. *Technologies* **2020**, *8*, 38. [CrossRef]
28. Cruz, N.; Martins, M.I.; Domingos Santos, J.; Gil Mur, J.; Tondela, J.P. Surface Comparison of Three Different Commercial Custom-Made Titanium Meshes Produced by SLM for Dental Applications. *Materials* **2020**, *13*, 2177. [CrossRef] [PubMed]
29. Kuzin, V.V.; Grigoriev, S.N.; Fedorov, M.Y. Role of the thermal factor in the wear mechanism of ceramic tools. Part 2: Microlevel. *J. Frict. Wear* **2015**, *36*, 40–44. [CrossRef]
30. Khodabakhshi, F.; Gerlich, A.P. Potentials and strategies of solid-state additive friction-stir manufacturing technology: A critical review. *J. Manuf. Process.* **2018**, *36*, 77–92. [CrossRef]
31. Kalender, M.; Kilic, S.F.; Ersoy, S.; Bozkurt, Y.; Salman, S. Additive Manufacturing and 3D Printer Technology in Aerospace Industry. In Proceedings of the 9th International Conference on Recent Advances in Space Technologies (RAST), Istanbul, Turkey, 11–14 June 2019; IEEE: New York, NY, USA, 2019; pp. 689–695.

32. Camacho, D.D.; Clayton, P.; O'Brien, W.J.; Seepersad, C.; Juenger, M.; Ferron, R.; Salamone, S. Applications of additive manufacturing in the construction industry—A forward-looking review. *Autom. Constr.* **2018**, *89*, 110–119. [CrossRef]
33. Chekurov, S.; Salmi, M.; Verboeket, V.; Puttonen, T.; Riipinen, T.; Vaajoki, A. Assessing industrial barriers of additively manufactured digital spare part implementation in the machine-building industry: A cross-organizational focus group interview study. *J. Manuf. Technol. Manag.* **2021**. [CrossRef]
34. Land, P.; Crossley, R.; Branson, D.; Ratchev, S. Technology Review of Thermal Forming Techniques for use in Composite Component Manufacture. *SAE Int. J. Mater. Manuf.* **2016**, *9*, 81–89. [CrossRef]
35. Liu, J.; Jalalahmadi, B.; Guo, Y.B.; Sealy, M.P.; Bolander, N. A review of computational modeling in powder-based additive manufacturing for metallic part qualification. *Rapid Prototyp. J.* **2018**, *24*, 1245–1264. [CrossRef]
36. Bambach, M.; Sizova, I.; Sydow, B.; Hemes, S.; Meiners, F. Hybrid manufacturing of components from Ti-6Al-4V by metal forming and wire-arc additive manufacturing. *J. Mater. Process. Technol.* **2020**, *282*, 116689. [CrossRef]
37. Echsel, M.; Springer, P.; Huembert, S. Production and planned in-orbit qualification of a function-integrated, additive manufactured satellite sandwich structure with embedded automotive electronics. *CEAS Space J.* **2020**. [CrossRef]
38. Hafenstein, S.; Hitzler, L.; Sert, E.; Öchsner, A.; Merkel, M.; Werner, E. Hot Isostatic Pressing of Aluminum–Silicon Alloys Fabricated by Laser Powder-Bed Fusion. *Technologies* **2020**, *8*, 48. [CrossRef]
39. Salman, O.O.; Funk, A.; Waske, A.; Eckert, J.; Scudino, S. Additive Manufacturing of a 316L Steel Matrix Composite Reinforced with CeO2 Particles: Process Optimization by Adjusting the Laser Scanning Speed. *Technologies* **2018**, *6*, 25. [CrossRef]
40. Saroia, J.; Wang, Y.; Wei, Q.; Lei, M.J.; Li, X.P.; Guo, Y.; Zhang, K. A review on 3D printed matrix polymer composites: Its potential and future challenges. *Int. J. Adv. Manuf. Technol.* **2020**, *106*, 1695–1721. [CrossRef]
41. Protasov, C.E.; Khmyrov, R.S.; Grigoriev, S.N.; Gusarov, A.V. Selective laser melting of fused silica: Interdependent heat transfer and powder consolidation. *Int. J. Heat Mass Transf.* **2017**, *104*, 665–674. [CrossRef]
42. Hartmann, C.; Lechner, P.; Himmel, B.; Krieger, Y.; Lueth, T.C.; Volk, W. Compensation for Geometrical Deviations in Additive Manufacturing. *Technologies* **2019**, *7*, 83. [CrossRef]
43. Shulunov, V.R. Several advantages of the ultra high-precision additive manufacturing technology. *Int. J. Adv. Manuf. Technol.* **2016**, *85*, 1941–1945. [CrossRef]
44. Okunkova, A.; Peretyagin, P.; Vladimirov, Y.; Volosova, M.; Torrecillas, R.; Fedorov, S.V. Laser-beam modulation to improve efficiency of selecting laser melting for metal powders. *Proc. SPIE* **2014**, *9135*, 913524.
45. Gusarov, A.V.; Grigoriev, S.N.; Volosova, M.A.; Melnik, Y.A.; Laskin, D.V.; Kotoban, D.V.; Okunkova, A.A. On productivity of laser additive manufacturing. *J. Mater. Process. Technol.* **2018**, *261*, 213–232. [CrossRef]
46. Metel, A.S.; Stebulyanin, M.M.; Fedorov, S.V.; Okunkova, A.A. Power Density Distribution for Laser Additive Manufacturing (SLM): Potential, Fundamentals and Advanced Applications. *Technologies* **2019**, *7*, 5. [CrossRef]
47. Canaday, H. Making 3D-printed parts for Boeing 787s. *Aerospace Am.* **2018**, *56*, 18–21.
48. Jelaca, M.S.; Boljevic, A. Critical Success Factors and Negative Effects of Development—The Boeing 787 Dreamliner. *Strateg. Manag.* **2016**, *21*, 30–39.
49. Rutkowski, M. Safety as an Element of Creating Competitive Advantage among Airlines Given the Example of The Airbus A350 XWB and The Boeing 787 Dreamliner Aircraft. *Sci. J. Sil. Univ. Technol. Ser. Transp.* **2020**, *108*, 201–212. [CrossRef]
50. Giannis, S. Testing and Analysis Building Block Approach: Evaluation of the Performance of the Integrated Lattice Fuselage Section. *SAMPE J.* **2016**, *52*, 22–33.
51. Kelkar, R.; Andreaco, A.; Ott, E.; Groh, J. Alloy 718: Laser Powder Bed Additive Manufacturing for Turbine Applications. In *Minerals Metals & Materials Series, Proceedings of the 9th International Symposium on Superalloy 718 & Derivatives: Energy, Aerospace, And Industrial Applications, Champion, PA, USA, 17–21 September 2000*; Ott, E., Liu, X., Andersson, J., Bi, Z., Bockenstedt, K., Dempster, I., Groh, J., Heck, K., Jablonski, P., Kaplan, M., et al., Eds.; Springer International Publishing AG: Cham, Switzerland, 2018; pp. 53–68.
52. Schanz, J.; Hofele, M.; Hitzler, L.; Merkel, M.; Riegel, H. Laser Polishing of Additive Manufactured AlSi10Mg Parts with an Oscillating Laser Beam. *Adv. Struct. Mater.* **2016**, *61*, 159–169.
53. Yang, T.; Liu, T.T.; Liao, W.H.; MacDonald, E.; Wei, H.L.; Chen, X.Y.; Jiang, L.Y. The influence of process parameters on vertical surface roughness of the AlSi10Mg parts fabricated by selective laser melting. *J. Mater. Process. Technol.* **2019**, *266*, 26–36. [CrossRef]
54. Krawczyk, M.B.; Królikowski, M.A.; Grochała, D.; Powałka, B.; Figiel, P.; Wojciechowski, S. Evaluation of Surface Topography after Face Turning of CoCr Alloys Fabricated by Casting and Selective Laser Melting. *Materials* **2020**, *13*, 2448. [CrossRef] [PubMed]
55. Texier, D.; Copin, E.; Flores, A.; Lee, J.; Terner, M.; Hong, H.U.; Lours, P. High temperature oxidation of NiCrAlY coated Alloy 625 manufactured by selective laser melting. *Surf. Coat. Technol.* **2020**, *398*, 126041. [CrossRef]
56. Antanasova, M.; Kocjan, A.; Hocevar, M.; Jevnikar, P. Influence of surface airborne-particle abrasion and bonding agent application on porcelain bonding to titanium dental alloys fabricated by milling and by selective laser melting. *J. Prosthet. Dent.* **2020**, *123*, 491–499. [CrossRef]
57. Yu, J.; Kim, D.; Ha, K.; Jeon, J.B.; Lee, W. Strong feature size dependence of tensile properties and its microstructural origin in selectively laser melted 316L stainless steel. *Mater. Lett.* **2020**, *275*, 128161. [CrossRef]

58. Zhao, Y.Z.; Sun, J.; Guo, K.; Li, J.F. Investigation on the effect of laser remelting for laser cladding nickel based alloy. *J. Laser Appl.* **2019**, *31*, UNSP 022512. [CrossRef]
59. Jeyaprakash, N.; Yang, C.H. Microstructure and Wear Behaviour of SS420 Micron Layers on Ti-6Al-4V Substrate Using Laser Cladding Process. *Trans. Indian Inst. Met.* **2020**, *73*, 1527–1533. [CrossRef]
60. Kotoban, D.; Grigoriev, S.; Okunkova, A.; Sova, A. Influence of a shape of single track on deposition efficiency of 316L stainless steel powder in cold spray. *Surf. Coat. Technol.* **2017**, *309*, 951–958. [CrossRef]
61. Smolenska, H.; Konczewicz, W.; Bazychowska, S. The Impact of Material Selection on Durability of Exhaust Valve Faces of a Ship Engine—A Case Study. *Adv. Sci. Technol. Res. J.* **2020**, *14*, 165–174.
62. du Plessis, A.; Yadroitsev, I.; Yadroitsava, I.; Le Roux, S.G. X-Ray Microcomputed Tomography in Additive Manufacturing: A Review of the Current Technology and Applications. *3D Print. Addit. Manuf.* **2018**, *5*, 227–247. [CrossRef]
63. Matache, G.; Vladut, M.; Paraschiv, A.; Condruz, R.M. Edge and corner effects in selective laser melting of IN 625 alloy. *Manuf. Rev.* **2020**, *7*, 8. [CrossRef]
64. Bashevskaya, O.S.; Bushuev, S.V.; Poduraev, Y.V.; Mel'nichenko, E.A.; Shcherbakov, M.I.; Garskov, R.V. Use of Infrared Thermography for Evaluating Linear Dimensions of Subsurface Defects. *Meas. Tech.* **2017**, *60*, 457–462. [CrossRef]
65. Bashevskaya, O.S.; Bushuev, S.V.; Nikitin, A.A.; Romash, E.V.; Poduraev, Y.V. Assessment of Surface Roughness Using Curvature Parameters of Peaks and Valleys of the Profile. *Meas. Tech.* **2017**, *60*, 128–133. [CrossRef]
66. Valente, E.H.; Gundlach, C.; Christiansen, T.L.; Somers, M.A.J. Effect of Scanning Strategy During Selective Laser Melting on Surface Topography, Porosity, and Microstructure of Additively Manufactured Ti-6Al-4V. *Appl. Sci.* **2019**, *9*, 5554. [CrossRef]
67. Luongo, A.; Falster, V.; Doest, M.B.; Ribo, M.M.; Eiriksson, E.R.; Pedersen, D.B.; Frisvad, J.R. Microstructure Control in 3D Printing with Digital Light Processing. *Comput. Graph. Forum* **2020**, *39*, 347–359. [CrossRef]
68. Johnson, A.R.; Procopio, A.T. Low cost additive manufacturing of microneedle masters. *3D Print. Med.* **2019**, *5*, 2. [CrossRef]
69. Covarrubias, E.E.; Eshraghi, M. Effect of Build Angle on Surface Properties of Nickel Superalloys Processed by Selective Laser Melting. *JOM* **2018**, *70*, 336–342. [CrossRef]
70. Bashevskaya, O.S.; Bushuev, S.V.; Ilyukhin, Y.V.; Kovalskiy, M.G.; Mel'nichenko, E.A.; Romash, E.V.; Poduraev, Y.V. Comparative Analysis of Thermal Deformations in Structural Elements of Measurement Stands and Supports. *Meas. Tech.* **2015**, *58*, 760–765. [CrossRef]
71. Chen, Y.; Sun, H.; Li, Z.; Wu, Y.; Xiao, Y.; Chen, Z.; Zhong, S.; Wang, H. Strategy of Residual Stress Determination on Selective Laser Melted Al Alloy Using XRD. *Materials* **2020**, *13*, 451. [CrossRef]
72. Wan, H.Y.; Luo, Y.W.; Zhang, B.; Song, Z.M.; Wang, L.Y.; Zhou, Z.J.; Li, C.P.; Chen, G.F.; Zhang, G.P. Effects of surface roughness and build thickness on fatigue properties of selective laser melted Inconel 718 at 650 degrees C. *Int. J. Fatigue* **2020**, *137*, 105654. [CrossRef]
73. Jamshidi, P.; Aristizabal, M.; Kong, W.; Villapun, V.; Cox, S.C.; Grover, L.M.; Attallah, M.M. Selective Laser Melting of Ti-6Al-4V: The Impact of Post-processing on the Tensile, Fatigue and Biological Properties for Medical Implant Applications. *Materials* **2020**, *13*, 2813. [CrossRef] [PubMed]
74. Grigoriev, S.N.; Metel, A.S.; Tarasova, T.V.; Filatova, A.A.; Sundukov, S.K.; Volosova, M.A.; Okunkova, A.A.; Melnik, Y.A.; Podrabinnik, P.A. Effect of Cavitation Erosion Wear, Vibration Tumbling, and Heat Treatment on Additively Manufactured Surface Quality and Properties. *Metals* **2020**, *10*, 1540. [CrossRef]
75. Gola, A.M.; Ghadamgahi, M.; Ooi, S.W. Microstructure evolution of carbide-free bainitic steels under abrasive wear conditions. *Wear* **2017**, *376*, 975–982. [CrossRef]
76. Bankowski, D.; Spadlo, S. Vibratory Machining Effect on the Properties of the Aluminum Alloys Surface. *Arch. Foundry Eng.* **2017**, *17*, 19–24. [CrossRef]
77. Niemczewski, B. A Comparison of Ultrasonic Cavitation Intensity in Liquids. *Ultrasonics* **1980**, *18*, 107–110. [CrossRef]
78. Landau, L.D.; Lifshitz, E.M. Hydrodynamic Fluctuations. *Soviet Phys. JETP-USSR* **1957**, *5*, 512–513.
79. Endo, H. Thermodynamic Consideration of the Cavitation Mechanism in Homogeneous Liquids. *J. Acoust. Soc. Am.* **1994**, *95*, 2409–2415. [CrossRef]
80. Wang, Q. Local energy of a bubble system and its loss due to acoustic radiation. *J. Fluid Mech.* **2016**, *797*, 201–230. [CrossRef]
81. Pelekasis, N.A.; Tsamopoulos, J.A. Bjerknes Forces between 2 Bubbles.1. Response to a Step Change in Pressure. *J. Fluid Mech.* **1993**, *254*, 467–499. [CrossRef]
82. Makarov, P.V. Mathematical theory of evolution of loaded solids and media. *Phys. Mesomech.* **2008**, *11*, 213–227. [CrossRef]
83. Gusev, A.I.; Shveikin, G.P. Energy of Elastic Lattice Deformation in Formation of Solid-Solutions of Transition-Metal Carbides and Nitrides. *Inorg. Mater.* **1976**, *12*, 1283–1286.
84. Alekseev, A.A.; Strunin, B.M. Change of Elastic Energy of Crystal during Its Plastic-Deformation. *Fizika Tverdogo Tela* **1975**, *17*, 1457–1459.
85. Akhmedzhanov, R.A.; Zelenskii, I.V.; Gushchin, L.A.; Nizov, V.A.; Nizov, N.A.; Sobgaida, D.A. Observation of Coherent Population Trapping in Ensembles of Diamond NV-Centers under Ground-State Level Anticrossing Conditions. *Opt. Spectrosc.* **2019**, *127*, 260–264. [CrossRef]
86. Metel, A.S.; Grigoriev, S.N.; Tarasova, T.V.; Filatova, A.A.; Sundukov, S.K.; Volosova, M.A.; Okunkova, A.A.; Melnik, Y.A.; Podrabinnik, P.A. Influence of Postprocessing on Wear Resistance of Aerospace Steel Parts Produced by Laser Powder Bed Fusion. *Technologies* **2020**, *8*, 73. [CrossRef]

87. Shmakov, V.A. Surface Quality of Small Components after Ultrasonic Abrasive Machining. *Russ. Eng. J.* **1976**, *56*, 33–34.
88. Isobe, H.; Tsuji, S.; Hara, K.; Ishimatsu, J. Improvement of Removal Rate of Tape Lapping by Applying Fluid with Ultrasonic Excited Cavitation. *Int. J. Autom. Technol.* **2021**, *15*, 65–73. [CrossRef]
89. Bolmatov, D.; Soloviov, D.; Zhernenkov, M.; Zav'yalov, D.; Mamontov, E.; Suvorov, A.; Cai, Y.Q.; Katsaras, J. Molecular Picture of the Transient Nature of Lipid Rafts. *Langmuir* **2020**, *36*, 4887–4896. [CrossRef] [PubMed]
90. Caupin, F.; Anisimov, M.A. Thermodynamics of supercooled and stretched water: Unifying two-structure description and liquid-vapor spinodal. *J. Chem. Phys.* **2019**, *151*, 034503. [CrossRef] [PubMed]
91. Lyashchenko, A.K.; Zasetskii, A.Y. Structural transition to electrolyte-water solvent and changes in the molecular dynamics of water and properties of solutions. *J. Struct. Chem.* **1998**, *39*, 694–703. [CrossRef]
92. Tan, K.L.; Yeo, S.H. Surface finishing on IN625 additively manufactured surfaces by combined ultrasonic cavitation and abrasion. *Addit. Manuf.* **2020**, *31*, 100938. [CrossRef]
93. Wang, J.; Zhu, J.; Liew, P.J. Material Removal in Ultrasonic Abrasive Polishing of Additive Manufactured Components. *Appl. Sci.* **2019**, *9*, 5359. [CrossRef]
94. Tan, K.L.; Yeo, S.H. Surface modification of additive manufactured components by ultrasonic cavitation abrasive finishing. *Wear* **2017**, *378–379*, 90–95. [CrossRef]
95. Grechnikov, F.V.; Surudin, S.V.; Erisov, Y.A.; Kuzin, A.O.; Bobrovskiy, I.N. Influence of Material Structure Crystallography on its Formability in Sheet Metal Forming Processes. *IOP Conf. Ser. Mater. Sci. Eng.* **2018**, *286*, UNSP 012021. [CrossRef]
96. Dong, G.; Marleau-Finley, J.; Zhao, Y.F. Investigation of electrochemical post-processing procedure for Ti-6Al-4V lattice structure manufactured by direct metal laser sintering (DMLS). *Int. J. Adv. Manuf. Technol.* **2019**, *104*, 3401–3417. [CrossRef]
97. Rotty, C.; Mandroyan, A.; Doche, M.-L.; Monney, S.; Hihn, J.Y.; Rouge, N. Electrochemical Superfinishing of Cast and ALM 316L Stainless Steels in Deep Eutectic Solvents: Surface Microroughness Evolution and Corrosion Resistance. *J. Electrochem. Soc.* **2019**, *166*, C468–C478. [CrossRef]
98. Hryniewicz, T.; Rokosz, K.; Rokicki, R. Electrochemical and XPS studies of AISI 316L stainless steel after electropolishing in a magnetic field. *Corros. Sci.* **2008**, *50*, 2676–2681. [CrossRef]
99. Ni, X.; Zhang, L.; Wu, W.; Song, J.; He, B.B.; Zhu, D.X. Improved Surface Properties for Nanotube Growth on Selective Laser Melted Porous Ti6Al4V Alloy via Chemical Etching. *Int. J. Electrochem. Sci.* **2019**, *14*, 5679–5689. [CrossRef]
100. Xu, W.; Ma, L.; Chen, Y.; Liang, H. Mechano-oxidation during cobalt polishing. *Wear* **2018**, *416*, 36–43. [CrossRef]
101. Chirkov, A.M.; Rybalko, A.P.; Rogal'skij, J.I.; Sedoj, E.A.; Merkukhin, A.V.; Borisov, N.V. Method of Laser-Plasma Polishing of Metallic Surface. RU Patent 2 381 094, 10 February 2010.
102. Marinin, E.A.; Chirkov, A.M.; Gavrilov, G.N.; Fetisov, G.P.; Chernyshov, D.A.; Kurganova, Y.A. Experimental Evaluation of the Methods of Laser Cementation of Low-Alloy Tool Steels. *Russ. Metall.* **2018**, *13*, 1259–1263. [CrossRef]
103. Koval, N.N.; Teresov, A.D.; Ivanov, Y.F.; Petrikova, E.A. Pulse Electron-Beam Metal Product Surface Polishing Method. RU Patent 2 619 5434, 16 May 2017.
104. Teresov, A.D.; Ivanov, Y.F.; Petrikova, E.A.; Koval, N.N. Structure and Properties of VT6 Alloy Obtained by Layered Selective Sintering of a Powder. *Russ. Phys. J.* **2017**, *60*, 1367–1372. [CrossRef]
105. Uglov, V.V.; Krutsilina, E.A.; Shymanski, V.I.; Kuleshov, A.K.; Koval, N.N.; Ivanov, Y.F. Heat Transfer in Surface Layer of a T15k6 Heterogeneous Hard Alloy under Pulsed High-Energy Irradiation. *Russ. Phys. J.* **2020**, *63*, 693–698. [CrossRef]
106. Lober, L.; Flache, C.; Petters, R.; Kuhn, U.; Eckert, J. Comparison of different post processing technologies for SLM generated 3161 steel parts. *Rapid Prototyp. J.* **2013**, *19*, 173–179. [CrossRef]
107. Gordon, E.R.; Shokrani, A.; Flynn, J.M.; Goguelin, S.; Barclay, J.; Dhokia, V.A. Surface Modification Decision Tree to Influence Design in Additive Manufacturing. In *Smart Innovation Systems and Technologies, Proceedings of the 3rd International Conference on Sustainable Design and Manufacturing (SDM), Chania, Greece, 4–6 April 2016*; Setchi, R., Howlett, R.J., Liu, Y., Theobald, P., Eds.; Springer: Berlin/Heidelberg, Germany, 2016; Volume 52, pp. 423–434.
108. Gatto, A.; Bassoli, E.; Denti, L.; Sola, A.; Tognoli, E.; Comin, A.; Porro, J.A.; Cordovilla, F.; Angulo, I.; Ocana, J.L. Effect of Three Different Finishing Processes on the Surface Morphology and Fatigue Life of A357.0 Parts Produced by Laser-Based Powder Bed Fusion. *Adv. Eng. Mater.* **2019**, *21*, 1801357. [CrossRef]
109. Kudrya, V.P.; Maishev, Y.P. Applications of the Technology of Fast Neutral Particle Beams in Micro-and Nanoelectronics. *Mikroelektronika* **2018**, *47*, 51–63. [CrossRef]
110. Vlcak, P.; Fojt, J.; Drahokoupil, J.; Brezina, V.; Sepitka, J.; Horazdovsky, T.; Miksovsky, J.; Cerny, F.; Lebeda, M.; Haubner, M. Influence of surface pre-treatment with mechanical polishing, chemical, electrochemical and ion sputter etching on the surface properties, corrosion resistance and MG-63 cell colonization of commercially pure titanium. *Mater. Sci. Eng. C* **2020**, *115*, 111065. [CrossRef] [PubMed]
111. Grigoriev, S.; Metel, A. Plasma-and Beam-Assisted Deposition Methods. In *Nanostructured Thin Films and Nanodispersion Strengthened Coatings*; NATO Science Series II: Mathematics, Physics and Chemistry; Voevodin, A.A., Shtansky, D.V., Levashov, E.A., Moore, J.J., Eds.; Springer: Dordrecht, The Netherlands, 2004; Volume 155, pp. 147–154. [CrossRef]
112. Isakova, Y.I.; Prima, A.I.; Pushkarev, A.I. A Conical Ion Diode with Self-Magnetic Insulation of Electrons. *Instrum. Exp. Tech.* **2019**, *62*, 506–516. [CrossRef]
113. Ghyngazov, S.A.; Zhu, X.P.; Pushkarev, A.I.; Egorova, Y.I.; Matrenin, S.V.; Kostenko, V.A.; Zhang, C.C.; Lei, M.K. Surface Modification of ZrO2-3Y(2)O(3) with Highintensity Pulsed N2+ Ion Beams. *Russ. Phys. J.* **2020**, *63*, 176–179. [CrossRef]

114. Metel, A.; Bolbukov, V.; Volosova, M.; Grigoriev, S.; Melnik, Y. Source of metal atoms and fast gas molecules for coating deposition on complex shaped dielectric products. *Surf. Coat. Technol.* **2013**, *225*, 34–39. [CrossRef]
115. Grigoriev, S.N.; Melnik, Y.A.; Metel, A.S.; Panin, V.V. Broad beam source of fast atoms produced as a result of charge exchange collisions of ions accelerated between two plasmas. *Instrum. Exp. Tech.* **2009**, *52*, 602–608. [CrossRef]
116. Suhara, M.; Matsuzaka, N.; Fukumitsu, M.; Okumura, T. Characterization of argon fast atom beam source and application to mesa etching process for GaInP/GaAs triple-barrier resonant tunneling diodes. In Proceedings of the 18th International Microprocesses and Nantechnology Conference, Tokyo, Japan, 26–28 October 2005; Institute of Pure Applied Physics: Tokyo, Japan, 2006; Volume 45, pp. 5504–5508.
117. Grigoriev, S.N.; Melnik, Y.A.; Metel, A.S.; Panin, V.V.; Prudnikov, V.V. A compact vapor source of conductive target material sputtered by 3-keV ions at 0.05-Pa pressure. *Instrum. Exp. Tech.* **2009**, *52*, 731. [CrossRef]
118. Musil, J.; Jaroš, M. Plasma and floating potentials in magnetron discharges. *J. Vac. Sci. Technol. A* **2017**, *35*, 060605. [CrossRef]
119. Anders, A. Tutorial: Reactive high power impulse magnetron sputtering. *J. Appl. Phys.* **2017**, *121*, 171101. [CrossRef]
120. Berlin, E.V.; Grigoriev, V.Y. Features of super-high-speed deposition of copper by a magnetron operating in target vapors on dielectric substrates. In Proceedings of the 11th International conference of a "Films and Coatings—2013", Saint Petersburg, Russia, 6–8 May 2013; pp. 104–106.
121. Shandrikov, M.V.; Artamonov, I.D.; Bugaev, A.S.; Oks, E.M.; Oskomov, K.V.; Vizir, A.V. Deposition of Cu-films by a planar magnetron sputtering system at ultra-low operating pressure. *Surf. Coat. Technol.* **2020**, *389*, 125600. [CrossRef]
122. Shandrikov, M.V.; Bugaev, A.S.; Oks, E.M.; Vizir, V.; Yushkov, G.Y. Ion mass-to-charge ratio in planar magnetron plasma with electron injections. *J. Phys. D Appl. Phys.* **2018**, *51*, 415201. [CrossRef]
123. Markov, A.B.; Yakovlev, E.V.; Shepel', D.A.; Petrov, V.I.; Bestetti, M. Liquid-Phase Surface Alloying of Copper with Stainless Steel Using Low-Energy, High-Current Electron Beam. *Russ. Phys. J.* **2017**, *60*, 1455–1460. [CrossRef]
124. Volosova, M.A.; Grigor'ev, S.N.; Kuzin, V.V. Effect of titanium nitride coating on stress structural inhomogeneity in oxide-carbide ceramic. Part 4. Action of heat flow. *Refract. Ind. Ceram.* **2015**, *56*, 91–96. [CrossRef]
125. Maishev, Y.P.; Shevchuk, S.L.; Kudrya, V.P. Formation of fast neutral beams and their using for selective etching. *Proc. SPIE* **2014**, *9440*, UNSP 94400K.
126. Metel, A.S.; Grigoriev, S.N.; Melnik, Y.A.; Panin, V.V. Filling the vacuum chamber of a technological system with homogeneous plasma using a stationary glow discharge. *Plasma Phys. Rep.* **2009**, *35*, 1058–1067. [CrossRef]
127. Metel, A.S.; Grigoriev, S.N.; Melnik, Y.A.; Bolbukov, V.P. Characteristics of a fast neutral atom source with electrons injected into the source through its emissive grid from the vacuum chamber. *Instrum. Exp. Tech.* **2012**, *55*, 288–293. [CrossRef]
128. Mesyats, G.A.; Proskurovsky, D.I.; Yankelevich, E.B.; Tregubov, V.F. Observation of micro-tip regeneration and the cathode polishing at nanosecond pulses of explosive emission current. *Rep. USSR Acad. Sci.* **1976**, *227*, 1335–1337.
129. Zhong, Z.W. Advanced polishing, grinding and finishing processes for various manufacturing applications: A review. *Mater. Manuf. Process.* **2020**, *35*, 1279–1303. [CrossRef]
130. Sagbas, B. Post-Processing Effects on Surface Properties of Direct Metal Laser Sintered AlSi10Mg Parts. *Met. Mater. Int.* **2020**, *26*, 143–153. [CrossRef]
131. Lazarenko, B.R.; Lazarenko, N.I. Electric Spark Machining of Metals in Water and Electrolytes (Elektroiskrovaya Obrabotka Metallov V Vode I Elektrolitakh). *Surf. Eng. Appl. Electrochem. (Elektronnaya Obrabotka Materialov)* **1980**, *1*, 5–8.
132. Danilov, I.; Hackert-Oschätzchen, M.; Zinecker, M.; Meichsner, G.; Edelmann, J.; Schubert, A. Process Understanding of Plasma Electrolytic Polishing through Multiphysics Simulation and Inline Metrology. *Micromachines* **2019**, *10*, 214. [CrossRef] [PubMed]
133. Afanasiev, Y.V.; Chichkov, B.N.; Demchenko, N.N.; Isakov, V.A.; Zavestovskaya, I.N. Ablation of metals by ultrashort laser pulses: Theoretical modeling and computer simulations. *Proc. SPIE* **2000**, *3885*, 266–274.
134. Sobol', O.V.; Andreev, A.A.; Grigoriev, S.N.; Gorban', V.F.; Volosova, M.A.; Aleshin, S.V.; Stolbovoy, V.A. Physical characteristics, structure and stress state of vacuum-arc tin coating, deposition on the substrate when applying high-voltage pulse during the deposition. *Probl. Atom. Sci. Technol.* **2011**, *4*, 174–177.
135. Aleshin, N.P.; Grigor'ev, M.V.; Shchipakov, N.A.; Krys'ko, N.V.; Krasnov, I.S.; Prilutskii, M.A.; Smorodinskii, Y.G. On the Possibility of Using Ultrasonic Surface and Head Waves in Nondestructive Quality Checks of Additive Manufactured Products. *Russ. J. Nondestruct.* **2017**, *53*, 830–838. [CrossRef]
136. Metel, A.; Bolbukov, V.; Volosova, M.; Grigoriev, S.; Melnik, Y. Equipment for deposition of thin metallic films bombarded by fast argon atoms. *Instrum. Exp. Tech.* **2014**, *57*, 345–351. [CrossRef]

MDPI AG
Grosspeteranlage 5
4052 Basel
Switzerland
Tel.: +41 61 683 77 34

Technologies Editorial Office
E-mail: technologies@mdpi.com
www.mdpi.com/journal/technologies

Disclaimer/Publisher's Note: The title and front matter of this reprint are at the discretion of the Guest Editors. The publisher is not responsible for their content or any associated concerns. The statements, opinions and data contained in all individual articles are solely those of the individual Editors and contributors and not of MDPI. MDPI disclaims responsibility for any injury to people or property resulting from any ideas, methods, instructions or products referred to in the content.

www.ingramcontent.com/pod-product-compliance
Lightning Source LLC
LaVergne TN
LVHW072332090526

838202LV00019B/2405